GREEN POLYMERS AND ENVIRONMENTAL POLLUTION CONTROL

GREEN POLYMERS AND ENVIRONMENTAL POLLUTION CONTROL

Edited by
Moayad N. Khalaf

Apple Academic Press Inc. | Apple Academic Press Inc.
3333 Mistwell Crescent | 9 Spinnaker Way
Oakville, ON L6L 0A2 | Waretown, NJ 08758
Canada | USA

©2016 by Apple Academic Press, Inc.

First issued in paperback 2021

Exclusive worldwide distribution by CRC Press, a member of Taylor & Francis Group
No claim to original U.S. Government works

ISBN 13: 978-1-77463-553-7 (pbk)
ISBN 13: 978-1-77188-139-5 (hbk)

Typeset by Accent Premedia Services (www.accentpremedia.com)

Library and Archives Canada Cataloguing in Publication

Green polymers and environmental pollution control / edited by Moayad N. Khalaf.

Includes bibliographical references and index.
Issued in print and electronic formats.
ISBN 978-1-77188-139-5 (hardcover).--ISBN 978-1-4987-3249-9 (pdf)
1. Biopolymers. 2. Biopolymers--Industrial applications. 3. Polymerization. 4. Green technology. 5. Pollution prevention. I. Khalaf, Moayad N., author, editor

| TP248.65.P62G74 2015 | 572 | C2015-906478-3 | C2015-906479-1 |

Library of Congress Cataloging-in-Publication Data

Green polymers and environmental pollution control / [edited by] Moayad N. Khalaf.

pages cm
Includes bibliographical references and index.
ISBN 978-1-77188-139-5 (alk. paper)
1. Polymerization--Environmental aspects. 2. Renewable natural resources. 3. Source reduction (Waste management) I. Khalaf, Moayad N.

| TP156.P6G75 2015 | 668.9'2--dc23 | 2015035596 |

Apple Academic Press also publishes its books in a variety of electronic formats. Some content that appears in print may not be available in electronic format. For information about Apple Academic Press products, visit our website at **www.appleacademicpress.com** and the CRC Press website at **www.crcpress.com**

CONTENTS

LIST OF CONTRIBUTORS

S. Aisverya
Department of Chemistry, D.K.M. College for Women, Thiruvalluvar University, Vellore, Tamilnadu, India, Tel: (+91) 98429 10157

Agus Arsad
Enhanced Polymer Research Group (EnPRO), Faculty of Chemical Engineering, Universiti Teknologi Malaysia, 81310 UTM Johor Bahru, Johor, Malaysia, E-mail: agus@cheme.utm.my

Robin Augustine
International and Interuniversity Centre for Nanoscience and Nanotechnology, Mahatma Gandhi University, Priyadarshini Hills P.O., Kottayam – 686 560, Kerala, India

Aznizam Abu Bakar
Enhanced Polymer Research Group (EnPRO), Department of Polymer Engineering, Faculty of Chemical Engineering, Universiti Teknologi Malaysia, 81310 UTM Skudai, Johor Malaysia, Tel: +60 7 5537835; Fax: +60 7 5581 463

Hossein Baniasadi
Department of Chemical and Petroleum Engineering, Sharif University of Technology, Tehran, Iran

Deivasagayam Dakshinamoorthy
Department of Chemistry, Pennsylvania State University, New Kensington, PA 15068, USA

Jibrin Mohammed Danlami
Enhanced Polymer Research Group (EnPRO), Faculty of Chemical Engineering, Universiti Teknologi Malaysia, 81310 UTM Johor Bahru, Johor, Malaysia

Ana M. Diez-Pascual
Analytical Chemistry, Physical Chemistry and Chemical Engineering Department, Faculty of Biology, Environmental Sciences and Chemistry, Alcalá University, 28871 Alcalá de Henares, Madrid, Spain

Fariba Ghaderinezhad
Department of Chemical and Petroleum Engineering, Sharif University of Technology, Tehran, Iran

T. Gomathi
Department of Chemistry, D.K.M. College for Women, Thiruvalluvar University, Vellore, Tamilnadu, India, Tel: (+91) 98429 10157

A. K. Haghi
University of Guilan, Rasht, Iran

Azman Hassan
Enhanced Polymer Research Group (EnPRO), Department of Polymer Engineering, Faculty of Chemical Engineering, Universiti Teknologi Malaysia, 81310 UTM Skudai, Johor Malaysia, Tel: +60 7 5537835; Fax: +60 7 5581 463

Cintil Jose
Department of Chemistry, Newman College, Thodupuzha, Kerala, India

Jithin Joy
Department of Chemistry, Newman College, Thodupuzha, Kerala, India

Nandakumar Kalarikkal
School of Pure and Applied Physics, Mahatma Gandhi University, Priyadarshini Hills P.O., Kottayam – 686 560, Kerala, India; Department of Biotechnology, St. Joseph's College, Irinjalakuda, Thrissur – 680 121, Kerala, India; E-mail: nkkalarikkal@mgu.ac.in

Moayad N. Khalaf
Chemistry Department, College of Science, University of Basrah, P. O. Box 773, Basrah, Iraq

Stewart P. Lewis
Innovative Science Corp. (www.innovscience.com), Salem, VA 24153, USA; Visiting Scientist, Department of Chemistry, Pennsylvania State University, New Kensington, PA 15068, USA

Nuranassuhada Mahzam
Enhanced Polymer Research Group (EnPRO), Faculty of Chemical Engineering, Universiti Teknologi Malaysia, 81310 UTM Johor Bahru, Johor, Malaysia

Khaliq Majeed
Enhanced Polymer Research Group (EnPRO), Department of Polymer Engineering, Faculty of Chemical Engineering, Universiti Teknologi Malaysia, 81310 UTM Skudai, Johor Malaysia, Tel: +60 7 5537835; Fax: +60 7 5581 463

Robert T. Mathers
Department of Chemistry, Pennsylvania State University, New Kensington, PA 15068, USA

P. Lovely Mathew
Department of Chemistry, Newman College, Thodupuzha, Kerala, India

K. Nasreen
Department of Chemistry, D.K.M. College for Women, Thiruvalluvar University, Vellore, Tamilnadu, India, Tel: (+91) 98429 10157

R. Nithya
Department of Chemistry, D.K.M. College for Women, Thiruvalluvar University, Vellore, Tamilnadu, India, Tel: (+91) 98429 10157

S. A. Ahmad Ramazani
Department of Chemical and Petroleum Engineering, Sharif University of Technology, Tehran, Iran

Peter S. Shuttleworth
Institute of Polymer Science and Technology, ICTP-CSIC, Juan de la Cierva 3, 28006 Madrid, Spain

Davi Nogueira da Silva
Faculdade de Tecnologia, Universidade Federal do Amazonas, Av. General. Rodrigo O. J. Ramos, 3000, CEP 69077–000, Manaus – AM, Brazil; E-mail: davi.nogueira@am.senai.br

S. Snigdha
International and Interuniversity Centre for Nanoscience and Nanotechnology, Mahatma Gandhi University, Priyadarshini Hills P.O., Kottayam – 686 560, Kerala, India

P. N. Sudha
Department of Chemistry, D.K.M. College for Women, Thiruvalluvar University, Vellore, Tamilnadu, India, Tel: (+91) 98429 10157; E-mail: drparsu8@gmail.com

Sabu Thomas
International and Inter University Centre for Nanoscience and Nanotechnology; School of Chemical Sciences, Mahatma Gandhi University, Priyadarshini Hills P.O., Kottayam – 686 560, Kerala, India; E-mail: sabupolymer@yahoo.com

Bhavana Venugopal
International and Interuniversity Centre for Nanoscience and Nanotechnology, Mahatma Gandhi University, Priyadarshini Hills P.O., Kottayam – 686 560, Kerala, India

Adalena Kennedy Vieira
Faculdade de Tecnologia, Universidade Federal do Amazonas, Av. General. Rodrigo O. J. Ramos, 3000, CEP 69077–000, Manaus – AM, Brazil; E-mail: adalenakennedy@gmail.com

Raimundo Kennedy Vieira
Faculdade de Tecnologia, Universidade Federal do Amazonas, Av. General. Rodrigo O. J. Ramos, 3000, CEP 69077–000, Manaus – AM, Brazil; E-mail: maneiro01@ig.com.br

K. Vijayalakshmi
Department of Chemistry, D.K.M. College for Women, Thiruvalluvar University, Vellore, Tamilnadu, India, Tel: (+91) 98429 10157

P. Angelin Vinodhini
Department of Chemistry, D.K.M. College for Women, Thiruvalluvar University, Vellore, Tamilnadu, India, Tel: (+91) 98429 10157

LIST OF ABBREVIATIONS

1,4-CHD	1,4-cyclohexadiene
2fFCS	dual focus fluorescence correlation spectroscopy
AAD	aryl-alcohol dehydrogenases
AAO	arylalcohol oxidase
ABC	atomistic-based continuum
AcOH	glacial acetic acid
AFM	atomic force microscopy
AOPSC	acid treated oil palm shell charcoal
APS	ammonium peroxydisulfate
BA	Brønsted acid
BD	1,3-butadiene
BD	Brownian dynamics
BGL	β-glucosidases
BRF	brown-rot fungi
CBAM	chitosan blended alginate matrix
CBIC	Chamber of Construction Industry
CD	cyclodextrin
CHC	Cahn–Hilliard–Cook
CHD	1,4 or 1,3-cyclohexadiene
CHF	congestive heart failure
CHIT	chitosan
CMC	carboxymethyl chitin
CNTs	carbon nanotubes
CONAMA	National Council of Environment
CRV	carvedilol
CT	chain transfer
CTC	charge transfer complex
CVD	chemical vapor deposition
DA	degree of acetylation
DE	degree of esterification
DEX	dextran sulfate

DFT	dynamic density functional theory
DGEBA	diglycidyl ether of bisphenol A
DLS	dynamic light scattering
DMA	dynamic mechanical analysis
DMAEMA	dimethyl aminoethyl methacrylate
DPD	dissipative particle dynamics
DS	degree of substitution
DSC	differential scanning calorimetry
EA	electron acceptor
EB	emeraldine base
ECM	extracellular matrix
EMCMCR	ethylenediamine-modified crosslinked magnetic chitosan resin
ERM	effective reinforcing modulus
ESD	electrostatic discharge
ESEM	environmental scanning electron microscope
EWC	equilibrium water content
FCS	fluorescence correlation spectroscopy
FEM	finite element method
FITC	fluorescein isothiocyanate
FRP	fiber reinforced polymer
FT-NIR	Fourier transform near infrared
FTIR	Fourier-transform infrared spectroscopy
GO	graphene oxide
GSI	gigascale integration
H_2SO_4	sulfuric acid
HCl	hydrochloric acid
HE	heulandite
HEMA	hydroxyethyl methacrylate
HM	high methoxy
IB	isobutene
IBL	implantable bioartificial liver
IOP	iontophoresis
IP	isoprene
ISS	interfacial shear strength
ITO	indium tin oxide

IUPAC	International Union of Pure and Applied Chemistry
KMnO4	potassium permanganate
KPS	potassium peroxodisulfate
LA	Lewis acid
LB	lattice Boltzmann
LbL	layer-by-layer
LCST	lower critical solution temperature
LDPE	low density polyethylene
LJ	Lennard-Jones
LM	low methoxy
LMA	lauryl methacrylate
MAA	methacrylic acid
MAO	methylaluminoxane
MAP	modified atmosphere packaging
MAPE	maleic anhydride grafted polyethylene
MAPP	maleic anhydride grafted PP
MC	Monte Carlo
MD	molecular dynamics
MFI	melt flow index
MFR	melt mass flow rate
MH	multi-scale homogenization
MM	molecular mechanics
MMT	montmorillonite
MW	molecular weight
MWNTs	multi-wall carbon nanotubes
$NaNO_3$	sodium nitrate
NDD	nasal drug delivery
NFRPCs	natural fiber reinforced polymer composites
NG	nanogels
NMT	nano-magnetite
NR	natural rubber
O-CMCS	O-carboxymethyl chitosan
ODD	oral drug delivery
OPF	oil palm fiber
OSA	octenyl succinic anhydride
P(NiPAM-co-MAA)	poly(N-isopropylacrylamide-co-methacrylic acid)

PA-6	polyamide-6
PAA	poly(acrylic acid)
PAG	PANI/graphene particles
PAH	Poly(Allylamine Hydrochloride)
PANI	Polyaniline
pARG	poly L-arginine
pASP	poly L-aspartic acid
PCD	polycrystalline diamond tooling
PDACMAC	poly(diallyldimethylammonium chloride)
PDGFB	platelet-derived growth factor B
PE	polyelectrolytes
PE	polyethylene
PenG	penicillin G
PET	poly(ethylene terephthalate)
PF	phenol formaldehyde
PFLA	perfluoroarylated Lewis acid
PGA	poly l-glutamic acid
PGA	polyglycolide
PGLA	pectin/poly lactide-co-glycolide
PIB	polyisobutene
PLA	polylactic acid
PLL	poly L-lysine
PLLA	poly (L-lactic acid)
PMAA	polymethacrylate
PNCs	polymeric nanocomposites
PNiPAM	poly(N-isopropylacrylamide)
PP	polypropylene
PS	polystyrene
PVC	poly (vinyl chloride)
QR	quinone reductases
RCM	rate controlling membrane
RH	rice husk
rPE	recycled polyethylene
RVE	representative volume element
SA	sodium alginate
SANS	small angle neutron scattering

SCFC	sugarcane fiber cellulose
SDS	sodium dodecyl sulfate
SEM	scanning electron microscopy
SF	silk fibroin
SHP	sterically hindered pyridine
SI	swelling index
SSF	solid state fermentation
ST	styrene
SWNTs	single-wall nanotubes
TDDS	transdermal drug delivery system
TDGL	time-dependent Ginsburg–Landau theory
TEM	transmission electron microscopic
TEMED	tetramethylenediamin
TETA	triethylene tetramine
TGA	thermogravimetric analysis
TPP	sodium tripolyphosphate
UD	unidirectional fiber orientation
vdW	Van Der Waals
VE	vinyl ether
VESFA	vinyl ether soybean fatty acids
VGCFs	vapor-grown carbon fibers
VPT	volume phase transition
WCA	weakly coordinating anion
WPNC	wood polymer nanocomposite
WRF	white-rot fungi
XG	xanthan gum
XRD	X-ray diffraction

LIST OF SYMBOLS

B	benzenic-type rings
D_{001}	interlayer distance between clay layers
E	electric field strength
E	Young's modulus of the composite
E_f	Young's modulus of the filler
E_m	Young's modulus of the matrix
F^c_{ij}	conservative force of particle j acting on particle i, γ and σ are constants depending on the system
H(i) and H(j)	Hamiltonian associated with the original and new configuration, respectively
I	light intensity
K_B	Boltzmann constant
L	embedded length of the nanotube
n_0	refractive index of the sample
P	permeability coefficient (mL-mm/m^2/24 h/atm)
Pi	momentum of particle i
Q	quinonic-type rings
q	wave vector
t	film thickness under investigation (mm)
U^a	energies associated with truss elements that represent covalent bond stretching
U^b	energies associated with truss elements that represent bond-angle bending
U^c	energies associated with truss elements that represent van der Waals interactions
U_k	kinetic energy
U^r	energies associated with covalent bond stretching
U_v	Hookian potential energy
U^{vdw}	energies associated with van der Waals interactions
U^θ	energies associated with bond-angle bending
v_0	Poisson's ratio of the matrix
V_0	volume of dry scaffold
v_f	volume fractions of the fillers

v_m	volume fractions of the matrix
V_p	pore volume
W_0	weight of the dry scaffold
W_d	weight of dried film
W_{es}	weight of swollen films
W_{LDPE}	weight fraction of LDPE in the sample
x	pullout distance of the nanotube
X_c	degree of crystallinity

Greek Symbols

Γ	decay rate
	melting enthalpy of 100% crystalline polyethylene
Δp	partial pressure difference on two sides of the film (atm)
ΔU	change in the sum of the mixing energy and the chemical potential of the mixture
ε_0	permittivity of free space
ε_r	dielectric constant of the dispersion medium
$\zeta(t)$	Gaussian random noise term
η	dynamic viscosity of the dispersion medium
θ	half of the diffraction angle at the first peak
θ	the angle at which the detector is located with respect to the sample cell
λ	incident laser wavelength
λ	wavelength of the X-ray beam
λ_1	due to the electronic transition π to π^* band
λ_2	electronic transition of the polaronic band to π^* band in the benzenoid ring
λ_3	corresponds to the electronic transitions of the π band to the polaronic band
ρ	density of the scaffold
τ	delay time
υ	velocity of a dispersed particle
φ	porosity of the scaffold

PREFACE

Green polymers are those produced using green (or sustainable) chemistry, a term that appeared in the 1990s. According to the definition of the International Union of Pure and Applied Chemistry (IUPAC), green chemistry relates to the "design of chemical products and processes that reduce or eliminate the use or generation of substances hazardous to humans, animals, plants, and the environment." Thus, green chemistry seeks to reduce and prevent pollution at its source. Natural polymers are usually green. The polymer industry looks at alternatives to petrochemical sources to ensure a viable long-term future.

Green polymers are a crucial area of research and product development that continues to grow in its influence over industrial practices. Developments in these areas are driven by environmental concerns and interest in sustainability, desire to decrease our dependence on petroleum, and commercial opportunities to develop "green" products. Publications and patents in these fields are increasing as more academic, industrial, and government scientists become involved in research and commercial activities.

Green Polymers and Environmental Pollution Control examines the latest developments in producing conventional polymers from sustainable sources. The purpose of this book is to publish new work from a cutting-edge group of leading international researchers from academia, government, and industrial institutions.

Providing guidelines for implementing sustainable practices for traditional petroleum-based plastics, biobased plastics, and recycled plastics, green polymers and environmental pollution control explains what green polymers are, why green polymers are needed, which green polymers to use, and how manufacturing companies can integrate them into their manufacturing operations. The volume will be a vital resource for practitioners, scientists, researchers, and graduate students.

With the recent advancements in synthesis technologies and the finding of new functional monomers, research on green polymers has shown strong potential in generating better property polymers from renewable resources. This book, describing these advances in synthesis, processing, and technology of such polymers, not only provides the state-of-the-art information to researchers but also acts to stimulate research in this direction.

Green Polymers and Environmental Pollution Control offers an excellent source for researchers, upper-level graduate students, brand owners, environment and sustainability managers, business development and innovation professionals, chemical engineers, plastics manufacturers, agriculture specialists, biochemists, and suppliers to the industry to debate sustainable, economic solutions for polymer synthesis.

—Professor Moayad N. Khalaf

ABOUT THE EDITOR

Moayad N. Khalaf

Moayad N. Khalaf is a professor of polymer chemistry at the Department of Chemistry, College of Science, University of Basrah, Iraq. He received his BSc in chemistry science, MSc in physical-organic chemistry, and PhD in polymer chemistry from the University of Basrah in Iraq. Professor Khalaf has more than 27 years of professional experience in the petrochemical industry, earned while working with the company for Petrochemical Industries, Iraq. In 2005, he joined the Chemistry Department at the University of Basrah, where now he is lecturing on most polymer related subjects. Dr. Khalaf supervised more than 12 MSc and 4 PhD students. He has 19 Iraqi patents and more than 100 scientific papers published in peer-reviewed journals and conference proceedings. His research interests are:

- Modified polymer for corrosion inhibitor, demulsifier, additive for oil.
- Polymer additive (light stabilizer and antioxidant)
- Lubricant antioxidant
- Lubricant modifier
- Lubricant recycling
- Waste polymers recycling
- Lingnosulfonate for well drilling
- Mechanical properties of composite polymer
- Rheological properties of polymer and composite using nanofiller
- Water desalination
- Preparation of flocculent from waste polystyrene
- Using ground water as source for industrial water and agriculture
- Preparation of polymer for solar cell application
- Preparation of conductive polymer

He also works on modifying the chemical security and safety strategy at the Department of Chemistry of the University of Basrah through funds from CRDF Global. Dr. Khalaf also received funds totaling ($350.000US) so far from the Arab Science and Technology Foundation and the Iraqi Ministry of Higher Education and Scientific Research to support his research in the fields of polymer chemistry and water treatment.

CHAPTER 1

PREPARATION AND CHARACTERIZATION OF NOVEL CONDUCTIVE POROUS CHITOSAN-BASED NANOCOMPOSITE SCAFFOLDS FOR TISSUE ENGINEERING APPLICATIONS

HOSSEIN BANIASADI, S. A. AHMAD RAMAZANI, and FARIBA GHADERINEZHAD

Department of Chemical and Petroleum Engineering, Sharif University of Technology, Tehran, Iran

CONTENTS

ABSTRACT

In this chapter, graphene oxide was prepared via modified Hummer's method and then chemically reduced to graphene nanosheets. These graphene nanosheets were used to synthesize polyaniline/graphene nanocomposites with chemical oxidation of aniline monomer via in-situ emulsion polymerization in the presence of sodium dodecyl sulfate in acid media. Morphological studies confirmed that graphene nanosheets were prepared successfully with this chemical reduction method. Electrical conductivity of synthesized polymer and nanocomposites investigated using a standard four-point probe technique. Electrical conductivity, FTIR and UV-Vis investigation confirmed that conductive binary doped emeraldine salt polyaniline and its nanocomposites with graphene nanosheets should be synthesized via emulsion polymerization. Morphological studies illustrated that this conductive powder, which was designated as PAG, has almost spherical shape with size of about 10–20 nm. We used it to fabricate novel porous-conductive scaffolds composed of chitosan and gelatin via lyophilization method. Conductivity measurements of the scaffolds revealed that with low amount of PAG (~2.5 wt.%) the conductivity reached close to 10^{-3} S.cm^{-1}, which was suitable for tissue engineering applications. Pore and swelling behavior studies showed porous scaffolds with porosity more than 50% and the pore size between 10–70 μ, with ability of absorbing more than 200% water, had successfully produced by freeze-drying process. The main mechanical properties, such as tensile strength, elongation at break point and tensile modulus of the scaffolds were examined in both dry and hydrated states. The results indicated that with increasing PAG content the scaffold showed relatively stiff behavior especially for the scaffolds with more than 5 wt.% PAG. However, many prepared scaffolds exhibited the desired mechanical strength for some kinds of tissue engineering applications.

1.1 INTRODUCTION

The natural polymers have the advantage to be biologically distinguished. They also can be supported by adhesion property and cellular performance but are weak in mechanical properties. Lots of them have some limitations in their production and hence are expensive. Chitin and its deacetylated derivative (chitosan) are of the polymers that show the necessary properties for biological applications. As a linear polysaccharide, chitosan is a cationic natural biopolymer and the second most abundant biosynthesized material. It is partially N-deacetylated derivative of chitin (an important constituent of the exoskeleton in animals). As shown in Figure 1.1, the only difference of chitin and chitosan is presence of two acetamide groups in chitosan instead of two hydroxyl groups in chitin. Chitosan is unique in terms of its abundance, accessibility, and cheapness and it is especially due to the nitrogen-containing functional group [1–8]. Chitosan which has a crystalline structure and several hydrogen bonds is a copolymer composed of b-(1,4)-2-acetamido-2-deoxy- D-glucopyranosyl and b-(1,4)-2-amino-2-deoxy-D-glucopyranosyl units [6, 9].

Chitosan is insoluble in aqua, alcoholic, and most mineral acidic systems but it is soluble in diluted acetic acid, formic acid, and lactic acid. In presence of a little amount of acid, chitosan is also soluble in the systems of water-methanol, water-ethanol, and water-acetone. Also, chitosan has a free amine group, which makes it to be a positive polyelectrolyte in the range of 2 to 6 of pH while the other polysaccharides will be charged negatively or neutrally; this is effective in chitosan solubility. Nevertheless the

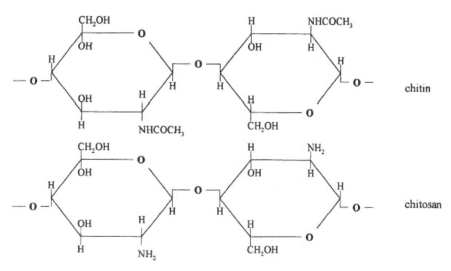

FIGURE 1.1   Schematically structure of chitin and chitosan.

chitosan solution is very viscous, and the formation of hydrogen binds in a three-dimensional network prevents of the polymeric chains in the applied electric field [1, 10].

Chitosan has the wide applications in a broad range as biomaterial and especially as the supporting matrix or delivery system for tissue repair and regeneration.

Due to the outstanding properties of chitosan in terms of low cost, large-scale availability, biocompatibility, biodegradability, non-antigenicity, low toxicity, good mechanical properties, swelling capacity, ease of preparation in various forms, and antibacterial activity, it has become one of the most important biomaterials for diverse applications in drug release, nutrition supplements, wound healing agents, drug carriers, membrane filter for water treatment, biodegradable coating or film for food packaging, tissue repair and regeneration [2–8, 11–14]. Chitosan scaffolds which marginally support biological activity of diverse cell types can be obtained with various geometries, pore sizes, and pore orientation. They are highly brittle with a strain at break of 40–50% in the wet state.

To improve the mechanical and biological properties of chitosan, natural macromolecules like gelatin have been extensively used due to their inherent advantages such as biocompatibility and biodegradability [2, 15]. Gelatin contains Arg–Gly–Asp (RGD)-like sequence that promotes cell adhesion and migration, and also forms a polyelectrolyte complex, that is the reason why it can improve the biological activity of chitosan [2].

Gelatin is a heterogeneous mixture of hot water-soluble proteins of high-average molecular weight obtained by partially hydrolyzing of collagen. Collagen is a major component of the ECM in skin, white connective tissue, and bones of animal. Due to the animal origin of collagen, it shows antigenicity whereas gelatin has relatively low antigenicity compared to its precursor. However gelatin still keeps some of the information signals which may promote cell adhesion, differentiation and proliferation, such as the Arg–Gly Asp (RGD) sequence of collagen. Gelatin whose prime property is the Sol–Gel transition under aqueous condition is consisting of microcrystallites interconnected network with amorphous regions of randomly coiled segments and it has the properties like heat reversibility [5–7, 15–17]. Gelatin contains free carboxyl groups on its chain backbones and has the potential to blend with chitosan because of the ability of forming hydrogen bonding. It is composed of a repeating sequence of amino acids glycine- X-Y triplets, where X and Y frequently are proline and hydroxyproline. Figure 1.2 shows the formula of gelatin. Gelatin as a biodegradable polymer which is widely used in biomedical and pharmaceutical fields has lots of attractive properties, such as excellent biocompatibility, plasticity and adhesiveness, promotion of cell adhesion and growth, nonantigenicity, and low cost. Hence, gelatin has been selected as an appropriate candidate blended with chitosan [3, 4, 6, 17]. Pulieri et al. [18] prepared dehydro-thermally crosslinked chitosan/gelatin blend films, in which the gelatin amount affected the physicochemical properties significantly. In comparison to chitosan, gelatin shows poor mechanical properties and very high swelling capacity in aqueous solutions [5]. To promote cell adhesion, migration, differentiation and proliferation of chitosan scaffolds, they have been mixed with gelatin [15]. Recent research revealed that, gelatin could exhibit activation of macrophages and high hemostatic effects. It is completely resorbable in vivo, and because of the existence of many functional groups its physicochemical properties can be suitably modulated.

As mentioned before, at the suitable pH value (in acidic environments), free amino groups of chitosan have positive charge, so they can be ionically bound to the negatively charged moieties of gelatin in an aqueous solution. Therefore, Gelatin can form a polyelectrolyte complex with chitosan [4, 17].

To modulate a broad spectrum of characteristics of chitosan-based biomaterials for tissue repair, crosslinking is another effective approach. Gelatin–chitosan scaffolds have been formed without or with crosslinkers

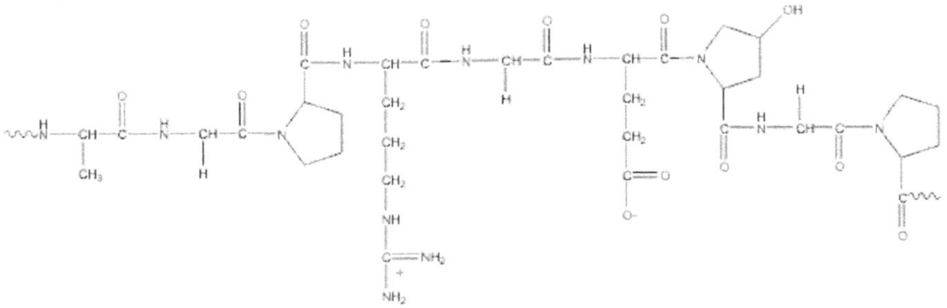

FIGURE 1.2 Schematically structure of gelatin.

such as glutaraldehyde or enzymes and tested in regenerating various tissues including skin, cartilage, and bone [2]. Lack of shape and mechanical stability of the gelatin-chitosan non-crosslinked gels impose using the crosslinking agents [5]. The size and type of crosslinker agent and the functional groups of chitosan mainly influence the crosslinking reaction. The small molecular size of crosslinker has faster crosslinking reaction due to its easier diffusion. Depending on the nature of the crosslinker, covalent or ionic bonds are the main interactions forming the network. In covalently crosslinked hydrogels, crosslinking degree has been presented as the main parameter which influences significant properties such as mechanical strength, swelling and drug release. Such gels generally exhibit pH-sensitive swelling and drug release by diffusion through their porous structure [19]. Ionic and covalent are two types of crosslinking used for the chitosan porous scaffolds. It is also well known that the covalent crosslinkers like formaldehyde, glutaraldehyde and genipin present a higher toxicity than ionic crosslinkers such as sodium sulfate, magnesium sulfate, sodium tripolyphosphate (TPP), hyaluronic acid, and alginate [5]. In comparison with the non-crosslinked chitosan scaffolds, the crosslinked scaffolds are of higher mechanical strength and resistance to enzymatic degradation [20]. Figure 1.3 schematically shows crosslinking reaction of chitosan-gelatin by glutaraldehyde as the crosslinking agent.

Conducting polymers which are extensively conjugated molecules have π electron delocalization along their backbone which giving them unique optical and electrical properties. This characteristic causes them to act as a semiconductor or a conductor. These molecules have alternating single and double bonds and electrons are able to move from one end to the other through the extended p-orbital system [22]. Some conductive polymers which are well known in the biomedical and tissue engineering applications in terms of their chemical and physical properties include polyaniline, polythiophene and polypyrrole. They are used as electroactive substrates for the culture of electrically excitable cells, such as neuronal or muscle cells [23, 24]. Polyaniline (PANI) due to its good environmental, thermal, and chemical stability, tunable conductivity switching between insulating and semiconducting materials, low operational voltage, facile synthesis, potentiality for practical applications, low cost, and ease of synthesis, is one of the most attractive conducting polymers among the known

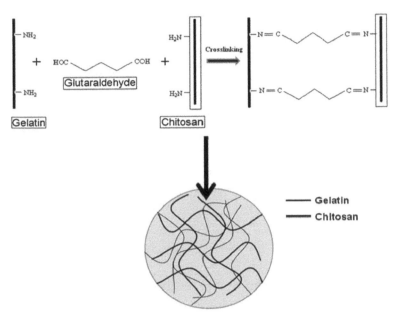

FIGURE 1.3 Schematic presentation of crosslinking between chitosan and gelatin polymer precursors using glutaraldehyde [21].

conducting polymers [25, 26]. PANI was initially found in 1834 by Runge, and referred to as aniline black. Following this, in order to analyze this substance, some researches were carried out in 1862. Finally in 1912, it was discovered that PANI is a mixed oxidation state polymer composed of reduced benzoid units and oxidized quinoid units [22]. The leucoemeraldine oxidation state, the emeraldine oxidation state, and the pernigraniline oxidation state are the different forms of PANI describe below:

(i) Leucoemeraldine base: the fully reduced form of non-doped PANI. It is composed solely of reduced units (Figure 1.4a).

(ii) Pernigraniline base: the fully oxidized form of non-doped PANI. It is composed solely of oxidized base unites (Figure 1.4b).

(iii) Emeraldine base: the intermediate oxidation state of PANI. It is composed of equal amounts of alternating reduced base and oxidized base units (Figure 1.4c) [27].

Volumetric titration methods using $TiCl_3$ can be used to determine the oxidation state of PANI. Qualitative information about its average

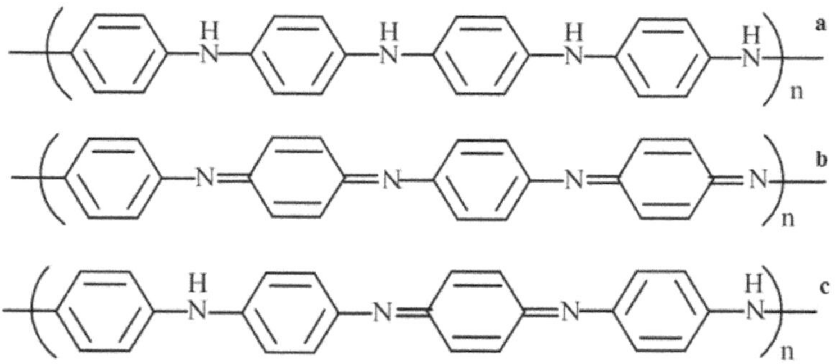

FIGURE 1.4   (a) Leucoemeraldine base, (b) Pernigraniline base, and (c) Emeraldine base [27].

oxidation state can also be tested by spectroscopic methods including FTIR, Raman, and UV/Vis [27].

Recently, it has been shown that tunable electroactivity of PANI has some applications in diverse biological activity such as for biosensors or as scaffolds in tissue engineering. It is also shown that PANI is biocompatible in vitro and in long-term animal studies in vivo [23]. Some extensively studied subject about PANI is related to many potential applications including secondary battery electrodes, supercapacitors, electromagnetic shielding devices, anticorrosion coating, gas separation membranes, light-emitting diodes, conducting molecular wires, sensors, switchable membranes, and so on. On one hand, PANI has the flexibility and processability of conventional polymers and on the other hand has the similar electronic, magnetic, and optical properties to metals and its doping level can be readily controlled through an acid-doping/base-dedoping process [26, 28–31]. Some characteristics of PANI including easy-synthesized, light-weight polymer exhibiting high conductivity, low operational voltage and high stress make it a good option for the development of actuators. In spite of all these properties, the main problem in the successful utilization of PANI is its poor mechanical properties and processability that the insoluble nature of it in common organic solvents has led to it. Incorporation of polar functional groups or long and flexible alkyl chains in the polymer backbone such as substituting one or more hydrogens by an alkyl, an alkoxy, an aryl hydroxyl, an amino group, or halogen group in an aniline nucleus is a common method that makes PANI soluble in water and/or organic solvents [27, 32–34]. Formation of polymer matrix/PANI composites is another approach to this problem. In these composites, the polymer brings specific properties such as better solubility or mechanical properties and the PANI brings conducting properties. They may also respond to the different stimuli such as pH or electric stimuli [35]. To apply PANI and PANI derivative composites and blends in biomedical realm, their matrices should be biocompatible and biodegradable. Thus, only limited synthetic or natural polymers are applicable [34].

Graphene, a single layer of carbon atoms in a hexagonal lattice, has recently attracted much attention due to its novel electronic and mechanical properties [36]. Graphene is usually prepared by the reduction of its precursor graphene oxide [37]. It is a typical pseudo-two-dimensional oxygen-containing solid in bulk form and possesses functional groups including hydroxyls, epoxides, and carboxyls [38–41]. Both graphene and graphene oxide sheets show very high mechanical properties with well biocompatibility, and they have potential applications as biomaterials [42–44]. The chemical groups of graphene oxide have been found to be a feasible and effective means of improving the dispersion of graphene. Additionally, functional side groups bound to the surface of graphene oxide or graphene sheets may improve the interfacial interaction between graphene oxide/graphene and the matrix [45].

Currently, nanomaterials are used to improve physicochemical properties of materials due to their structural features. In this regard, discovery of graphene and graphene-based polymer nanocomposites play a key role in modern science and technology. In comparison to the neat polymer, polymer/graphene nanocomposites show enhanced characteristics especially in mechanical, thermal, gas barrier, electrical and flame retardant properties. It is also reported that in comparison to the clay or other carbon filler-based polymer nanocomposites, graphene-based polymer nanocomposites have better mechanical and electrical properties. These hybrid nanomaterials at a very low filler loadings show considerable improvement in properties that cannot normally be achieved using conventional composites or virgin polymers. It should be mentioned that this improvement is directly related to dispersion of the nanofillers in the polymer matrix [46].

The first aim of the present study is to synthesize polyaniline/graphene nanocomposites with appropriate morphology and electrical properties second, to prepare porous conductive chitosan-based scaffolds, which are expected to have a well-defined pore structure, acceptable mechanical properties and a desired conductivity. Since PANI generally is non-flexible and non-processable, only a very low amount of PANI can be used for this kind of scaffold also the size of polymer particles should be small enough to easily excrete through circulatory system after the matrix of the scaffold has fully degraded. For this purpose, first we produced conductive PANI powders with chemical oxidation via emulsion polymerization of aniline in the presence of graphene nanosheets. In the next step, the conductive chitosan-based porous scaffolds were successfully prepared with mixing chitosan/gelatin matrix and conductive PANI/graphene powder. Combination of chitosan and gelatin brings very good biocompatibility and biodegradability while incorporation small amount of PANI/graphene makes scaffolds electrical conductive. Finally, it is expected that the produced scaffolds be appropriate for some tissue engineering applications.

1.2 EXPERIMENTAL PART

1.2.1 MATERIALS

The materials used in this study include: (a) natural flake graphite and sodium dodecyl sulfate (SDS) from Sigma Aldrich, (b) the aniline monomer was obtained from Aldrich and purified by vacuum distillation and kept under nitrogen in a refrigerator at about 4°C prior to use, (c) medium molecular weight chitosan powder with $M_w = 480$ KDa and DD = 75–85% and ammonium peroxydisulfate (APS) from Aldrich, (d) glacial acetic acid (AcOH), hydrochloric acid (HCl), sulfuric acid (H_2SO_4), sodium nitrate ($NaNO_3$), potassium permanganate ($KMnO_4$) and sodium tripolyphosphate (TPP) from Merck, and (e) deionized water.

1.2.2 SYNTHESIS OF GRAPHENE OXIDE AND REDUCTION TO GRAPHENE NANOSHEETS

Graphene oxide (GO) was prepared via the modified Hummers' method [46]. Briefly, 2 g of graphite powder and 1 g of $NaNO_3$ were mixed with 46 mL of concentrated H_2SO_4 at 0°C in ice bath. The mixture was stirred for 30 min. Then 6 g of $KMnO_4$ was slowly added into the mixture. The temperature should be lower than 20°C. The stirring was continued for 2 h. Subsequently the temperature was increased to 35°C and 65°C while stirring was continued for 30 and 40 min, respectively. After that, 92 mL deionized water was added to the mixture and the temperature was increased up to 100°C. The concentration of the solution was reduced by adding 280 mL distillated water and hydrogen peroxide (30 wt.%). Finally graphene oxide was isolated from the mixture by centrifugation and washed with diluted hydrochloric acid (HCl 5 wt.%) to eliminate metal ions, then washed with deionized water to remove the extra amount of HCl. The mixture was filtered and obtained solid was dried at 60°C for 2 days.

In order to prepare graphene, the precursor graphene oxide was first dispersed in water followed by the addition of an aqueous KOH solution. It is known that KOH, a strong base, can confer a large negative charge through reactions with the reactive hydroxyl, epoxy and carboxylic acid groups on the graphene oxide sheets, resulting in extensive coating of the sheets with negative charges and K^+ ions [46]. After that, to produce a homogeneous suspension of reduced graphene, hydrazine monohydrate was added to KOH-treated graphene oxide (Figure 1.5).

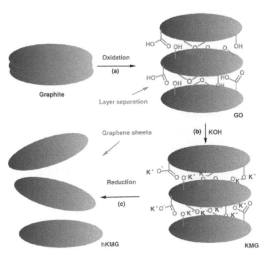

FIGURE 1.5   Schematic presentation of reduction of graphene oxide to graphene in KOH medium via hydrazine monohydrate [47].

1.2.3 SYNTHESIS OF POLYANILINE

To synthesize conducting polymers, there are a number of methods including electrochemical oxidation of the monomers, chemical synthesis and some less common one such as enzyme-catalyzed and photochemically initiated polymerization [22]. PANI is usually synthesized by chemical or electrochemical polymerization in the aqueous acid media. The synthesized polymer is called an emeraldine salt. For bulk production, the chemical oxidation of aniline is the more feasible method. The limitation of this method is the poor processability of the obtained polymer due to its insolubility in common solvents, although it can be improved by using different dopants [25, 27]. There are some descriptions about these two methods as follows:

(a) Electrochemical Oxidation Synthesis
In this case, PANI is synthesized in acidic media by constant potential and current and on an inert metallic electrode, for example, Pt or conducting Indium Tin Oxide (ITO) glass.

(b) Chemical Synthesis
In this method, monomer (aniline) is synthesized in aqueous solution containing oxidant, for example, ammonium peroxydisulfate and acid, for example, hydrochloric. In this type of synthesis, the monomer is converted directly to conjugated polymer by a condensation process. However, an excess of the oxidant lead to materials that are essentially intractable is one of its disadvantages. By progressing the oxidative condensation of aniline, the color of solution turns to black which probably is due to the soluble oligomers. The nature of the medium and the concentration of the oxidant are the effective parameters on the intensity of coloration. The major effective parameters on the course of the reaction and on the nature of the final product are as follows: nature and temperature of medium, concentration of the oxidant, and duration of the reaction. To obtain desirable results, some factors as low ionic strength, volatility, and non-corrosive nature of the medium should be controlled although no medium satisfies all of these requirements [27].

1.2.3.1 Polymerization Mechanism

There is a close similarity in the electrochemically or chemically polymerization mechanism of aniline. The following mechanism proceeds in both cases:

The first step is the formation of the aniline radical cation which has several resonant forms and is formed by transferring the electron from the 2s energy level of the nitrogen atom.

Among the different resonance forms shown in Figure 1.6, form (c) is the more reactive one due to its important substituent inductive effect and its absence of steric hindrance. The next step is the dimer formation between the radical cation and its resonant form (most probably form (c)) by the so-called "head-to-tail" reaction in acidic medium.

Then, as shown in Figure 1.7, a new radical cation dimer is formed by oxidizing the dimer. This radical has two possible reactions, either

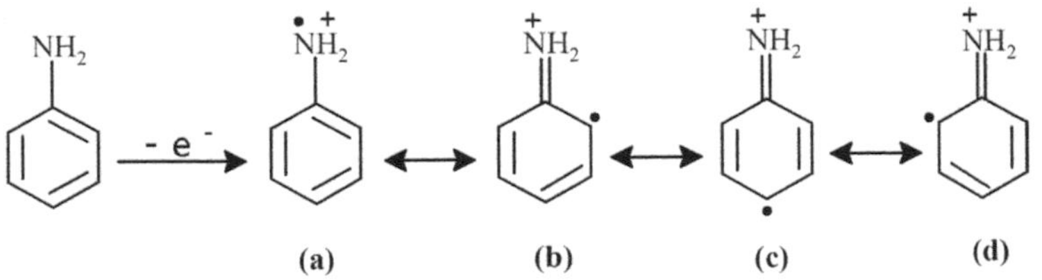

(a)   (b)   (c)   (d)

FIGURE 1.6   The formation of the aniline radical cation and its different resonant structures [27].

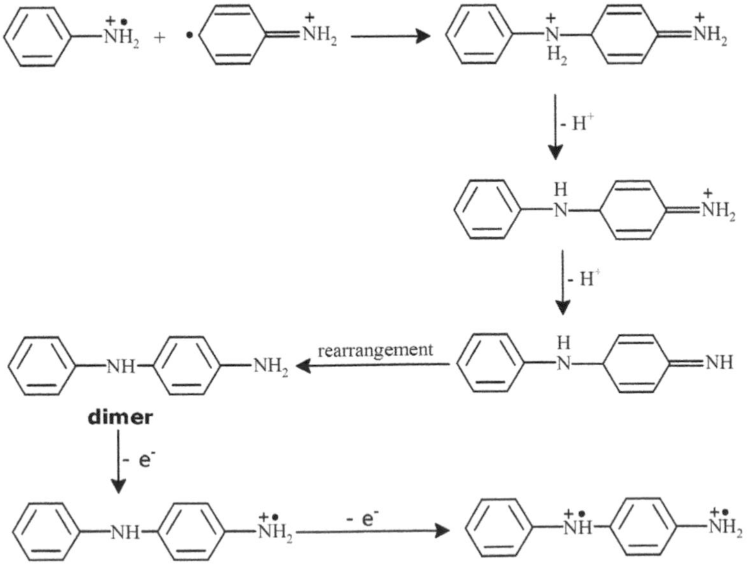

FIGURE 1.7   Formation of the dimer and its corresponding radical cation [27].

with the radical cation monomer or with the radical cation dimer to form, respectively, a trimer or a tetramer. By continuation of above steps, the PANI polymer is formed (Figure 1.8).

In addition to the synthesis reaction of p-coupled PANI, there are some side reactions including (Figure 1.9):

- coupling of aniline and its oligomers in "ortho" position;
- formation of benzidine groups ("tail-to-tail" coupling);

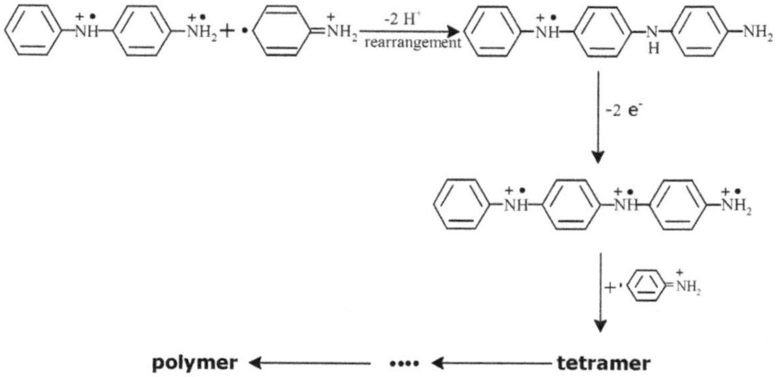

FIGURE 1.8   One possible way of PANI polymer formation [27].

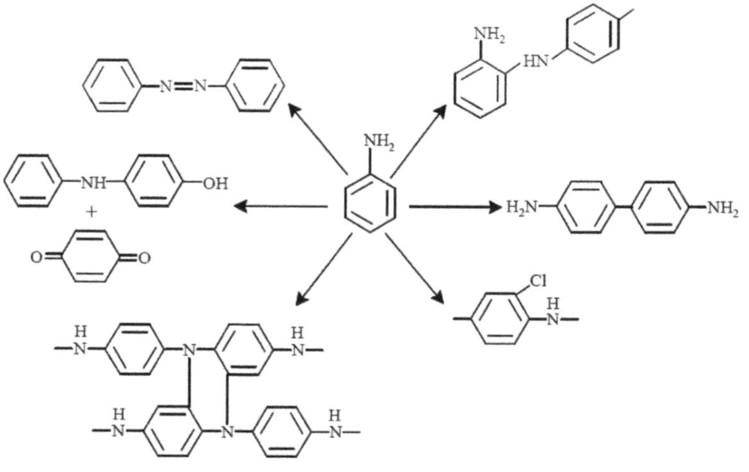

FIGURE 1.9   Side reaction occurring during PANI synthesis [27].

- chlorine substitution in aromatic ring (in systems with HCl and LiCl or NaCl);
- formation of N=N bonds (azo groups);[2] formation of N-CAr grafting bridges between chains;
- polymer hydrolysis (=O and -OH groups).

All these side reactions considered as chain defects introduce undesirable elements to the structure of PANI [27].

1.2.3.2 Protonic Acid Doping of Polyaniline

It is known that only emeraldine state of PANI can be used for non-redox doping process to produce conductive polymer. Charge transfer is a kind of a doping process where the number of electrons of the polymer remains unchanged. Angelopoulos et al. [27] for the first time converted emeraldine base form of PANI to highly conducting metallic regime by this doping method. They did this doping process by treating emeraldine base with aqueous protonic acids as shown in Figure 1.10. It is known that the conductivity of PANI which doped with this method is ~9 to 10 orders of magnitude greater than that of non-doped polymer.

Earlier studies indicate that the doped polymer is a stable polysemiquinone radical cation as shown in Figure 1.11 [27].

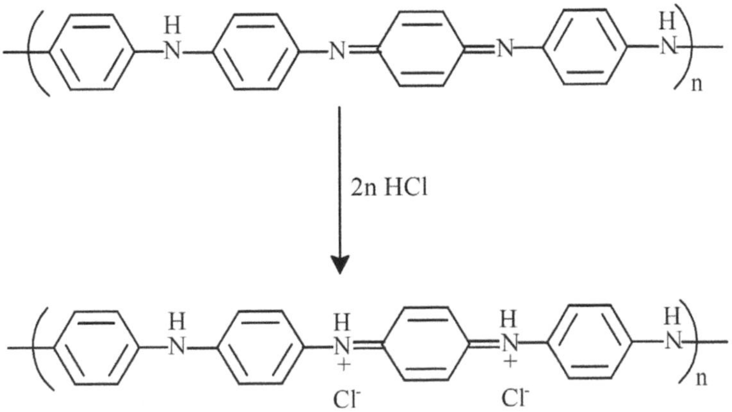

FIGURE 1.10 Protonic acid doping of PANIs [27].

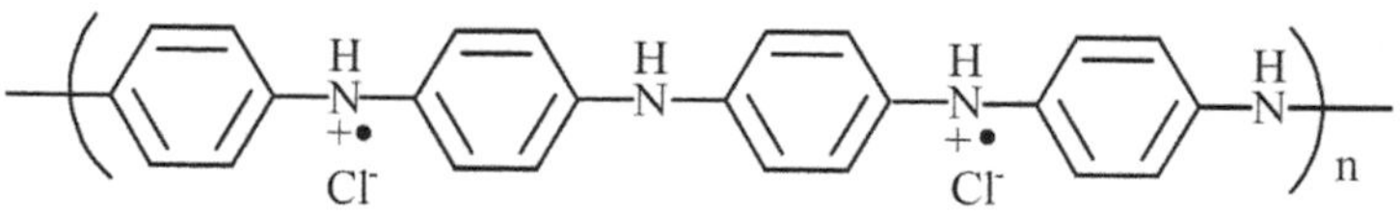

FIGURE 1.11 A stable polysemiquinone radical cation [27].

Here we synthesized PANI (emeraldine salt, ES) by chemical oxidation via two methods: conventional emulsion polymerization to produce binary-doped PANI and homogeneous solution polymerization to produce single-doped PANI.

In conventional emulsion system 5.768 g SDS was dispersed in 40 mL HCl (1 M) in a two necked round bottom flask, then 0.745 g aniline in 10 mL HCl (1 M) was introduced to the mixture with vigorous stirring at room temperature under nitrogen atmosphere to obtain a uniform solution. After 30 min, 10 mL HCl (1 M) including 0.923 mL APS as an oxidant were added drop-wise into 100 mL of reaction mixtures during 20–30 min. After the induction period of about 20–40 min (including time of APS addition), the homogeneous transparent reaction mixtures were turned into bluish tint and the coloration was pronounced as polymerization proceeded. The stirring was continued for 6 h and then the reaction was allowed to proceed without agitation for 24 h at room temperature. Finally dark green colored PANI dispersions were obtained without any precipitation. In our experiments, the molar ratios of APS to aniline and SDS to aniline were kept 0.5 and 2.5, respectively. Excess amount of methanol was added into the SDS-HCl binary doped PANI dispersion to precipitate PANI powder by breaking the hydrophilic–lipophilic balance of the system and to stop the reaction. After that, the solution was centrifuged for 20 min at, 8000 rpm. The precipitation was washed sometimes with methanol, acetone and water to remove unreacted chemicals, aniline oligomers and SDS. The obtained binary-doped emeraldine salt PANI cakes were dried in a vacuum oven at 50°C for 48 h. Figure 1.12 schematically shows what happens during oxidation of aniline monomer with APS in acidic media.

With the same molar ratio of oxidant to monomer, a homogeneous solution polymerization was also carried out for comparison. The solution polymerization was stirred for 24 h at room temperature under nitrogen atmosphere, and the obtained PANI was also centrifuged with the same procedure described above. The obtained single-doped emeraldine salt

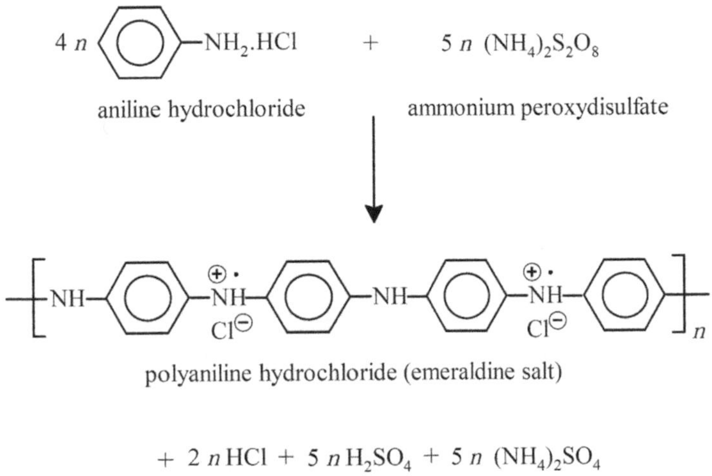

$$4n \ \text{<benzene>}-NH_2 \cdot HCl \quad + \quad 5n \ (NH_4)_2S_2O_8$$

aniline hydrochloride   ammonium peroxydisulfate

$$\left[-NH-\text{<benzene>}-\overset{\oplus \cdot}{NH}-\text{<benzene>}-NH-\text{<benzene>}-\overset{\oplus \cdot}{NH}-\text{<benzene>} \right]_n$$
$$Cl^{\ominus} \qquad\qquad\qquad\qquad Cl^{\ominus}$$

polyaniline hydrochloride (emeraldine salt)

$$+ \ 2n \, HCl + 5n \, H_2SO_4 + 5n \ (NH_4)_2SO_4$$

FIGURE 1.12   Oxidation of aniline hydrochloride with APS [27].

PANI was dried in a vacuum oven at 50°C for 48 h. Emeraldine base (EB) PANI also was prepared as a control by suspending prepared PANI-ES with constant stirring in 100 mL of NH$_4$OH (24%) solution in order to convert the PANI hydrochloride (emeraldine salt) to PANI (emeraldine base) as shown in Figure 1.13.

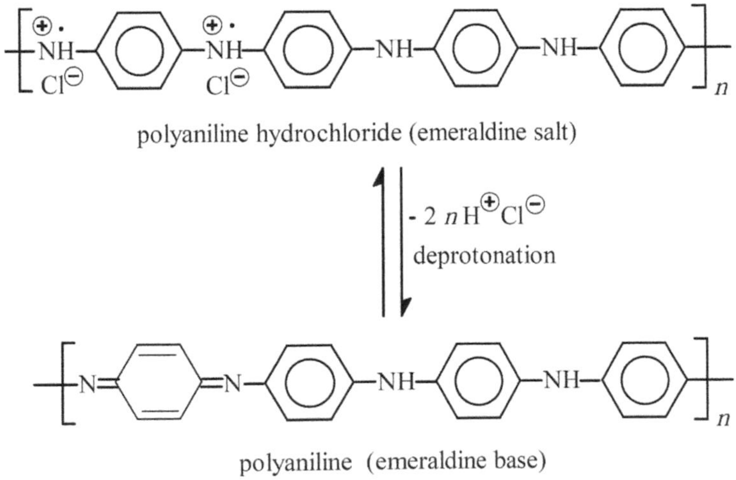

$$\left[\overset{\oplus \cdot}{NH}-\text{<benzene>}-\overset{\oplus \cdot}{NH}-\text{<benzene>}-NH-\text{<benzene>}-NH-\text{<benzene>} \right]_n$$
$$Cl^{\ominus} \qquad\quad Cl^{\ominus}$$

polyaniline hydrochloride (emeraldine salt)

$$- \ 2n \, H^{\oplus} Cl^{\ominus}$$
deprotonation

$$\left[N=\text{<benzene>}=N-\text{<benzene>}-NH-\text{<benzene>}-NH-\text{<benzene>} \right]_n$$

polyaniline (emeraldine base)

FIGURE 1.13   PANI emeraldine salt is deprotonated in the alkaline medium to PANI emeraldine base [27].

1.2.4 SYNTHESIS OF POLYANILINE/GRAPHENE NANOCOMPOSITES

The PANI/graphene nanocomposites were prepared with similar method described for aniline with the difference that prescribed amounts of graphene were dispersed in 1 M HCl solution of aniline monomer with sonication and the monomer-graphene dispersion were used for both emulsion and solution polymerization. Figure 1.14, shows schematically preparation of PANI/graphene nanocomposite. The graphene amounts were 0, 0.1, 0.2, 0.3, 0.4, 0.5, 0.7, and 1 wt.% according to the monomer net weight.

It should be mentioned that, here 1 M concentration was selected as an optimum concentration of HCl for doping according to the Thanpitcha et al. report [31]. Briefly, the enhancement of the electrical conductivity with increasing HCl concentration (from 0.1 To 1 M) is due to the increasing degree of protonation of the imine group of PANI. At higher HCl concentrations (2–6 M HCl), a decrease in electrical conductivity occurs. This result is probably due to the over protonation of PANI chains causing a decrease in the delocalization length of PANI.

1.2.5 SYNTHESIS OF POROUS CONDUCTIVE SCAFFOLDS

To determine the optimum loading of graphene for preparation porous conductive scaffolds, electrical conductivity of synthesized PANI (emulsion and solution polymerization samples) containing various amount of graphene was measured. After determining the percolation threshold of

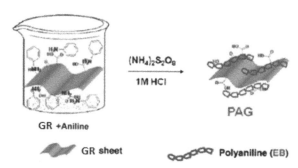

FIGURE 1.14   Schematic process of preparing PANI/graphene nanocomposites [47].

each samples (see Section 1.4.1), the emulsion synthesized PANI with 0.5 wt.% graphene was chosen.

In order to prepare composite scaffolds, chitosan was dissolved in aqueous acetic acid (2% v/v) for about 6 h in the beaker under stirring. Then gelatin was introduced into the system while beaker was heated in 40°C. After 3 h, clear solution of chitosan-gelatin was obtained. In these experiments 1:1 mass ratio of chitosan to gelatin proportion was selected to reach the optimum mechanical properties [48]. PAG powder was dispersed in aqueous acetic acid (2% v/v) and sonicated for half an hour and then specific volume of the suspension was added to the chitosan-gelatin solution. Stirring was continued for 3 h to obtain homogenous dark green mixture. The mixture was cast into a plastic dish and introduced into a vacuum oven to evaporate at a controlled rate at ambient temperature. So, a gelatinous membrane obtained right at the end of this process. Crosslinking of gelatinous membrane was done by dipping in a 5 wt.% acidic solution of sodium tripolyphosphate with pH 3 (*see* Section 1.4.3). The crosslinked membrane was then immersed into 4.0 wt.% NaOH aqueous solution for 1h to completely neutralize remained acetic acid. After this, the solid-like membrane was exhaustively washed with deionized water until neutrality was achieved and frozen at −40°C for 24 h and then lyophilized at −40°C by freeze dryer (Lyotrap-Plus). In all prepared scaffolds, the amount of both chitosan and gelatin was fixed at 2 wt.% and PAG content was 0, 2.5, 5, 7.5 and 10 wt.% of total weight of chitosan and gelatin. They were designated as C2G2, C2G2_PAG2.5, C2G2_PAG5, C2G2_PAG7.5, and C2G2_PAG10, respectively.

Non-porous chitosan-based membrane with different PAG contents were also prepared via the similar method, and used as controls. The only difference is that the samples were dried at ambient temperature after neutralizing by NaOH and washing by deionized water.

1.3 CHARACTERIZATION

1.3.1 *FOURIER-TRANSFORM INFRARED SPECTROSCOPY (FTIR)*

The FTIR spectra were recorded using a Shimadzu FTIR spectrometer. All spectra were the average of 64scans at a resolution of 4 cm^{-1}, from 500 cm^{-1} to, 4000 cm^{-1} which were done at 25°C.

1.3.2   WIDE-ANGLE X-RAY DIFFRACTION (XRD)

The XRD patterns were recorded using a X'pert PRO MRD, Philips, Netherlands) using CuKa radiation ($\lambda = 1.5406$ Å) at a generator voltage of 40 kV and generator current of 40 mA. Scanning was in 0.02° at a rate of 1°/s. The interlayer spacing (d_{001}) of graphene oxide was calculated in accordance with Bragg equation: $2d \sin \theta = \lambda$.

1.3.3   TRANSMISSION ELECTRON MICROSCOPIC (TEM)

The TEM images of graphene sheets were taken by Philips CM200 with an accelerating voltage of 200 kV. The samples were redispersed in deionized water and sonicated for 5 min with an ultrasonic bath cleaner, then a droplet of graphene dispersion was cast onto a TEM copper grid and the solvent was evaporated overnight at room temperature.

1.3.4   ELECTRICAL CONDUCTIVITY

The conductivity measurements were performed using a standard four-point probe method at a constant current of 0.5 mA and ambient temperature. About 0.05 g of dried PANI salt (and nanocomposites) powders was compressed into a disk pellet of 13 mm in diameter with a hydraulic pressure at, 3000 psi to measure the conductivity. Each measurement was repeated three times and the average values reported as the result.

1.3.5   UV-VISIBLE

UV-Visible absorption spectra of samples dissolved in N-methylpyrrolidone were recorded with a UV-Vis spectrophotometer UV2450 from Shimadzu Corporation.

1.3.6   SCANNING ELECTRON MICROSCOPY (SEM)

For SEM analysis, samples were redispersed into acetone and sonicated for 5 min. Then it was dropped on the microscopic Lam surface with micro-needle and dried at ambient temperature. All samples sputter coated with gold layers and images were taken with scanning electron micro-scope TESCAN, VEGA series, 2007.

1.3.7 THERMOGRAVIMETRIC ANALYSIS (TGA)

TGA was done with a PL-1500 thermoanalyzer. The temperature range studied was 30–900°C at a heating rate of 10°C/min under a nitrogen atmosphere.

1.3.8 DIFFERENTIAL SCANNING CALORIMETRY (DSC) ANALYSIS

DSC analysis was performed using a Perkin–Elmer, Pyris 1DSC model differential scanning calorimeter under nitrogen atmosphere. For DSC experiments, each sample was heated from 30 to 250°C at 10°C/min.

1.3.9 SWELLING BEHAVIOR

The porous and non-porous scaffold was weighed initially (W_d) and then immersed into the 250 mL phosphate buffer (pH 7.4) at 37°C for adequate time. The swollen weight (Ws) was also obtained by gently removing the surface water with blotting paper. Swelling index (SI) was then calculated using the following formula:

$$SI(\%) = \frac{W_s - W_d}{W_d} \times 100 \qquad (1)$$

The equilibrium water content (EWC) was also calculated by using the formula:

$$EWC(\%) = \frac{W_{es} - W_d}{W_d} \times 100 \qquad (2)$$

where, W_d is the weight of dried film, and W_{es} is the weight of swollen films (soaked for 48h in phosphate buffer, pH 7.4) after removing the surface water with blotting paper.

1.3.10 THE PORE VOLUME

A flotation method was employed to determine the density of the dry scaffold [49]. The scaffold was dried in a vacuum until a constant weight was reached prior to the density measurement. Carbon tetrachloride (with the density of 1.582 g.cm^{-3}) and heptanes (with the density of 0.683 g.cm^{-3}) were used as the solvents. Replicate measurements were made and the density of the scaffold (ρ) was taken as an average with an uncertainty of $\pm$ 0.002 g.cm^{-3}.

The pore volume (V_p) was studied by monitoring the weight gain of the scaffold immersed in a given solvent as reported method by Wan and et al. [50]. The dry scaffold was placed into a flask which was already fully filled with cyclohexane (with the density of 0.778 g/cm^{-3}) for 48 h at 23°C. The volume of cyclohexane absorbed by the scaffold was used to estimate the porosity. The porosity of the scaffold (φ) was represented via the following formula:

$$\varphi(\%) = \frac{V_P}{V_0 + V_P} \times 100\% \tag{3}$$

where, V_0 is the volume of dry scaffold and was calculated by the following formula:

$$V_0 = \frac{W_0}{\rho} \tag{4}$$

where, W_0 and ρ are the weight and density of the dry scaffold, respectively.

1.3.11 MECHANICAL PROPERTIES MEASUREMENT

The mechanical properties of dry crosslinked scaffolds were measured using an INSTRON universal testing machine (model, 4206) at 25°C with relative humidity of 50%. A strain rate of 2 mm.min^{-1} and gauge length of 20 mm was employed. The scaffolds were rectangular (60×20 mm) and contained different amount of PAG. The thickness of each strip was measured using a common micrometer. The Young's modulus, ultimate tensile strength and elongation at break of the scaffolds were directly obtained from tensile tests. Replicate measurements were made and the average values were quoted as the results. The hydrated samples were prepared by immersing the dry samples in deionized water for a period of 30 min prior to testing and cut into specimens with the same dimensions (60×20 mm²). The thickness of each specimen was also measured with common micrometer.

1.4 RESULTS AND DISCUSSION

1.4.1 CHARACTERIZATION OF SYNTHESIZED GRAPHENE NANOSHEETS

Figure 1.15 shows FTIR spectra of graphite, graphene oxide and chemically reduced graphene. It is clearly seen that no significant peak is detectable in graphite while, the presence of different type of oxygen functionalities

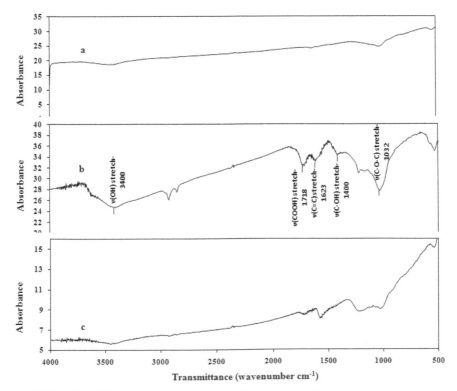

FIGURE 1.15   FTIR spectra of (a) graphite, (b) graphene oxide, and (c) graphene.

in graphene oxide is confirmed at, 3400 cm^{-1} (O-H stretching vibrations), at, 1718 cm^{-1} (stretching vibrations from C=O), at, 1623 cm^{-1} (skeletal vibrations from unoxidized graphitic domains), at, 1220 cm^{-1} (C-OH stretching vibrations), and at, 1032 cm^{-1} (C-O stretching vibrations) [51]. For chemically reduced graphene FTIR peaks exhibits that O-H stretching vibrations observed at, 3400 cm^{-1} was significantly reduced due to deoxygenation. However, stretching vibrations from C=O at, 1718 cm^{-1} and C-O at, 1032 cm^{-1} were still observed, which were caused by remaining carboxyl groups even after hydrazine reduction [52].

The XRD patterns of the graphite, graphene oxide, and chemically reduced graphene are illustrated in Figure 1.16. As can be seen, there is a strong and sharp peak at $2\theta = 6.8°$ corresponds to the interlayer spacing of 0.83 nm in the XRD pattern of graphene oxide which is within the range of values that had been previously reported [53]. Also, chemically reduced graphene shows a broad peak at $2\theta = 23\sim26°$ which has good agreement with others reports [54, 55].

High transparency of graphene to the electron beam in comparison to the thin amorphous carbon has been investigated using TEM image. As can be seen in Figure 1.17a the appearance of stable and transparent graphene sheets in the TEM image indicates the presence of single layer graphene in other words; graphene was fully exfoliated into individual sheets by ultrasonic treatment. Figure 1.17b shows TEM image of graphene with different number of layers. Transparent location in this figure indicates monolayer graphene, while dark location in the edge of the graphene that folded back shows multi layer graphene.

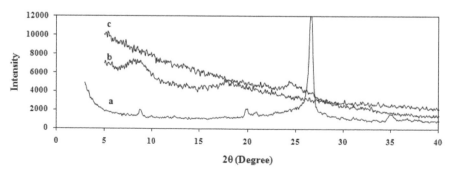

FIGURE 1.16   X-ray diffractograms of (a) graphite, (b) graphene oxide, and (c) graphene.

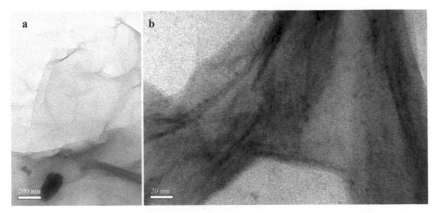

FIGURE 1.17   TEM images of prepared graphene sheet.

1.4.2   ELECTRICAL CONDUCTIVITY AND UV-VIS STUDIES

The room temperature DC conductivity of doped polyaniline powder synthesized via emulsion and solution polymerization with different amount of graphene is shown in Figures 1.18a and 1.18b. It shows a sudden increase in conductivity (in both case of emulsion and solution method), in a relatively narrow concentration range around the so-called percolation threshold. The percolation threshold calculated by fitting the data was at low loading of graphene ($\sim$ 0.4 wt.%). This was attributed to the high aspect ratio, large specific surface area and homogeneous dispersion of the graphene nanosheets in the PANI matrix. It was also seen that the electrical conductivity of binary-doped synthesized samples was much higher than single-doped synthesized one. According to this result, 0.5 wt.% loading of graphene and the emulsion polymerization method were chosen as the optimum conditions of synthesizing PANI/graphene nanocomposite and we designated it as PAG.

Figure 1.19 presents the UV-Vis spectra of single and binary-doped PANI and also PAG. In single-doped PANI spectrum, three peaks at $\lambda_1 = 380$ nm, $\lambda_2 = 425$ nm and $\lambda_3 = 800$ nm, are observed. These three peaks are characteristic of doped PANI (emeraldine salt) [56] where λ_1 is due to the electronic transition π to π^* band, λ_2 is the electronic transition of the polaronic band to π^* band in the benzenoid ring and λ_3 corresponds to the electronic transitions of the π band to the polaronic band. The first two

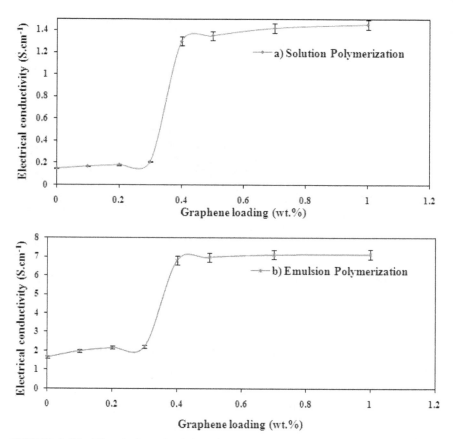

FIGURE 1.18   Electrical conductivity of PANI/graphene nanocomposites: (a) solution polymerization, and (b) emulsion polymerization method (each data is the average of 3 times measuring).

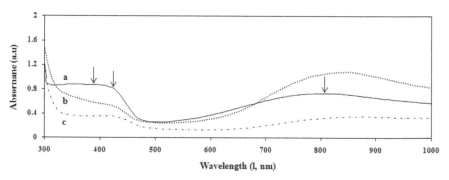

FIGURE 1.19   UV-Vis spectra of (a) single and (b) binary doped PANI, and (c) PAG.

bands are often combined into a flat or distorted single peak with a local maximum between 360 and 420 nm. The binary-doped PANI spectrum shows similar peak than single-doped one, indicating that the doped stated in binary-doped PANI is kept. Nevertheless, the position of peak in λ_3 presents a slight shift at larger wavelength compared to the λ_3 position of the single-doped PANI. Here, the shift of λ_3 peak was found from 800 nm to 830 nm and this was reflected in the conductivity of the binary-doped PANI which also confirmed via conductivity measurements. Similar behavior of λ_3 was found in nanostructures of PANI doped with four different inorganic acids [57]. In the spectrum of PAG all three absorption bands are observed, however, presence of graphene nanosheets appeared to merged the $\pi-\pi^*$ transition of the benzenoid rings.

1.4.3   FTIR SPECTRA OF SYNTHESIZED POLYANILINE AND POLYANILINE/GRAPHENE NANOCOMPOSITES

FTIR spectra of PANI-EB, single and binary doped PANI-ES and PAG are illustrated in Figure 1.20. PANI-EB powder (Figure 1.20a) shows the following five major vibrational bands of, 1597, 1500, 1300, 1145, and 820 cm^{-1}. They are in excellent agreement with previously published values [47, 58, 59]. In addition, it is important to note that there are two bands associated with N-H stretching vibrations- a major broad band at ~3385 cm^{-1} and a minor sharp band at ~3394 cm^{-1}. The bands close to 820 cm^{-1} are characteristics of the p-substituted chains. The band close to, 1145 cm^{-1} is described as being characteristic of the conducting polymer due to the delocalization of electrical charges caused by deprotonation and it can be attributed to the band characteristics of B-NH-Q or B-NH-B, where B refers to the benzenic-type rings and Q to the quinonic-type rings [60]. In the region close to, 1300 cm^{-1}, the peaks are attributed to the presence of aromatic amines. The bands around, 1500–1600 cm^{-1} are related to the stretching of the C-N bonds of the benzenics and quinonics rings, respectively. The intensity of these bands gives an idea of the oxidation state of polyaniline when they present similar intensities; the polyaniline is in the emeraldine form. The appearance of wider band instead of a peak is due to the presence of a high concentration of the mentioned groups in the sample.

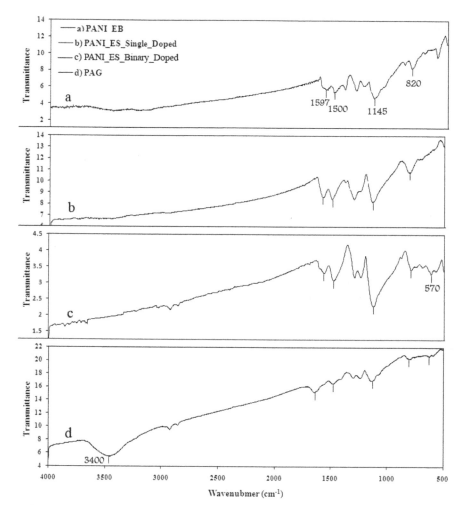

FIGURE 1.20 FTIR spectra of (a) PANI-EB, (b) single and (c) binary doped PANI-ES, and (d) PAG.

The characteristic bands of the doped PANI-ES in single and binary-doped PANI are observed in Figures 1.20b and 1.20c, respectively. The position of the bands of the binary-doped PANI's remains unchanged into the single-doped PANI. Both spectra exhibit the clear presence of benzoid and quinoid ring vibrations at, 1500 cm^{-1} and, 1600 cm^{-1}, respectively, thereby indicating the oxidation state of emeraldine salt polyaniline. The very weak and broad band near, 3400 cm^{-1} is assigned to the N-H stretching mode. The strong band at, 1150 cm^{-1} was described by Mc. Diarmid et al. [59] as the "electronic-like band" and is considered to be a criterion for the degree of delocalization of electrons and thus it is characteristic peak of PANI conductivity. The peak of SO^{-3} group from SDS dopant agents appears at 570 cm^{-1} in binary doped sample as reported before [61] and it is in agreement with notable increase of conductivity of these binary-doped PANI's compared with single-doped PANI.

The FTIR spectrum of PAG nanocomposite (Figure 1.20d) illustrates several clear differences from the spectrum of the neat PANI-ES (binary doped, Figure 1.20c). The nanocomposite spectrum exhibits an inverse, 1600/1500 cm^{-1} intensity ratio compared to that of the emeraldine salt without graphene nanosheets. These data reveal that the PANI in the nanocomposite is richer in quinoid units than the pure PANI-ES. This fact may suggest that graphene-PANI interactions promote and/or stabilize the quinoid ring structure. The π-bonded surface of the carbon nanosheets might interact strongly with the conjugated structure of polyaniline, especially through the quinoid ring. Aromatic structures, in general, are known to interact strongly with the basal plane of graphitic surfaces via π-stacking [62].

A striking difference between the two spectra in Figure 1.20 (c and d) is found in the N-H stretching region near, 3400 cm^{-1}. This signal is broad and strong in the nanocomposite samples while it was very weak in the pure ES polyaniline spectrum. Although it is not clear where the source of the N-H peak intensity difference is, the interaction between polyaniline and graphene may result in "charge transfer" as suggested previously [63]. In this case, the carbon nanosheets sp^2 carbons compete with the chloride ion and thus perturb the H-bonding environment and increase the N-H stretch intensity. The intensity of other peaks in the nanocomposite spectrum is also increased and shifted relative to the peaks of the pure emeraldine salt (Figure 1.20c). For example, the intensity of the signal at, 1145 cm^{-1} increased and shifted to, 1128 cm^{-1}. This dramatic increase of the designated "electronic-like absorption" peak defined as (-N=quinoid=N-) agrees well with the increased conductivity measurements. It appears that the interaction between PANI and graphene increases the effective degree of electron delocalization, and thus enhances the conductivity of the polymer chains. The strong interaction may result in carbon nanosheets functioning as a chemical dopant for PANI conductivity.

1.4.3   MORPHOLOGICAL STUDIES

It is known that surfactant concentration and any additives which added into the solution are affected on the polymer shape and size in emulsion polymerization. Generally, as surfactant concentration increases, micellar shape changes from sphere to cylinder, hexagonal and lamellar structure successively. For SDS, the CMC (critical micelle concentration) was reported to be 8.0 mM concentration in aqueous solution without any additives at 20°C [64]. Micellar size and aggregation number at CMC are approximately 6 nm (in diameter) and 62, respectively, at 20°C [64]. At this concentration, the shape of micelle is assumed to be spherical. In the later, if solubilizate, salt, or any other additives are added into the solution, the system becomes so complex that the micellar shape, size and morphology could not be predicted without experimental investigation. When hydrochloride acid is used it is expected the it influenced of the on the surfactant micelle probably trough adsorption onto the micellar surface and therefore reduce electric repulsion between hydrophilic head groups of the ionic surfactant, thus increase micellar size and aggregation number and also affect intermicellar interaction as depressing electrical double layer.

In our experiments the concentrations of monomer and SDS have been selected 0.09 and 0.15 M, respectively. Figure 1.21 shows SEM

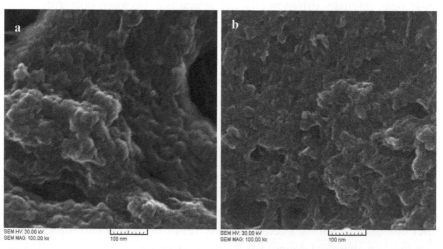

FIGURE 1.21   SEM micrographs of (a) binary doped PANI, and (b) PAG.

micrographs of binary doped PANI and PAG. Although, due to the high concentrations more than CMC and also existence of some additive into to reaction media it is not expected to have exactly spherical morphology for our synthesized PANI, but it can be seen from Figure 1.21a some spherical nanoparticles with the size of 10–20 nm are recognizable. For PAG (Figure 1.21b) the morphology of the polymer somewhat has changed probably duo to incorporation of graphene nanosheets into polymeric matrix, nevertheless no phase separation cannot be seen. On the other hand, although presence of graphene nanosheets into matrix has changed the morphology of polymer, but good dispersion of nanosheets into matrix as well as proper interaction between them has been achieved which is also confirmed with other tests like as TGA.

1.4.4 THERMAL BEHAVIOR OF SYNTHESIZED POLYANILINE

Figure 1.22 presents the TGA thermograms of the PANI-EB, single and binary-doped PANI and PAG. As be seen, single-doped PANI curve shows three main weight loss steps; it is well known that the first (110°C) is due to the volatilization of water molecules and oligomers, as well as unreacted monomer elimination; the second (110–300°C) indicates the mass loss of the protonic acid component of the polymer and the third (>300°C) is attributed to the complete decomposition of the polymer chain and can lead to production of gases such as acetylene and ammonia [62, 65]. Emeraldine-base form showed after water loss almost a straight

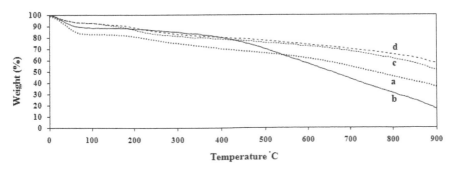

FIGURE 1.22   TGA thermograms of (a) single doped, (b) EB and (c) binary doped PANI, and (d) PAG.

line up to its decomposition onset. Gradual weight loss over the wide temperature in the binary-doped PANI can be attributed to a good thermal stability of the PANI main chain. This improvement also can be seen for PAG probably due to can high thermal stability of graphene nanosheets, good dispersion in polymeric matrix, and proper interaction between graphene and PANI chains.

DSC thermograms of samples are illustrated in Figure 1.23. As can be seen all samples shows two peaks, namely, an endothermic peak at 90–120°C, and an exothermic peak at 150–220°C. The PANI powder had discernible moisture content therefore; the endothermic peaks were most likely due to the vaporization of water. This was in agreement with the TGA results. The water influence on the polymer crystal structure is poorly studied. Freitas [66] suggested that water can even be bound to the PANI lattice and Lubentsov et al. [67] concluded that water influences the crystal structure of PANI and thus changes its conductivity. The chemical process related to the exothermic peak was due to crosslinking reaction. This crosslinking reaction resulted from a coupling of two neighboring -N=Q=N- groups (where Q represents the quinoid ring), to give two -NH-B-NH- groups (where B represents the benzenoid ring) through a link of the N with its neighboring quinoid ring [27]. As can be seen the exothermic peak in EB PANI is stronger than the others probably due to existence more nitrogen atoms which have potential to make link as described while in single and binary doped PANI the number of these nitrogen atoms are less. Surprisingly, this peak becomes again strong for PAG likely in

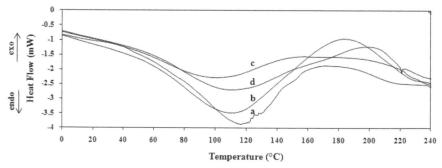

FIGURE 1.23   DSC thermograms of (a) single doped, (b) EB and (c) binary doped PANI, and (d) PAG.

order to presence of graphene nanosheets and good interaction between PANI chains and nanosheets. DSC analysis were not performed successfully in PANI's glass transition temperature (T_g point) detecting due to the usual difficulties of DSC technique, such as the lack of accuracy for determining this temperature.

1.4.5 ELECTRICAL CONDUCTIVITY OF SCAFFOLDS

Although a dry chitosan membrane is nearly non-conducting, it is known as a weak cationic polyelectrolyte and some of its amino groups could be partially protonated when it is hydrated, hence a fully swollen chitosan membrane may show a somewhat semi-conductive property [67, 68]. In addition, incorporation of gelatin into chitosan matrix doesn't change significantly electrical property of chitosan. In the present case, we found that all the types of porous chitosan/gelatin scaffolds which did not contain any PANI particles only showed a conductivity of less than 10^{-7} S.cm^{-1}, even though they were fully hydrated in deionized water. Therefore, to enhance the conductivity of scaffolds, it is necessary to use the appropriate conductive fillers. As stated earlier, the conductivity of a pressed disc of PAG containing 0.5 wt.% graphene nanosheets (PAG) was found to be around 5.50 (±0.31) S.cm^{-1}. Hence, it seems that PAG conductive powder is suitable to enhance electrical property of chitosan/gelatin scaffolds. Figure 1.24 represents the

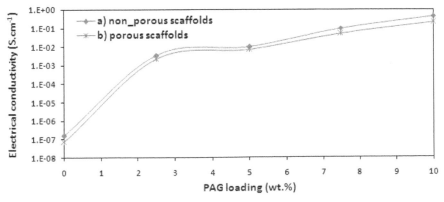

FIGURE 1.24 Electrical conductivity of (a) non-porous and (b) porous scaffolds as a function of PAG loading (each data is the average of 5 times measuring).

dependence of the conductivity of porous and non-porous scaffolds on the PAG content. It is clearly seen that the electrical conductivity of the scaffolds increases significantly with increasing PAG content. This is a direct result of the uniform dispersion of highly conductive PAG into chitosan/gelatin matrix. At the same PAG content, the conductivity of porous scaffold is noticeably lower than that of the non-porous one. These results are understandable if some details are inspected further. In the case of the porous scaffolds, some conductive paths located in the integrated conductive network built by the PAG particles will be inevitably disconnected by numerous pores, leading to a decrease in the conductivity of porous scaffolds.

The choice of conductivity for conductive tissue engineering scaffolds is mainly governed by the possible applications of these scaffolds. In general, a physiological electrical potential should be limited within an order of millivolt (mV). As an example, a 100 mV electrical potential is a common value for physiological stimulation. Meanwhile, an electrical current in the range of 0.6 to 400 mA has been demonstrated to be biologically effective in both in vitro and in vivo [69, 70]. Accordingly, only porous chitosan-based scaffolds with conductivity higher than 1.0×10^{-3} S.cm^{-1} can be considered as candidates. Therefore, chitosan-based porous scaffold with a PAG content of around 2.5 wt.% (Figure 1.24) is an appropriate option for the tissue engineering applications.

1.4.6   SWELLING BEHAVIOR

In order to obtain optimum crosslinking time, non-porous chitosan/gelatin membranes were immersed into solution of TPP (5 wt.%) from 15 to 90 min with 15 min interval. For each period, swelling index (SI) of samples was calculated by using Eq. (3) as a function of time; the results are shown in Figure 1.25. The equilibrium water content (EWC) of each crosslinked sample which was calculated by using Eq. (4) is also shown in Figure 1.26. As it can be seen in Figure 1.26 swelling index of each sample has a growing trend and over the time it becomes constant. It can also be seen that with increasing crosslinking time the SI and the amount of adsorbed water decreases. This behavior confirms that TTP has crosslinked chitosan/gelatin matrix due to formation internal network in polymeric matrix [71] which prevents excess swelling of polymer, it also

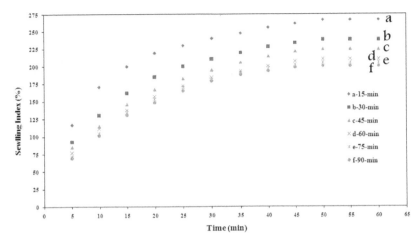

FIGURE 1.25 Swelling index of non-porous chitosan/gelatin membranes which crosslinked for (a) 15, (b) 30, (c) 45, (d) 60, (e) 75, and (f) 90 min.

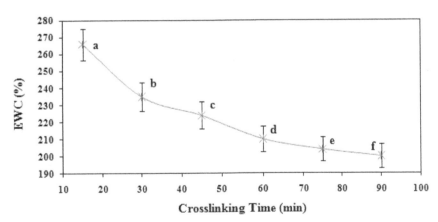

FIGURE 1.26 Equilibrium water content (24 h) of chitosan/gelatin membranes which crosslinked for (a) 15, (b) 30, (c) 45, (d) 60, (e) 75, and (f) 90 min.

shows that increasing crosslinking time causes the formation of more network structure inside polymeric matrix so that the SI decreases with the time. As seen in Figure 1.26, with increasing crosslinking time from 60 to 90 min, the EWC does not change considerably; therefore, the time 60 min was chosen as a suitable crosslinking time for the following study.

The equilibrium water content of porous and non-porous scaffolds as a function of PAG content are shown in Figure 1.27. In both porous and non-porous scaffolds incorporation of PAG to chitosan/gelatin matrix reduces EWC of scaffolds especially, after adding more than 5 wt.% PAG. So, the EWC dropped 100% with introducing 10% PAG to the scaffold. This behavior is probability due to the hydrophobicity nature of synthesized PANI. The amount of EWC for porous scaffolds is slightly higher than non-porous one. This increase in water uptake of the freeze-dried samples is probably the consequence of its more open and porous microstructure. The system operates like a sponge, with water being retained within the polymeric matrix and as free water within the pores [72].

1.4.7 PORE STUDY OF SCAFFOLDS

In this section, the effect of PAG content on pore parameters of prepared scaffolds is investigated. It was found that the water content, evaporation rate, and evaporation time of solvent (acetic acid aqueous solution) significantly affected the pore parameters of the scaffolds. Hence, all scaffolds were remained in the water for 24 h to reach the equilibrium swelling then freeze dried with the same evaporation rate and time, the amount of PAG was the only variable. Figure 1.28 shows SEM

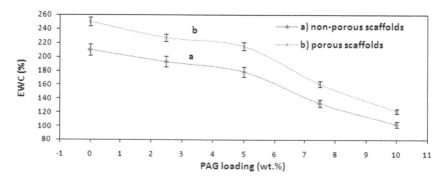

FIGURE 1.27 Equilibrium water content (24 h) of crosslinked (60 min) (a) non-porous and (b) porous scaffolds as a function of PAG loading (each data is the average of 5 times measuring).

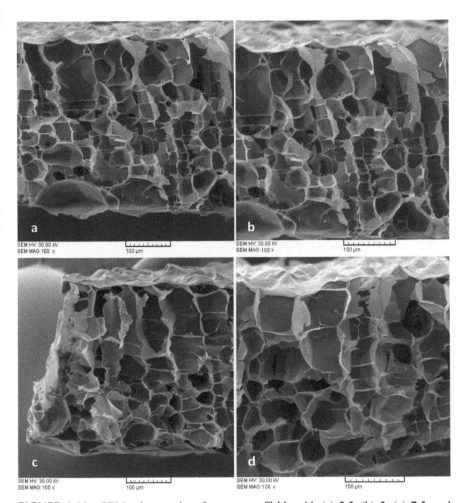

FIGURE 1.28 SEM micrographs of porous scaffolds with (a) 2.5, (b) 5, (c) 7.5, and (d) 10 wt.% PAG.

micrograph of porous scaffolds with different amount of PAG. Some calculated results for the pore parameters of the scaffolds are given in Table 1.1. It can be seen that the pore volume, porosity and pore size of the scaffolds decreases with increasing PAG amount. As seen in swelling behavior section, with increasing PAG, the equilibrium water content decreases due to hydrophobicity nature of prepared PAG. Reduction in the amount of water adsorbed by the scaffolds reduces the pore volume after freeze drying.

TABLE 1.1 Pore Parameters of Porous Scaffolds*

Sample code	PAG (wt.%)	Pore volume (V_p) (ml.g^{-1})	Porosity (φ) (%)	Pore size** (μm)
C2G2-PAG2.5	2.5	2.05	61.48±3.3	10–25
C2G2-PAG5	5	1.89	59.32±2.74	20–40
C2G2-PAG7.5	7.5	1.64	55.21±2.66	45–60
C2G2-PAG10	10	1.11	49.94±2.94	55–70

*The value for V_p and φ are the average value from three time measurements for each sample and the crosslinking time for all scaffolds was kept constant at 60 min.

**The pore size of scaffolds was estimated from their SEM images in the area of cross-section.

1.4.5 MECHANICAL PROPERTIES

The mechanical properties of porous scaffolds are one of the important parameters for many tissue-engineering applications, such as articular cartilage grafts, conduits for peripheral nerve regeneration and tissue transplantation. Thus, in the present study, the mechanical properties of the porous chitosan- based scaffolds in dry and hydrated states are mainly investigated and the relevant results are listed in Table 1.2. It is seen that C2G2 crosslinked dry porous scaffold has more elongation than scaffolds with PAG at break point but its tensile modulus and tensile strength are dramatically lower than those with PAG. C2G2scaffold also shows a well-defined elasticity and ductility, but introducing PAG displays a relatively stiff behavior especially for the scaffolds with more than 5 wt.% PAG. It is probably due to brittle and stiff nature of polyaniline and graphene. The same behavior is seen for the hydrated porous scaffolds, but the effect of PAG on change of mechanical behavior from elastic-ductile to stiff-brittle can be observed on higher content of PAG (more than 7.5 wt.%). It is also evident from Table 1.2 that elongation at break point increases in all hydrated samples while the tensile modulus and tensile strength decreases. It can be attributed to the porous characteristics and swollen property of the scaffolds. Under a tensile strain, the chitosan chains inside the hydrated porous scaffolds, which permit the scaffolds to reach more extensible deformation and give a larger elongation at fracture are easily aligned with the tensile axis. In addition, a higher pore volume and a larger pore size inside the hydrated scaffold also provide a scaffold with a larger

TABLE 1.2 Mechanical Properties of All Prepared Scaffolds

Samples Code	Thickness (µm)	Tensile Strength (MPa)	Tensile Modulus (MPa)	Elongation at Break (%)
	Crosslinked dry porous scaffolds			
C2G2	35±2.04	5.83±0.28	7021±3.52	29.21±1.17
C2G2_PAG2.5	37±2.01	6.69±0.31	80.25±3.96	26.25±1.1
C2G2_PAG5	40±2.02	7.42±0.35	93.21±4.84	22.32±0.92
C2G2_PAG7.5	39±2.1	9.66±0.41	109.75±5.52	18.20±0.69
C2G2_PAG10	44±1.95	11.02±0.49	120.88±6.02	12.65±0.49
	Crosslinked hydrated porous scaffolds			
C2G2	185±8.21	0.71±0.035	5.21±0.18	63.32±2.85
C2G2_PAG2.5	161±7.01	0.84±0.043	6.21±0.22	55.24±2.41
C2G2_PAG5	157±6.54	0.88±0.047	7.12±0.26	46.32±2.12
C2G2_PAG7.5	125±6.85	1.02±0.054	8.23±0.27	32.21±1.47
C2G2_PAG10	121±6.12	1.29±0.061	9.02±0.31	23.47±1.01
	Crosslinked dry non-porous scaffolds			
C2G2	39±1.77	40.81±2.25	1950±117	35.41±1.24
C2G2_PAG2.5	40±1.97	48.25±2.54	2450±141	30.22±1.1
C2G2_PAG5	42±1.62	57.96±3.21	2680±157	27.54±0.95
C2G2_PAG7.5	41±1.74	75.41±4.02	3100±186	20.55±0.72
C2G2_PAG10	47±2.1	91.24±5.12	3450±201	15.21±0.49
	Crosslinked hydrated non-porous scaffolds			
C2G2	81±5.22	7.61±0.22	75.21±3.72	57.41±2.58
C2G2_PAG2.5	75±4.29	8.92±0.25	82.25±4.21	51.95±2.30
C2G2_PAG5	70±5.06	10.59±0.21	99.32±5.01	41.21±1.91
C2G2_PAG7.5	59±4.31	16.75±0.51	114.54±5.7	32.21±1.52
C2G2_PAG10	55±4.84	17.52±0.51	129.35±6.51	27.14±1.21

*The values are the average values for three samples.

deformable space and a much lower chain-tangling density, thus, a corresponding larger breaking elongation for this kind of hydrated scaffolds is expected. Reduction in tensile strength and tensile modulus of the hydrated scaffold is probably due to the disappearance of microcrystalline domains in polymer. It is noticeable in Table 1.2 that all porous scaffolds exhibit a remarkable decrease in their tensile strength and modulus by an amount of around 10-fold in comparison to non-porous scaffolds. Certainly, it is due to the existence of pore and free volume inside porous scaffolds structure which causes a decrease in mechanical properties.

In addition, it was also found that scaffolds containing 7.5 wt.% PAG or more are somewhat brittle and rigid. These results show that the C2G2_PAG7.5 and C2G2_PAG10 are not suitable candidates for tissue engineering applications.

1.5 CONCLUSION

Single and binary doped conductive polyaniline have been prepared with chemical oxidation polymerization via solution and emulsion polymerization methods. Binary doped samples were used to prepare conductive nanocomposites with chemically reduced graphene nanosheets. Spectroscopic and morphological investigations on graphene such as FTIR, TEM, and XRD confirmed that graphene nanosheets successfully prepared via modified Hummer's method followed by chemical reduction from graphite. Conductivity measurements showed that conductive doped PANI was obtained and also incorporation small amount of graphene nanosheets into polymeric matrix increased conductivity up to 7.5 S.cm^{-1}. UV-Vis and FTIR studies also confirmed preparation of conductive samples. The SEM micrographs showed that almost spherical morphology of PANI nanoparticles was synthesized due to presence of SDS emulsifier in reaction environment, on other words, SDS plays as emulsifier component and also binary doped agent. Also, SEM analysis indicated that incorporation of graphene nanosheets into PANI matrix changes spherical shapes of particles nevertheless good dispersion and also proper interaction between PANI and graphene was achieved. This good interaction and also the effect of doping type were investigated with thermal analysis and the results showed that binary doped samples degraded softer than single doped ones. Novel porous –conductive scaffolds were designed and prepared by incorporating highly conductive PANI/graphene particles (PAG) and employing a phase separation technique to build pores inside the scaffolds. Electrical conductivity measurement with FTIR spectra confirmed that PANI-ES was synthesized successfully via emulsion polymerization. It also showed the interaction between PANI and graphene increased the effective degree of electron delocalization, and thus enhanced the conductivity of the polymer chains. The conductivity results of prepared scaffolds showed that a conductivity close to 10^{-3} S.cm^{-1} was achieved for some scaffolds with a low PAG loading of around 2.5 wt.%. Pore and swelling behavior studies confirmed that incorporation of PAG into chitosan-based scaffolds decreased their porosity as well as equilibrium water content probably due to the hydrophobicity nature of prepared polyaniline. Nonetheless, the porosity and EWC of scaffolds remained more than 50% and 100% respectively, even after adding 10% PAG. It was also found that the incorporation of PAG into chitosan/gelatin scaffolds bring brittle properties for them due to the brittle nature of polyaniline and graphene. Nevertheless, scaffolds with less than 5 wt.% PAG were also suitable candidates for tissue engineering applications.

KEYWORDS

- **chitosan**
- **conductivity**
- **gelatin**
- **polyaniline**
- **porous scaffolds**

REFERENCES

1. Khor, E., Lim. Implantable applications of chitin and chitosan. Biomaterials (2003), 24, 2339.
2. Huang, Y., Onyeri, S., Siewe, M., Moshfeghian, A., Sundararajan, V. In vitro characterization of chitosan–gelatin scaffolds for tissue engineering. Biomaterials (2005), 26, 7616–7627.
3. Cheng, M., Deng, J., Yang, F., Gong, Y., Zhao, N., Zhang, X. Study on physical properties and nerve cell affinity of composite films from chitosan and gelatin solutions. Biomaterials (2003), 24, 2871–2880.
4. Kim, S., Nimni, M.E, Yang Zh, Han, B. Chitosan/Gelatin-Based Films Crosslinked by Proanthocyanidin. J Biomed Mater Res Part B: Appl Biomater (2005), 75B, 442–450.

5. Jatariu, A. N., Popa, M., Curteanu, S., Peptu, C. A. Covalent and ionic co-crosslink-ing—An original way to prepare chitosan–gelatin hydrogels for biomedical applications. J Biomed Mater Res Part A (2011), DOI: 10.1002.

6. Chang, Y., Xiao, L., Tang, Q. Preparation and Characterization of a Novel Thermo-sensitive Hydrogel Based on Chitosan and Gelatin Blends. J Appl Polym Sci (2009), 113, 400–407.

7. Mirzaei, E., Ramazani SA A, Shafiee, M., Alemzadeh I and Ebrahimi, H. Modeling and Comparison of Different Simulations for Release of Amoxicillin from Chitosan Hydrogels, Polymer-Plastics Technology and Engineering (2013), 52, 1147–1153.

8. Mirzaei, E., Ramazani, S.A A, Shafiee, M., Danaei, M. Studies on Glutaraldehyde Crosslinked Chitosan Hydrogel Properties for Drug Delivery Systems. International Journal of Polymeric Materials and Polymeric Biomaterials (2013), 62, 605–611.

9. Nagahama, H., Maeda, H., Kashiki, T., Jayakumar, R., Furuike, T., Tamura, H. Preparation and characterization of novel chitosan/gelatin membranes using chitosan hydrogel. Carbohydrate Polymers (2009), 76, 255–260.

10. Armentano, I., Dottori, M., Fortunati, E., Mattioli, S., Kenny, J.M. Biodegradable polymer matrix nanocomposites for tissue engineering: A review. Polymer Degrada-tion and Stability (2010), 95, 2126–2146.

11. Chandy, T., Sharma, C. P. Biomaterials, artificial cells, and artificial organs (1990), 18, 1.

12. Ge, Z. G., Baguenard, S., Lim, L. Y., Wee, A., Khor, E. Hydroxyapatite–chitin mate-rials as potential tissue engineered bone substitutes. Biomaterials (2004), 25, 1049.

13. Muzzarelli, R. A. A. Carboxymethylated chitin and chitosan. Carbohydrate Polymers (1988), 8, 11–21.

14. Muzzarelli, R. A. A., Muzzarelli, C. Native and modified chitins in the biosphere. In: Stankiewicz, B. A., van Bergen, P. F. (eds) Nitrogen-containing macromolecules in the bio- and geosphere. ACS Symposium Series (1998), 707, 148.

15. Peter, M., Ganesh, N., Selvamurugan, N., Nair, S. V, Furuike, T., Tamura, H., Jaya-kumar, R. Preparation and characterization of chitosan–gelatin/nanohydroxyapatite composite scaffolds for tissue engineering applications. Carbohydrate Polymers (2010), 80, 687–694.

16. Young, B. R., Pitt, W. G., Cooper, S. L. Protein adsorption on polymeric biomaterials, I. Adsorption isotherms. Journal of Colloid and Interface Science (1988), 124, 28.

17. Mao, J., Zhao, L., Yao, K. D., Shang, Q., Yang, G., Cao, Y. Study of novel chitosan-gelatin artificial skin in vitro. J Biomed Mater Res 64A (2003), 301–308.

18. Pulieri, E., Chiono, V., Ciardelli, G., Vozzi, G., Ahluwalia, A., Domenici, C., Vozzi, F., Giusti, P. Chitosan/gelatin blends for biomedical applications. Journal of Bio-medical Materials Research Part A (2008), 86, 311.

19. Gonçalves, V. L., Laranjeira, M. C. M., Fávere, V. T. Effect of Crosslinking Agents on Chitosan Microspheres in Controlled Release of Diclofenac Sodium. Ciência e Tecnologia (2005), 15, 6–12.

20. Lin, H. Y., Yeh, C. T. Alginate-crosslinked chitosan scaffolds as pentoxifylline deliv-ery carriers. Journal of Materials Science: Materials in Medicine (2010), 5, 21, 1611.

21. Kathuria, N., Tripathi, A., Kar, K. K., Kumar, A. Synthesis and characterization of elastic and macroporous chitosan–gelatin cryogels for tissue engineering. Acta Bio-materialia (2009), 5, 406–418.

22. Molapo, K. M., Ndangili, P. M., Ajayi, R. F., Mbambisa, G., Mailu, S. M., Njomo, N., Masikini, M., Baker, P., Iwuoha, E. I. Review Paper Electronics of Conjugated Polymers (I): Polyaniline. J. Electrochem. Sci. (2012), 7, 11859–11875.

23. Lia, M., Guob, Y., Weib, Y., MacDiarmidc, A. G., Lelkesa, P. I. Electrospinning polyaniline-contained gelatin nanofibers for tissue engineering applications. Biomaterials(2006), 27, 2705–2715.

24. Bidez, P. R., Li Sh., Macdiarmid, A. G., Venancio, E. C., Yenwei, Lelekes, P. I. Polyaniline, an electroactive polymer, supports adhesion and proliferation of cardiac myoblasts. J. Biomater. Sci. Polymer Edn (2006), 17, 199–212.

25. Sinha, S., Bhadra, S., Khastgir, D. Effect of Dopant Type on the Properties of Polyaniline. J Appl Polym Sci (2009), 112, 3135–3140.

26. Zhao, M., Wu, X., Cai Ch. Polyaniline Nanofibers: Synthesis, Characterization, and Application to Direct Electron Transfer of Glucose Oxidase. J. Phys. Chem. C (2009), 113, 4987–4996.

27. Yilmaz, F. Polyaniline: Synthesis, characterization, solution properties and composites, Doctor of Philosophy Thesis, 2007.

28. Tan, C. K., Blackwood, D. J. Interactions between polyaniline and methanol vapor Sensors and Actuators B: Chemical (2000), 71, 184–191.

29. Karami, H., Mousavi, M. F., Shamsipur, M. A new design for dry polyaniline rechargeable batteries. Journal of Power Sources (2003), 117, 255–259.

30. Sengupta, P. P. S., Barik, S., Adhikari, B. Mater. Manuf. Processes, 2006, 21, 263–270.

31. Thanpitcha, T., Sirivat, A., Jamieson, A. M., Rujiravanit, R. Preparation and characterization of polyaniline/chitosan blend film. Carbohydrate Polymers (2006), 64, 560–568.

32. Tiwari, A. Gum Arabic Graft Polyaniline: An Electrically Active Redox Biomaterial for Sensor Applications. Journal of Macromolecular Science, Part A Pure and Applied Chemistry (2007), 44, 735–745.

33. Yavuz, A. G., Uygun, A., Bhethanabotla, V. R. Substituted polyaniline/chitosan composites: Synthesis and characterization. Carbohydrate Polymers (2009), 75, 448–453.

34. Yavuz, A. G., Uygun, A., Can, H. K. The effect of synthesis media on the properties of substituted polyaniline/chitosan composites. Carbohydrate Research (2011), 346, 2063–2069.

35. Xu, X. H., Ren, G. L., Cheng, J., Liu, Q., Li, D. G., Chen, Q. J. Self-assembly of polyaniline-grafted chitosan/glucose oxidase nanolayered films for electrochemical biosensor applications. Journal of Materials Science (2006), 41, 4974–4977.

36. Ramanathan, T., Abdala, A. A., Stankovich, S., Dikin, D. A., Herrera-Alonso, M., Piner, R. D. Functionalized graphene sheets for polymer nanocomposites. Nature Nanotechnology (2008), 3(6), 327–331.

37. Chen, W. F., Yan, L. F., Bangal, P. R. Preparation of graphene by the rapid and mild thermal reduction of graphene oxide induced by microwaves. Carbon, 2010, 48(4), 1146–1152.

38. McAllister, M. J., Li, J. L., Adamson, D. H., Schniep, H. C., Abdala, A. A., Liu, J. Single sheet functionalized graphene by oxidation and thermal expansion of graphite. Chemistry of Materials (2007), 19(18), 4396–4404.

39. Niyogi, S., Bekyarova, E., Itkis, M. E., McWilliams, J. L., Hamon, M. A., Haddon R C Solution properties of graphite and graphene. Journal of the American Chemical Society (2006), 128(24), 7720–7721.

40. Stankovich, S., Dikin, D. A., Piner, R. D., Kohlhaas, K. A., Kleinhammes, A., Jia, Y. Synthesis of graphene-based nanosheets via chemical reduction of exfoliated graphite oxide. Carbon (2007), 45(7), 1558–1565.

41. Vickery, J. L., Patil, A. J., Mann, S. Fabrication of graphene–polymer nanocomposites with higher-order three-dimensional architectures. Advanced Materials (2009), 21(21), 2180–2184.

42. Stankovich, S., Dikin, D. A., Dommett, G. H. B., Kohlhaas, K. M., Zimney, E. J., Stach E A. Graphene-based composite materials. Nature (2006), 442(7100), 282–286.

43. Wakabayashi, K., Pierre, C., Dikin, D. A., Ruoff, R. S., Ramanathan, T., Brinson L C. Polymer–graphite nanocomposites: Effective dispersion and major property enhancement via solid-state shear pulverization. Macromolecules (2008), 41(6), 1905–1908.

44. Wang, S. R., Tambraparni, M., Qiu, J. J., Tipton, J., Dean, D. Thermal expansion of graphene composites. Macromolecules (2009), 42(14), 5251–5255.

45. Coleman, J. N., Cadek, M., Blake, R., Nicolosi, V., Ryan, K. P., Belton, C. High-performance nanotube-reinforced plastics: Understanding the mechanism of strength increase. Advanced Functional Materials (2004), 14(8), 791–798.

46. Kuilla, T., Bhadra, S., Yao, D., Kim, N. H., Bose, S., Lee, J. H. Recent advances in graphene based polymer composites. Progress in Polymer Science (2010), 35, 1350–1375.

47. Harada, I., Furukawa, Y., Ueda, F. Vibrational spectra and structure of polyaniline and related compounds. Synthetic Metals (1989), 29, 303.

48. Yu, S. H., Wu, S. J., Wu, J. Y., Peng, C. K., Mi, F. L. Tripolyphosphate Crosslinked Macromolecular Composites for the Growth of Shape and Size Controlled Apatites. Molecules (2013), 18, 27–40.

49. Quarshi, M. T., Blair, H. S., Allen, S. J. Studies on modified chitosan membranes. I. Preparation and characterization. Journal of Applied Polymer Science(1992), 46, 255.

50. Wan, Y., Huang, W., Wang, Z., Zhu, X. X. Preparation and characterization of high loading porous crosslinked poly(vinyl alcohol) resins. Polymer (2004), 45, 71.

51. Xu, Y., Bai, H., Lu, G., Li Ch and Shi, G. Flexible Graphene Films via the Filtration of Water-Soluble Noncovalent Functionalized Graphene Sheets. Journal of the American Chemical Society (2008), 130, 5856.

52. Li, D., Müller, M. B., Gilje, S., Kaner RB and Wallace, G. G. Processable aqueous dispersions of graphene nanosheets. Nature Nanotechnology (2008), 3, 101.

53. Bissessur, R., Liu, P. K. Y., White, W., and Scully S F. Encapsulation of polyanilines into graphite oxide. Langmuir (2006), 22(4), 1729–1734.

54. Fan, H., Wang, L., Zhao, K., Li, N., Shi, Z., Ge, Z., and Jin, Z. Fabrication, Mechanical Properties, and Biocompatibility of Graphene-Reinforced Chitosan Composites. Biomacromolecules (2010), 11, 2345–2351.

55. Ju, H. M., Choi SH and Huh, S. H. X-ray Diffraction Patterns of Thermally reduced Graphenes. Journal of the Korean Physical Society (2010), 57, 6, 1649–1652.

56. Huang, J., Virji, S., Weiller BH and Kaner, R. B. Polyaniline Nanofibers: Facile Synthesis and Chemical Sensors Journal of the American Chemical Society (2003), 125, 314.

57. Zhang, Z., Wei Z and Wan, M. Nanostructures of Polyaniline Doped with Inorganic Acids. Macromolecules (2002), 35, 5937.

58. Zheng, W., Angelopoulos, M., Epstein, A. J., MacDiarmid, A. G. Experimental Evidence for Hydrogen Bonding in Polyaniline: Mechanism of Aggregate Formation and Dependency on Oxidation State. Macromolecules (1997), 30, 2953.
59. Quillard, S., Louarn, G., Lafrant, S., MacDiarmid, A. G. Vibrational analysis of polyaniline: A comparative study of leucoemeraldine, emeraldine, and pernigraniline bases. Physical Review B (1994), 50, 12496.
60. Campos, T. L. A., Kersting, D. F., Ferreira, C. A. Chemical synthesis of polyaniline using sulphanilic acid as dopant agent into the reactional medium. Surface and Coatings Technology (1999), 122, 3.
61. Hino, T., Namiki, T., Kuramoto, N. Synthesis and characterization of novel conducting composites of polyaniline prepared in the presence of sodium dodecylsulfonate and several water soluble polymers. Synthetic Metals (2006), 156, 1327.
62. Chen, R. J., Zhang, Y., Wang, D., Dai, H. Noncovalent Sidewall Functionalization of Single-Walled Carbon Nanotubes for Protein Immobilization. Journal of American Chemical Society (2001), 123, 3838.
63. Cochet, M., Maser, W. K., Benito, A. M., Callejas, M. A., Martinez, M. T., Benoit, J. M., Schreiber, J., Chauvet, O. Synthesis of a new polyaniline/nanotube composite: "in-situ" polymerization and charge transfer through site-selective interaction. Chemical Communications (2001), 1450.
64. Van Os, N. M., Haak, J. R., Rupert, L. A. Physico-Chemical Properties of Selected Anionic, Cationic and Nonionic Surfactants, Elsevier, Amsterdam, 1993.
65. Kim, B. J., Oh, S. G., Han MG and Im, S. S. Synthesis and characterization of polyaniline nanoparticles in SDS micellar solutions. Synthetic Metals, 2001, 122, 297.
66. Freitas, P. S. Síntese da polianilina em escala piloto e seu Processamento. PhD Thesis. Universidade Estadual de Campinas, UNICAMP, Sao Paulo, Brazil, 2000. 156p.
67. Lubentsov, B., Timofeeva, O., Saratovskikh, S., Kirnichnyi, V., Pelekh, A., Dmiternko, V. Study of conducting polymer interaction with gaseous substances. IV. The water content influence on polyaniline crystal structure and conductivity. Synthetic Metals (1992), 47(2), 187–192.
68. Phillip, B., Dautzenberg, H. K., Linow, J., Kotz, J. Polyelectrolyte complexes recent developments and open problems. Progress in Polymer Science (1989), 14, 91.
69. Wan, Y., Creber, K. A. M., Peppley, B., Bui, V. T. Ionic conductivity of chitosan membranes. Polymer (2003), 44, 1057.
70. Kerns, J. M., Fakhouri, A. J., Weinri, H. P., Freeman, J. A. Electrical stimulation of nerve regeneration in the rat: The early effects evaluated by a vibrating probe and electron microscopy. Neuroscience (1991), 40, 93.
71. Kow, L. M., Pfaff, D. W. Estrogen effects on neuronal responsiveness to electrical and neurotransmitter stimulation: an in vitro study on the ventromedial nucleus of the hypothalamus. Brain Research (1985), 347, 1.
72. Tiwary, A. K., Rana, V. Crosslinked Chitosan Films: Effect of Crosslinking Density on Swelling Parameters. Pakistan Journal of Pharmaceutical Sciences (2010), 23, 4, 443–448.
73. Mao, J., Kondu, S., Ji, H. F., McShane, M. J. Study of the Near-Neutral pH-Sensitivity of Chitosan/Gelatin Hydrogels by Turbidimetry and Microcantilever Deflection. Biotechnology and Bioengineering (2006), DOI: 10.1002/bit.20755.

CHAPTER 2

MORPHOLOGICAL, THERMAL AND MECHANICAL PROPERTIES OF GREEN COMPOSITE BASED ON RECYCLED POLYETHYLENE/ POLYAMIDE-6/KENAF COMPOSITES

AGUS ARSAD, NURANASSUHADA MAHZAM, and
JIBRIN MOHAMMED DANLAMI

Enhanced Polymer Research Group (EnPRO), Faculty of Chemical Engineering, Universiti Teknologi Malaysia, 81310 UTM Johor Bahru, Johor, Malaysia, E-mail: agus@cheme.utm.my

CONTENTS

ABSTRACT

The objective of this study is to investigate the effect of different kenaf concentration (5–15 phr) on the morphological, thermal and mechanical properties of recycled polyethylene (rPE) and polyamide-6 (PA-6). Kenaf fiber was used to enhance the properties of the rPE/PA-6. NaOH treatment for kenaf and polyethylene grafted maleic anhydride (PE-g-MAH) were used to improve the adhesion between rPE/PA-6 with kenaf. Tensile, flexural and DMA test were conducted to study the mechanical properties of the composites. The mechanical properties such as Young's modulus, flexural modulus, and flexural strength were successfully enhanced by the addition of kenaf. However, some problem occurred which affected and lowered the tensile strength and elongation at break. Thermogravimetric analysis (TGA) and scanning electron microscope (SEM) tests were also performed for its thermal and morphological properties. The kenaf orientation and dispersion in rPE/PA-6 blend was observed to be well dispersed in the matrix and oriented with flow direction.

2.1 INTRODUCTION

Polymer waste has been increasing at city landfills tremendously [1]. Thus the best solution in overcoming the buildup of plastic solid waste is recycling. The demand for recycling has increased further, especially for expensive engineering resins such as polyamide (PA-6), because it possesses a lot of excellent properties. Previous studies have shown that unmodified PA-6s are excellently suitable for utilization in mechanical recycling processes [2].

Polyamide-6 (PA-6) and recycled polyethylene (rPE) were used in this study and polymer blending techniques have been used to improve the performance of polymeric materials by combining the best characteristics of each constituent. In recent years, there has been an increasing interest in incorporating rPE as new materials for packaging in the plastics industry [3]. Recycling mixed waste plastics in the form of blends is an economically desirable way of reusing recycled polymeric compounds with satisfactory cost, or performance, and application potentials [2, 3].

Unfortunately, PA-6 and rPE are incompatible due to the different polarity and different crystalline structure, which are thermodynamically immiscible, leading to incompatible blends with low mechanical properties. Nonetheless, one technical problem associated with plastics waste such as PA-6 and rPE is its heterogeneous composition, immiscibility and poor interfacial adhesion between dispersed phase and matrix, resulting in poor physical-mechanical properties such as low tensile strength and impact toughness [4]. Thus a compatibilizing agent is needed to overcome this problem.

In a recently study conducted for PA-6 and PP blend, PP-g-MAH (MAH = maleic anhydride) has been used as a compatibilizing agent and the best process was a single step [4]. Ha and co-workers [5] stated that the effect of PP-g-MAH on PA-6 and PP polymer blends was to increase the structural stability and morphology by the in-situ reaction of anhydride groups with the amino end groups of PA-6. Therefore, blends of PA-6 and rPE were compatibilized by PE-g-reactive reacted with the amino groups of PA-6.

The blends of rPE with PA-6 will not achieve a desired level of high mechanical strength due to the recycled aspect of PE. Therefore the use of natural fibers was expected to improve this drawback of the blends. Natural renewable fibers with low density, allow high volume of filling in the composite, and are nonabrasive to nature since they are biodegradable [5–8]. However, the high moisture absorption of the natural fibers and their low microbial resistance are great disadvantages that need to be considered. These disadvantages were minimized when encapsulated in the plastics and also by using these composites in applications where such drawbacks are not of prime consideration [6].

In this study, kenaf fiber was used as reinforcement filler for PA-6/rPE blends. The ratio of the blend to the kenaf is the most vital in this study. This is to reveal the most appropriate ratio in having the most tremendous impact on the properties of the blend. In the previous work, it has been reported that tensile and flexural strength as well as the elastic modulus of the kenaf fiber-reinforced polylactic acid (PLA) composites increased linearly[9] with the incorporation of kenaf into the composites. For these reasons, this study is aimed at studying the effects that kenaf content has on two different blends of PA-6 and rPE in terms of their mechanical, thermal and morphological properties.

2.2   EXPERIMENTAL PART

2.2.1   MATERIALS

The rPE was collected from industrial waste with melt flow index (MFI) of 35.6 g/10 min and rPA6 was prepared by re-extruding virgin

PA6 to produce PA6 with recycled properties with MFI of 42.9 g/10 min. Virgin PA-6 (Amilan CM1017) was obtained from Toray Plastics, Malaysia. Both MFI values were obtained and carried out at 230°C with load weight, 2.16 kg. The kenaf were from the most part were planted in Kelantan, Malaysia with bulk density of 0.1986 g per cm^3 after pre-treatment with alkaline treatment. The PE-g-MAH was obtained from Shanghai Jiangqiao Plastics Ltd., China with MFI 1 g/10 min, and sodium hydroxide (NaOH) for alkaline treatment was supplied by Quality Reagent Chemical, Malaysia.

2.2.2   SAMPLE PREPARATION

Kenaf was treated by immersing it into 5 vol. % NaOH solution for 1 h at room temperature to remove some hemicellulose, lignin, wax and oils, thereby increasing surface roughness and reducing its hydrophilic nature. This method of treatment was adopted from the method of Mishra et al. [10] and Ray et al. [11]. The treatment's main purpose is to disrupt hydrogen bonding in the network structure that influence the fiber's mechanical behavior, especially its strength and stiffness [12]. Then fiber was thoroughly washed with running water and it allowed to dry at room temperature for 24 h. Treated fibers have lower average tensile strength than untreated fibers [13]. After which the kenaf was ground to a particle size of 100 µm. The extrusion was conducted at 80 rpm and the barrels temperature was maintained at 165–180°C from hopper to the die, respectively. Prior to extrusion, the pellets were dehumidified using a hopper dryer at 90°C for 24 h. A pelletizer machine was then used to pelletize the extruded material. A 2 kg sample was then made by mixing 80 wt.% of PA-6, 20 wt.% of rPE, and 5 phr (parts of hundreds of parts of recycled PE/PA6) of PE-g-MAH. This was extruded at 30rpm and where the barrel temperature was maintained at 190–220°C from hopper to the die, respectively. After extrusion, the blends were molded with an injection-molding machine. However, because of PA-6 high moisture absorption limitation, the pellets were dehumidified in an oven for 90°C for 24 h before molding in a JSW N100 BII injection molding machine. The barrel temperature of the injection machine was maintained at 200 to 240°C. The selected ratio was extruded with different phr of kenaf which was 5–15 phr. The blends were finally molded with the injection-molding machine to form specimen. Finally, the specimens were then characterized.

2.2.3 MECHANICAL ANALYSIS

ASTM D638, standard test method was used in order to determine the tensile properties such as Young's Modulus, tensile strength and extension at break of the samples. The samples were in the form of dumbbell-shape test specimens type IV. This test was performed at 10 mm/min of crosshead speed for each formulation.

Flexural test covered the determination of flexural properties such as flexural strength and modulus. ASTM D790, three point bending test was performed at the speed 3 mm/min.

2.2.4 DYNAMIC MECHANICAL ANALYSIS (DMA)

Dynamic Mechanical Analysis determines the elastic modulus (storage modulus), viscous modulus (loss modulus) and damping coefficient (Tan δ) as a function of temperature. The test specimens' dimension was 3 mm × 13 mm × 20 mm and was the same for those used in the Izod impact test but without a notch. The test specimens were clamped between the movable and stationary fixtures, and then enclosed in the thermal chamber. The frequency, amplitude, and a temperature range of 25–220°C were set-up for the material. The analyzer applied torsional oscillation to the test sample while slowly moving through the specified temperature range of 25–220°C.

2.2.5 THERMOGRAVIMETRIC ANALYSIS (TGA)

The specimens were scanned at the scanning rate of 20°C/min under nitrogen gas at the range of 25 to 600°C. This method measures the weight loss and onset temperature, T_o. The Perkin Elmer TGA 7 was then used for characterizing thermal transitions at Polymer Laboratory, UTM Skudai.

2.2.6   *MORPHOLOGICAL ANALYSIS*

The morphological properties of the samples were analyzed using fractured specimens of tensile test as the specimens for Scanning Electron Microscope (SEM) tests. The magnification used in this test was 500x. The morphology of the fracture surfaces were observed in a scanning electron microscope (SEM: XL 40 type PHILIPS) at an accelerating voltage of 30kV. The tests were conducted at Material Laboratory, Faculty of Mechanical Engineering, UTM Skudai.

2.3   RESULTS AND DISCUSSION

2.3.1   *TENSILE PROPERTIES ANALYSIS*

Figure 2.1 shows the trend of Young's modulus as the kenaf content is increasing. Young's modulus is a measure of the stiffness of an elastic material and is a quantity used to characterize materials. In addition, it also measures the resistance of a material to elastic deformation under load. The higher the Young's modulus the stiffer the material is and the

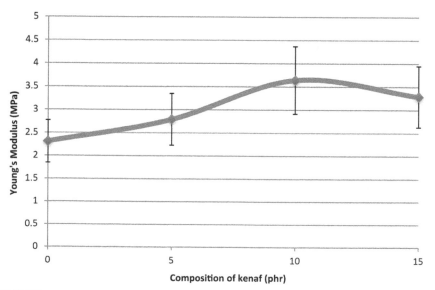

FIGURE 2.1   Young's modulus of rPE/PA-6 versus composition of kenaf.

less elastic it is. Figure 2.1 shows that modulus with 5phr of kenaf was higher in value (2.79 GPa) than without kenaf which was about 2.31 GPa. An increase in the kenaf content increases the Young's modulus significantly. However, at 15 phr kenaf content, Young's modulus for this blend decreases slightly. Based on the standard deviation of the data recorded, it can be concluded that addition of kenaf over 15 phr has no significant effect on the modulus. This is because kenaf concentration higher than this limit does not really improve the compatibility of PA6/rPE blends. Thus, incorporation of kenaf into the composites system achieved a maximum effect at 10phr of kenaf. This trend in Young's modulus is in contrast to the results of a study carried out for rPE/Bagasse natural fiber which showed that rPE modulus increased by about 50 wt.% when 30 wt.% of the bagasses were added [14]. The results were also supported by DMA studies and will be further discussed in DMA subsection.

Basically, tensile strength is the maximum stress that a material can withstand while being stretched before necking occurs (that is when the specimen's cross section starts to significantly contract). The tensile strength of the composites are shown in Figure 2.2. It is obvious that PA-6/rPE composites have a slight decrease in tensile strength with an increase in kenaf content. There was a decrease with the addition of kenaf at 5phr, and at 10phr of kenaf it gave a tensile strength of 141.1 MPa, which was

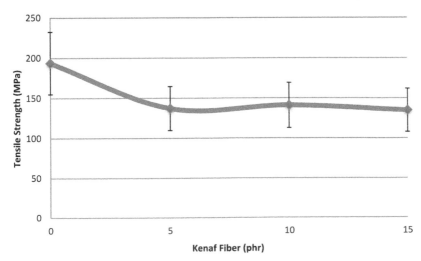

FIGURE 2.2   Tensile strength of rPE/PA-6 versus composition of kenaf.

lower than the sample without kenaf, 193.7 MPa. Kenaf is natural fiber and contains cellulosic material, which is hygroscopic in nature and has excellent moisture absorption, This influences the tensile strength of the composites. Thus, this disability (hydrophilic nature) of the kenaf caused a reduction in its strength and stiffness [14]. Besides that, PA-6 also has high moisture absorption and also provides a similar effect. The absorbed moisture formed voids in the matrix, which led to insufficient filling of matrix and consequently gave it a poor tensile strength [14]. It was concluded that 10 phr of kenaf gave the highest tensile strength as compared to 5 and 15phr. This trend was similar with Young's modulus trends, and it was concluded that when the amount of kenaf exceeds this amount the properties of the PA6/rPE composites do not improve. However, based on the error bars in the plot, the tensile strength was considered to be essentially a constant value with increasing in kenaf contents. Although kenaf is reinforcement for recycled materials, there are also some weaknesses in the compatibility between kenaf and the composite, which was a major problem. The decreased tensile strength may be attributed to insufficient filling of major resin [15]. A previous study has pointed out, the inherent polar and hydrophilic nature of the lignocellulosic fiber in kenaf, and the non-polar characteristics of the polymers, resulted in difficulties in compounding or blending of the fibers and matrix [13].

Figure 2.3 shows extension at break of the composites with the increase in kenaf content. Figure 2.3 showed the extension at break of the

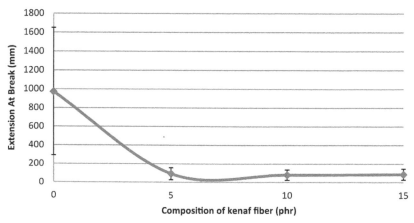

FIGURE 2.3   Extension at break of rPE/PA-6 versus composition of kenaf.

composites decreased tremendously by introduction of 5phr of the kenaf into the composites. The extension at break at 10phr of kenaf is lower than without kenaf. The decrease of extension at break is obviously attributed to the incorporation of kenaf in PA6/rPE blends, which did not improve the interaction of polymer chains because only microscale kenaf were found to fill the voids of the spaces between them. Thus, the flexibility of this polymer system to slide between the kenaf became harder which reduced the extension at break. Then, at 15phr of kenaf, the value of extension at break increased to 86.81 mm. In conclusion, the incorporation of kenaf has not really improved the elongation of the composites.

As shown in Figure 2.4, it can be observed that the impact energy or impact strength decreased as the composition of the kenaf was increased. Upon the incorporation of 5phr of kenaf in the PA6/rPE composites, the impact strength decreases to 130.53 MPa. Then at 10phr of kenaf, the impact strength slightly increases to 5 MPa and it is expected to decrease again to the lowest value when the amount of kenaf was beyond 15phr. The decrease in the impact strength, in the presence of kenaf may be attributed by the kenaf of PA6 and rPE through the chemical reaction between the free terminal of amine end group of PA6 and the anhydride function in PE-g-MAH. This indicates that more interaction between PA6 and rPE than with kenaf resulted in lower impact strength as well as poorer toughened behavior. However, this study is not in agreement with the previous

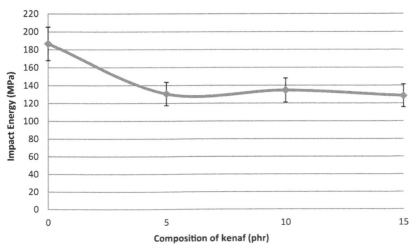

FIGURE 2.4   Impact properties of rPE/PA-6 versus composition of kenaf.

study [16], which shows an increase in impact strength of the polymer blend by increasing kenaf content.

2.3.2 FLEXURAL PROPERTIES ANALYSIS

Figure 2.5 shows the flexural modulus of rPE/PA-6 composites containing kenaf fibers from 0 to 15 phr. The result showed that the flexural modulus increased slightly as the kenaf content was increased. This was expected because kenaf acts as a reinforcing agent and it increases the flexural modulus as well as stiffness of the composites. Flexural modulus is a measure of stiffness during the initial part of the bending forces. These results are similar to the previous study, which has reported the flexural modulus of kenaf reinforced PLA composites, increases linearly with fiber content [17]. The maximum value of the flexural modulus was achieved at 10 phr of the kenaf content. The decrease in flexural modulus beyond this point indicates that the incorporation of kenaf did not provide any significant effects to the stiffness of the composites. From this result, the elasticity of the samples can be figured out as high flexural modulus means high stiffness. Though, high stiffness means low elasticity.

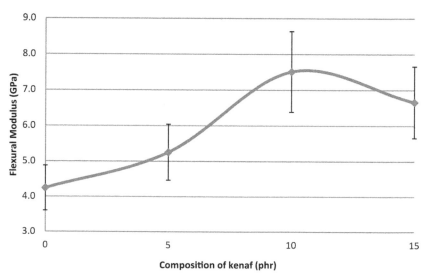

FIGURE 2.5   Flexural modulus of rPE/PA-6 versus composition of kenaf.

Figure 2.6 shows the flexural strength of rPE/PA-6 composites with different composition of the kenaf. Flexural strength is the ability of the material to withstand bending forces applied perpendicular to its longitudinal forces. This result shows that the addition of 10phr of the kenaf decreases its flexural strength. The flexural strength for PA-6/rPE blend should be increased proportionally with the kenaf content as expected. However, as mentioned in the tensile strength's result, the decrease of the flexural strength may due to lack of compatibility between the kenaf and the composites. The main problem, encountered during the addition of kenaf into polymer matrix was lack of good interfacial adhesion between the two components, which resulted in some poor properties in the final product [4, 18].

2.3.3 DMA PROPERTIES ANALYSIS

DMA was performed for virgin polyethylene and its composites to obtain further information on mechanical properties and molecular motions. DMA is a method that measures the stiffness and mechanical damping of a cyclically deformed material as a function of temperature. Identification of the glass transition temperature (Tg) is one of the most common uses of

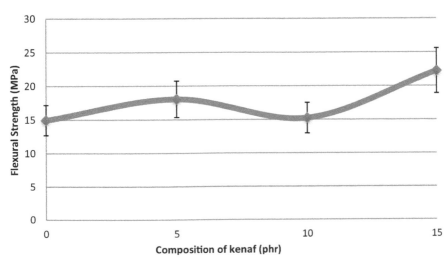

FIGURE 2.6   Flexural strength of rPE/PA-6 versus composition of kenaf.

DMA and generally *Tg* is determined as a maximum of tan delta curve and seen as a drop in storage modulus curve.

Figure 2.7 shows the loss modulus of PA-6/rPE blends with 5 phr of the kenaf was the highest at 50–60°C. Between these temperatures, the composite shows its liquid like behavior, followed by 15 and 10 phr of the kenaf contents, respectively. More so, as the temperature increased, the loss modulus for all PA6/rPE blends with kenaf slowly decreased. This indicates that by increasing the temperature, the composites tend to lose their liquid like behavior. This was due to the presence of heat providing enough energy for the kenaf fibers to vibrate, move, slip and rotate as much as polymer blends would have done and hence provided a lubricant effect to the composites. Consequently, the viscosity of PA6/rPE composite decreased with an increase in temperature. But for PA-6/rPE blend without kenaf, the loss modulus increased for the whole range of temperatures because an absence of lubricant effects by kenaf fibers. It was therefore concluded that the viscosity of PA6/rPE without kenaf was higher than PA6/rPE blend with kenaf.

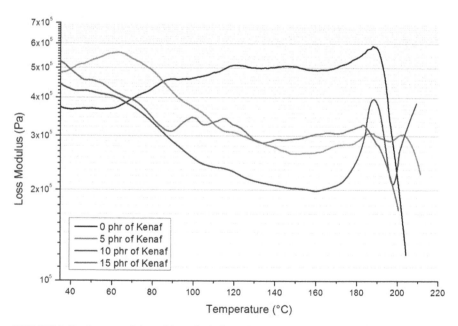

FIGURE 2.7   Loss modulus of kenaf reinforced rPE/PA-6 composites.

Figure 2.8 shows the storage modulus PE/PA-6 composites. The storage modulus of the composites decreased with an increase in temperature. This indicates that PA6/rPE blended with 10 phr of the kenaf has the lowest storage modulus compared to others. This shows that elasticity of this blend is lower. For 15 phr of the kenaf, the storage modulus is in between 10 phr and 0 phr of the kenaf. This was due to the fact that, the excess kenaf did not help much in improving the elasticity of the composites. This indicates that, the composites have a constant elasticity behavior regardless of the kenaf content. At 70°C storage modulus of PA-6 /rPE blends increased drastically while storage modulus for the composites constantly decreased. These composites can be categorized as liquid-like material, because the G" is greater than G' at all ranges of temperature studied regardless of the amount of kenaf in the systems. This phenomenon was attributed to the kenaf acting as a lubricant and therefore leading to polymer phase of PA6 and rPE to slip at the interphase and interfacial level. These results were similar to the study conducted on the mechanical properties of PA6/ABS blends with or without glass fiber by Arsad et al. [16].

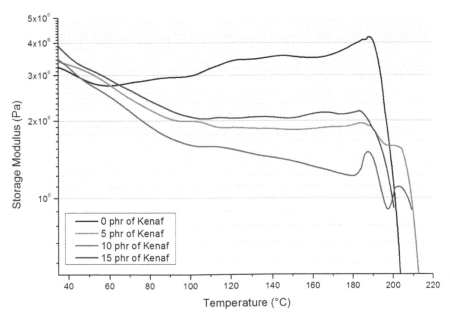

FIGURE 2.8   Storage modulus of kenaf reinforced rPE/PA-6 composites.

Figure 2.9 shows the tangent delta of the composites. Tan delta was calculated by loss modulus divided by storage modulus. The previous study shows that an increase in the content of the kenaf, leads to a decrease in tan delta of the composites. Tan delta is also very sensitive with structural change of the composites and viscoelastic behavior. This can be determined using the analysis of tan delta peak. Figure 2.9 shows that a peak occurred at 75°C which depends on the content of the kenaf in the PA6/rPE composites. It was also observed, that the unreinforced composite has a moderate tan delta, thereby indicating less elasticity than the reinforced composites. The elastic behavior became more prominent with the highest amount of the kenaf. In other words, the elasticity disappeared as the kenaf content was increased. Tan delta peak also indicates a solid liquid transition, leading the composites to undergo a transition with increasing temperature. At the transition point, tan delta was expected to be independent of temperature. The results of tan delta in this research were found to be similar to the study conducted on the mechanical properties of PA6/ABS blends with or without glass fiber [16]. Thus loss and storage modulus composites without kenaf increased with temperature

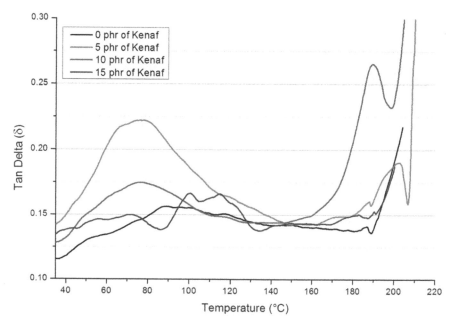

FIGURE 2.9   Tangent delta of kenaf reinforced rPE/PA-6 composites.

because MAH was affected during the increased in temperature. An addition of the kenaf helps in lubricating and thereby paving a space for the polymer and the kenaf to slip and move one another. In conclusion, the higher the temperature, the more the vibration of the polymer molecules and fiber particles and that led to a reduction in modulus of the composites.

2.3.4 THERMOGRAVIMETRIC ANALYSIS

Figure 2.10 shows the TGA weight loss curves for PA6/rPE composites with different kenaf contents. The purposes of the curves were to determine the maximum decomposition temperature (T_{max}). It shows a single step degradation process for all the composites. Based on the graph, the decomposition temperature was achieved at 500°C to 530°C and approximately 85–90% weight loss for all the samples was observed. The residue left was due to the impurities encountered in the samples during preparation the sample.

Previous studies assumed that degradation of the blends involved the mechanism of both PA-6 [19] and PA-66 [20], this research therefore took

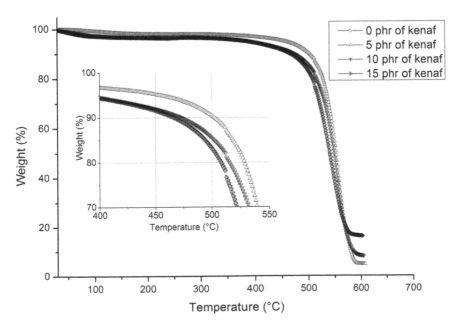

FIGURE 2.10   TGA curves of kenaf reinforced rPE/PA-6 composites.

both PA-6 and rPE polymer degradation mechanisms into account. They suggested that thermal decomposition of PA-6 starts with homolytic scission of the N-alkylamide bond, so that primary amide, nitrile, vinyl and alkyl chain ends are generated [20]. It is also reported that along amines, amides and nitriles, caprolactam is also a major product in the degradation of PA6 [21].

Furthermore, it is believed that rPE also starts to degrade by chain scission. From the above results, the degradation temperature approximately shifted towards higher temperatures after the introduction of kenaf for both PA-6 and rPE matrixes. This result has some similarity with previous research by Refs. [22, 23], where the degradation of PP was around 501.8°C while the kenaf reinforced PP was 524.1°C, which indicates that the composite has a higher thermal stability. The addition of kenaf resulted in an increase in degradation temperatures, which could be attributed to consolidation effects [22]. By contrast, the weight loss generally decreased as kenaf content increased, which is similar to results from previous researchers [23]. According to them, as the weight percent of the reinforcement in the blend increases, a gradual decrease in weight loss was noticed, indicating an increase in thermal stability. The weight losses were not 100%, which indicates that there were some residues. According to Lei et al. [14] the residue was due to the carbonization of the kenaf fiber and mainly derived from the fiber degradation. Thus, both composite systems showed similar T_o values, suggesting no significant difference in the thermal stability between them. This was in perfect agreement with previous researchers [19, 21, 24].

2.3.5 SEM PROPERTIES

Fractured tensile samples were used as SEM test samples, which were carried out at 500x of magnification. The objective of this test was to discuss the fracture surface morphology and the dispersion of PA-6/rPE blend reinforced by kenaf. Figure 2.11(a) micrographs showed a fine dispersion without the addition of the kenaf facilitates good adhesion between rPE/PA6 by the presence of PE-g-MAH compatibilizer in the composites. Huda et al. [7] reported that the coupling agent/compatibilizer causes a significantly better wetting of the kenaf through the polymer matrix.

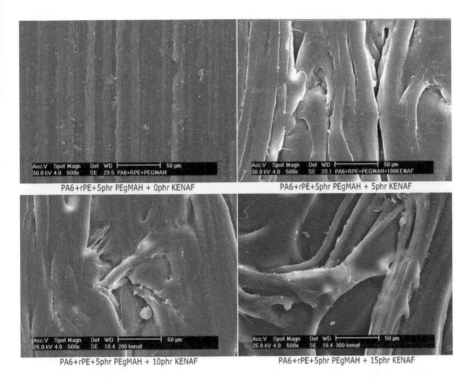

PA6+rPE+5phr PEgMAH + 0phr KENAF

PA6+rPE+5phr PEgMAH + 5phr KENAF

PA6+rPE+5phr PEgMAH + 10phr KENAF

PA6+rPE+5phr PEgMAH + 15phr KENAF

FIGURE 2.11   SEM micrograph of kenaf reinforced rPE/PA-6 composites.

Good dispersion avoids unoccupied area of the blending; hence it gave the desired improvement of thermal stability. This result has some similarities with the previous study whereby the addition of SEBS-g-MA improved the homogeneity of the blend of HDPE/PA-6 morphology [14].

The addition of 10phr of the kenaf in Figure 2.11(b) show that there were some void spaces in the composites fracture, meaning that 5phr of the kenaf was not enough to occupy all empty spaces of the composites. Hence, there were still some empty spaces in the composites, which reduced the efficiency of heat transfer to the composites, thereby lowering melting point and crystallization temperature of rPE. However, the occupied area of polymer matrix shows a very good adhesion and dispersions of the kenaf, it was therefore concluded that the use of kenaf was compatible with these polymers. It therefore states that good adhesion between polymer matrix and kenaf is shown with alkali treated kenaf. The large void spaces encountered made the tensile strength fluctuate a bit.

The voids were noticed to appear in Figure 2.11(c) as the content of the kenaf increased to 10phr. The furry-like shape was due to kenaf not being pulled out from the composites due to the breakage of the samples during tensile test. From this, it could be seen that the tensile strength was good. Figure 2.11(d) shows that there were many voids spaces compared to other composites at 15phr of kenaf in the blend. In spite of that, the voids spaces that occurred as kenaf was pulled out are randomly oriented which was similar with the result obtained by Liu et al. [25]. According to them, random orientation of kenaf in the matrix gave a very poor mechanical property, compared to the long unidirectional orientation. The long uni-directional orientation of the kenaf was seen in the Figure 2.11(d), which gave a better tensile strength of PA-6/rPE.

2.4 CONCLUSION

Generally, 5 phr of PE-g-MAH was seen to be the optimum concentration regarding the mechanical test. The mechanical properties for both PA-6/rPE blends with amount PE-g-MAH selected was enhanced successfully by kenaf as the Young modulus, flexural modulus and flexural strength increased proportionally as the kenaf content was increased. In addition, it was a good result because the extensions at break for PA-6/rPE blends were decreased by an increased in kenaf contents. However, there were some compatibility problems between kenaf and PA-6/rPE matrix, which led to insufficient matrix resin at some points and agglomeration on the kenaf at other place. This was seen in the SEM test where some fiber orientation lowered and bettered the mechanical properties. From the result obtained at 10phr of the kenaf was seen to be the optimum concentration for the mechanical and morphological properties. In conclusion, the effect or the function of PE-g-MAH on polymer blend was clearly seen in the morphological properties. The effect of the kenaf was clearly seen in the mechanical properties and thermal properties of the composites. While the TGA, clearly concludes that an increase in the kenaf content, decreases the loss weight of polymer composites and increases the thermal stability.

KEYWORDS

- **dynamic mechanical analysis**
- **green composites**
- **PA6/rPE/kenaf**
- **PE-g-MAH**

REFERENCES

1. Al-Salem, S. M., Lettieri, P., Baeyens, J., Recycling and recovery routes of plastic solid waste (PSW): A review. 29(10), 2625–2643 (2009).
2. Aziz, S. H., Ansell, M. P., The effect of alkalization and fiber alignment on the mechanical and thermal properties of kenaf and hemp bast fiber composites: Part 1 – polyester resin matrix. 64(9), 1219–1230 (2004).
3. Bertin, S., Robin, J.-J., Study and characterization of virgin and recycled LDPE/PP blends. 38(11), 2255–2264 (2002).
4. Edeerozey, A. M. M., Akil, H. M., Azhar, A. B., Ariffin, M. Z., Chemical modification of kenaf fibers. Mater Lett, 61(10), 2023–2025 (2007).
5. Ha, C.-S., Park, H.-D., Cho, W. J., Compatibilizers for recycling of the plastic mixture wastes. II. The effect of a compatibilizer for binary blends on the properties of ternary blends. J Appl Polym Sci, 76(7), 1048–1053 (2000).
6. Hong, S. M., Hwang, S. S., Jeon, B. H., Choi, J. S., Kim, H. B., Lim, S. T., Choi, H. J., Polypropylene/polyamide-6 blends based on commingled plastic wastes. J Mater Sci, 40(14), 3857–3859 (2005).
7. Huda, M. S., Drzal, L. T., Mohanty, A. K., Misra, M., Effect of fiber surface-treatments on the properties of laminated biocomposites from poly(lactic acid) (PLA) and kenafs. 68(2), 424–432 (2008).
8. Joshi, S. V., Drzal, L. T., Rothon, R., Are natural fiber composites environmentally superior to glass fiber reinforced composites?. 35(3), 371–376 (2004).
9. Karnani, R., Krishnan, M., Narayan, R., Biofiber-reinforced polypropylene composites. Polym Eng Sci, 37(2), 476–483 (1997).
10. Mishra, S., Misra, M., Tripathy, S. S., Nayak, S. K., Mohanty, A. K., Graft Copolymerization of Acrylonitrile on Chemically Modified Sisal Fibers. 286 (2), 107–113 (2001).
11. Ray, D., Sarkar, B. K., Rana, A. K., Bose, N. R., Effect of alkali treated jute fibers on composite properties. B Mater Sci, 24(2), 129–135 (2001).
12. Li, X., Tabil, L. G., Panigrahi, S., Chemical Treatments of Natural Fiber for Use in Natural Fiber-Reinforced Composites: A Review. 15, 25–33 (2007).

13. Kelar, K., Jurkowski, B., Preparation of functionalized low-density polyethylene by reactive extrusion and its blend with polyamide 6. Polymer, 41(3), 1055–1062 (2000).

14. Lei, Y., Wu, Q. L., Yao, F., Xu, Y. J., Preparation and properties of recycled HDPE/ natural fiber composites. Compos Part a-Appl S, 38(7), 1664–1674 (2007).

15. Akil, H. M., Omar, M. F., Mazuki, A. A. M., Safiee, S., Ishak, Z. A. M., Abu Bakar, A., Kenaf fiber reinforced composites: A review. Mater Design, 32(8–9), 4107–4121 (2011).

16. Arsad, A., Rahmat, A. R., Hassan, A., Iskandar, S. N., Mechanical and Rheological Properties PA6/ABS blends With or Without Short Glass Fiber. J Reinf Plast Comp, 29(18), 2807–2820 (2010).

17. Ochi, S., Mechanical properties of kenaf fibers and kenaf/PLA composites. Mech Mater, 40(4–5), 446–452 (2008).

18. Sukri, S. M., Suradi, N. L., Arsad, A., Rahmat, A. R., Hassan, A., Green composites based on recycled polyamide-6/recycled polypropylene kenaf composites: mechanical, thermal and morphological properties. J Polym Eng, 32(4–5), 291–299 (2012).

19. Zong, R., Hu, Y., Liu, N., Li, S., Liao, G., Investigation of thermal degradation and flammability of polyamide-6 and polyamide-6 nanocomposites. 104(4), 2297–2303 (2007).

20. Komalan, C., George, K. E., Varughese, K. T., Mathew, V. S., Thomas, S., Thermogravimetric and wide angle X-ray diffraction analysis of thermoplastic elastomers from nylon copolymer and EPDM rubber. 93(12), 2104–2112 (2008).

21. Pramoda, K. P., Liu, T., Liu, Z., He, C., Sue, H.-J., Thermal degradation behavior of polyamide 6/clay nanocomposites. 81(1), 47–56 (2003).

22. Manfredi, L. B., Rodríguez, E. S., Wladyka-Przybylak, M., Vázquez, A., Thermal degradation and fire resistance of unsaturated polyester, modified acrylic resins and their composites with natural fibers. 91(2), 255–261 (2006).

23. John, M. J., Bellmann, C., Anandjiwala, R. D., Kenaf–polypropylene composites: Effect of amphiphilic coupling agent on surface properties of fibers and composites. 82(3), 549–554 (2010).

24. Balakrishnan, H., Hassan, A., Isitman, N. A., Kaynak, C., On the use of magnesium hydroxide towards halogen-free flame-retarded polyamide-6/polypropylene blends. 97(8), 1447–1457 (2012).

25. Liu, H., Wua, Q., Zhang, Q., Preparation and properties of banana fiber-reinforced composites based on high density polyethylene (HDPE)/Nylon6 blends. Bioresource technology, 100(23), 6088–6097 (2009).

CHAPTER 3

EFFECT OF COMPATIBILIZER: FILLER RATIO ON THE TENSILE, BARRIER AND THERMAL PROPERTIES OF POLYETHYLENE COMPOSITE FILMS MANUFACTURED FROM NATURAL FIBER AND NANOCLAY

KHALIQ MAJEED, AZMAN HASSAN, and AZNIZAM ABU BAKAR

Enhanced Polymer Research Group (EnPRO), Department of Polymer Engineering, Faculty of Chemical Engineering, Universiti Teknologi Malaysia, 81310 UTM Skudai, Johor Malaysia,

Tel: +60 7 5537835; Fax: +60 7 5581 463

CONTENTS

ABSTRACT

Rice husk (RH)/montmorillonite (MMT) hybrid filler filled low-density polyethylene nanocomposite films were prepared by extrusion blown film. Maleic anhydride grafted polyethylene was used as compatibilizer and the compatibilizer to filler ratio was varied from 0.25:1 to 2:1. The results revealed that compatibilizer:filler ratio has a significant effect on the delamination of MMT and distribution of the filler, which in turn, define its tensile and barrier properties. X-ray diffraction analysis and scanning electron microscopy revealed that compatibilizer to filler ratio 1:1 enhanced the interlayer distance significantly and improved the filler distribution. Further, the tensile and barrier properties of RH/MMT/LDPE system could be significantly improved with the significant amount of the compatibilizer. The optimum compatibilizer:filler ratio for RH/MMT/ LDPE system is in the range of 0.75:1 to 1:1 as tensile and barrier results have showed. Melting point of the nanocomposite film was remained statistically unchanged, while the degree of crystallinity increased significantly for the same ratios.

3.1   INTRODUCTION

Plastic films are widely used for food-stuff and goods packaging; agriculture and merchandizing. They have played an important role in enhancing the shelf life of packaged foods and agricultural productivity [1]. Low density polyethylene (LDPE), a prime member of the polyolefin family, is predominantly used in film applications because of its high flexibility, easy process ability, easy to seal, impact toughness, stress crack resistance, microwaveability, recyclability, high resistant to moisture and fair gas barrier properties [2]. Each year, an estimated 500 billion to 1 trillion plastic bags are consumed worldwide and most of them, after serving their useful life as packaging film, finds their way to the land fill sites [3]. Unfortunately, the extensive and expanding use of these films will result in increased accumulation of plastic wastes because of their non-biodegradability.

Natural fibers are susceptible to microorganisms and their biodegradability is one of the most promising aspects of their incorporation in polymeric materials [4]. Among others, their renewability, abundance, lower cost, environmental friendliness, and relative no abrasiveness for the processing machinery has make natural fibers attractive candidate for reinforcement of polymers [5]. Rice husk (RH) is one of the widely available agricultural industrial residue materials. It would always be removed and separated from rice grain during the rice milling. Like many other agricultural by-products, the industrial applications of this biomass are very limited with little economic value. There have been studies on the biodegradation of natural fiber reinforced LDPE bio-composites and have produced encouraging results [6, 7]. But it has been reported that the performance properties of these biocomposites deteriorates with adding fiber [6–8]. A possible solution to enhance the performance properties of natural fiber reinforced composites can be represented by hybridization with nanoclays [9–12]. Nanoclays are considered one of the main particles for reinforcing polymeric matrices and promoting new and enhanced properties, such as enhanced toughness, stiffness, gas barrier and flame resistant characteristics [13, 14]. Among all, Montmorillonite (MMT) is most potentially promising nanoclay. The clay structure is formed by hundreds of platelets or sheets stacked into particles like pages of a book [15]. To take full advantage and realize large aspect ratio, the clays should be exfoliated and homogeneously dispersed in the matrix.

Research and development in materials coupled with appropriate filler, filler-matrix interaction and new formulation strategies to develop novel composites have many potential applications. As many conventional and present materials, such as metals, ceramics, plastics, polymer blends and composites cannot fulfill all requirements for this new era, it appears that these frontiers will not be realized solely by developing new materials, but by optimizing material combinations and taking advantage of their synergistic functions. In this context, hybridization, compounding of polymers with hybrid fillers, is a technique that can complement the drawbacks of conventional polymers and composites. Hybrid composites are an amalgamation of two or more fillers/fibers in a polymer matrix. They are attractive as they result in a balance between performance properties and the cost of the composite that cannot be obtained with a single kind of reinforcement [16–18]. In order to improve the performance properties and to meet varied demands, innovative and modified hybrid composites are being developed and optimized for potential commercial use. Though in principle, several fibers can be incorporated into a hybrid system, a combination of only two types of fiber would be the most beneficial [19]. There has been a renewed interest in finding high performance, cost effective and biodegradable material by using more than one filler in a single matrix. Work has been done by various researchers and recently published review on utilization of these hybrid materials in packaging has provided another dimension to the potential versatility of these materials [15].

In general, fiber to resin adhesion and delamination/dispersion of the clay platelets within the polymer matrix are the pertinent parameters to describe the performance properties of the composites. Generally, the performance properties of the filler reinforced composites are mainly influenced by the interfacial adhesion between the fillers and the matrix material; and dispersion of the fillers [20]. In this regard, many approaches have been described in the literature aiming to improve the interfacial adhesion between the fiber/filler and matrix material [21–24]. Among all, one of the most common and effective approach is the addition of maleic anhydride grafted polyolefin compatibilizer [23, 25, 26]. Zahedi, Pirayesh (9) studied the effect of MAPP compatibilizer content (4 and 6 wt.%) on the mechanical properties of walnut shell flour/montmorillonite filled polypropylene composites and reported that the properties with adding 6 wt.% MAPP are better than those prepared by adding 4 wt.% MAPP. Najafi et al. [27] compared the influence of 2 wt.% MAPP loading on the physical and morphological properties of PP/reed flour/MMT nanocomposites with the uncompatibilized composites. The authors concluded adding MAPP improved the mechanical properties and decreased thickness swelling. Further, the MMT interlayer spacing in the nanocomposites increased by adding MAPP. These studies indicate that different amounts of maleated polyolefin compatibilizer have been used in most of the published studies on natural fiber/nanoclay hybrid filler filled polyolefin composites and have achieved various level of success. It is also reported that the addition of these maleated polyolefin co polymer lowers the neat matrix properties due to their different molecular weights than the matrix and maleated groups that disrupt the crystallinity of the matrix [28, 29]. Further, their excessive usage negatively affects the downstream properties. In addition, costs of male-ated polyolefin are substantially higher compared to unmodified polyolefin. Hence, optimization studies are needed in the formulation of higher performance composites at lower cost.

The present study entails optimization of compatibilizer to filler ratio to obtain RH/MMT hybrid filler filled LDPE nanocomposite films with higher tensile and barrier properties. In addition, delamination of MMT layers and morphologies of fractured surfaces of the composite films with different compatibilizer:filler ratios are also studied.

3.2   EXPERIMENTAL PART

3.2.1   MATERIALS

Rice husk (RH) was obtained from BERNAS (Padiberas Nasional Berhad), Malaysia. Montmorillonite (MMT, 1.44P, modifier: Dioctadecyl Dimethyl Ammonium Chloride) was obtained from American Nanocor Corporation. Film grade low density polyethylene (LDPE, LDF200GG) used as the polymer matrix was obtained from Titan chemicals, Malaysia; Melt Mass Flow Rate (MFR, ASTM D1238) of 2 g/10 min and density of 0.922 g/cm^3. Maleic anhydride modified linear low density polyethylene (MAPE, OREVAC®18365) was used as a compatibilizer (MFR (190°C, 2.16 kg) of 2.5 g/10 min and density of 0.916 g/cm^3 at 23°C).

3.2.2   FIBER TREATMENT

The supplied RH was washed extensively with distilled water at room temperature to remove dust or other debris and dried in an air circulating oven at 100°C for 24 h to reduce moisture content to less than 4%. Drying washed rice husk will not only prevent microbial growth (fungi) but would facilitate particle size reduction as well. Pre-washed and dried RH was ground using grinder and RH flour having particle size less than 75 μm was separated using a sieve (Retsch test sieve, model AS200), for use as bio-filler.

3.2.3   COMPOSITE FILM PROCESSING

The RH flour was dried in an oven at 100°C overnight prior to processing. To provide a same matrix and filler for bridging and bonding of MAPE, the amount of LDPE, RH, and MMT was fixed to 92, 4, and 4 wt.%, respectively. LDPE and RH/MMT were melt compounded in a co-rotating twin extruder in the presence of maleic anhydride grafted polyethylene (MAPE) as compatibilizer; having compatibilizer:filler ratio of 0, 0.25, 0.5, 0.75, 1, 1.5 and 2. The extruded strands were air-dried in a cool air stream, pelletized and then film blown to prepare thin films.

3.2.4   EXPERIMENTAL TECHNIQUES

3.2.4.1   X-Ray Diffraction Analysis

Diffraction studies were carried out to investigate the delamination/exfo-liation of MMT in polymer matrix. A Bruker D8 Advance difractometer was used to measure the d-spacing of the hybrid filler-filled nanocomposite films. The diffraction patterns were obtained at room temperature in the range $2 < 2\theta < 10$ degree by step of $0.02°$. The X-ray beam Cu Kα radiation ($\lambda = 0.154$ nm), operated at 30 kV and 10 mA. The interlayer distance was determined by the peak, using the Braggs' equation

$$D_{001} = \frac{\lambda}{2 Sin\theta}$$

where, D_{001} = interlayer distance between clay layers; λ = wavelength of the X-ray beam, and θ = half of the diffraction angle at the first peak.

3.2.4.2   Morphological Analysis

The morphology of blown film samples was evaluated using scanning electron microscope (Philips XL 40). To investigate the surface topography, the film samples were mounted onto copper stubs using double sided sticky tapes and their surface was coated with a thin layer of gold by using BIO-RAD SEM coating system. For interfacial morphologies, the film samples were frozen in liquid nitrogen and broken into pieces. The fractured film samples were mounted onto copper stubs using double sided sticky tapes and sputter coated with gold to provide enhanced conductivity.

3.2.4.3 Tensile Properties

The tensile properties, including tensile modulus, tensile strength and elongation at break of the composite blown films were carried out by using Lloyds universal testing machine at room temperature, following the procedures described in ASTM D882. The composite blown films were cut into rectangular shaped specimen (102 x 15 mm²) along with their machine direction following the ASTM standard. Micrometer (Mitutoyo, Japan) with precision ±0.001 mm was used to measure thickness of the film samples. At least eight different samples were tested for each sample composition and values of tensile strength, elongation at break and Young's modulus were obtained from stress-strain curves at a crosshead speed of 50 mm/min.

3.2.4.4 Barrier Properties

The barrier properties of plastics are usually described by quantifying the amount of permeating gas/vapor molecules. The oxygen (O_2) and carbon dioxide (CO_2) permeability's for the composite blown films were measured at room temperature in a constant pressure/variable volume type permeation cell designed according to ASTM D1434–82 (Reapproved, 2009). Circular film samples of uniform thickness and 4.4 cm diameter were used to study the gas transmission rate (GTR, mL/m²/24hr). Micrometer (Mitutoyo, Japan) with precision ±0.001 mm was used to measure film sample thickness and permeability coefficient was calculated by using the relation given below.

$$P = t \frac{GTR}{\Delta p}$$

where: P = permeability coefficient (mL-mm/ m²/24hr/atm); t = film thickness under investigation (mm); Δp = partial pressure difference on two sides of the film (atm).

3.2.4.5   Thermal Properties

The thermal properties of composite films were examined using differential scanning calorimetry (DSC). DSC was carried out using a PerkinElmer DSC-7 over a temperature range of 30 to 180°C. The weight of the samples was about 5 to 10 mg and the scanning rate of 10°C per min was adopted. All measurements were performed under nitrogen environment. The degree of crystallinity (X_c) of LDPE/RH composite samples was obtained by the comparison of the heat of fusion with a 100% crystalline PE heat of fusion of (293 J/g) by using the following relation

$$Xc = \frac{\Delta H}{\Delta H_m . W_{LDPE}} 100$$

where, ΔH_m = Melting enthalpy of 100% crystalline polyethylene, 293 J/g [30]; W_{LDPE} = weight fraction of LDPE in the sample.

3.3   RESULTS AND DISCUSSION

3.3.1   X-RAY DIFFRACTION ANALYSIS

One of the keys to take full advantage of the reinforcement and/or tortuosity clay particles can provide to the nanocomposite is to obtain a satisfactory dispersion of the nanoclay in the polymer matrix and the insertion of polymer chains within the interlayer of the clay structure. In order to investigate the delamination of the MMT layers, XRD were performed on the composite film samples having different compatibilizer:filler ratios. Depending on the exfoliation and distribution of MMT layers, three different types of composites are achievable; tactoid or immiscible, intercalated or exfoliated. For a tactoid or immiscible composite structure, the filler diffraction peak should stay at the same position; for an intercalated structure, it should be shifted to lower 2θ value; and for an exfoliated structure it should disappear. Figure 3.1 compares the position of diffraction peaks of MMT and RH/MMT filled LDPE nanocomposite films, with or without MAPE.

The XRD patters of neat MMT shows a diffraction peak at about 3.46 2θ, representing an interlayer spacing of about 2.55 nm. For uncompatibilized

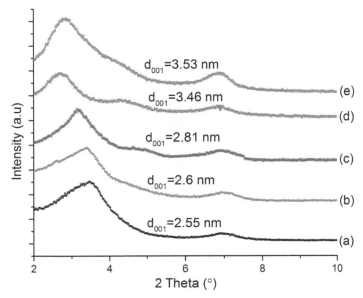

FIGURE 3.1   XRD pattern of (a) pristine MMT; and MMT/RH filled LDPE nanocomposite films having different compatibilizer:filler ratios, (b) 0, (c) 0.5, (d) 1, and (e) 1.5.

composite films, the corresponding peak slightly moved towards lower angle at about 3.40° 2θ, suggesting a small change in the interplanar distance. It is documented that very small or no intercalation is obtained by compounding polyolefin with clay in the absence of compatibilizer [30]. In the compatibilized composites, the diffraction peak of MMT was shifted to lower angles and the interlayer spacing becomes bigger than the neat MMT. These observations indicate that adding MAPE into the RH/MMT filled LDPE system improved interaction between filler and matrix materials; and the polymer chains entered into the MMT layers forming intercalated nanocomposites, without reaching complete exfoliation. The results suggest that overall effect of MAPE addition in interlayer spacing is quite significant; an increase of 10, 36 and 38%, for 0.5, 1 and 1.5 compatibilizer:filler ratio, respectively. Concerning optimum compatibilizer:filler ratio for MMT delamination, we can see that the acceptable delamination results were obtained for compatibilizer:filler ratio of 1 and higher. It is also worth mentioning that the ratios beyond 1 do not lead to a remarkable increase in the interlayer spacing. The results suggest that optimum intercalation occurs at a compatibilizer:filler ratio of 1 and further increase in the ratio is not beneficial.

3.3.2 MORPHOLOGICAL ANALYSIS

Mechanical properties of natural fiber reinforced composites are mainly influenced by the level of dispersion and interfacial adhesion between the matrix and fibers; and microscopy is one of the most effective methods for morphological observations. In this study, scanning electron microscope (SEM) was used to investigate the dispersion of natural fiber and interfacial morphologies; and the images are shown in Figures 3.2 (film surface) and 3.3 (cryo fractured surface). The SEM pictures show the differently shaped particles and the more or less spherical lignin-based particles.

Simple naked eye examination of the prepared composite films indicated that better dispersion of the filler was achieved in the films having higher compatibilizer:filler ratio. Figure 3.2a depicts the poor dispersion and agglomeration of filler particles in the matrix. This filler particles agglomeration decreases film homogeneity, resulting in reduction in mechanical properties [7]. Further, these aggregates provide an evidence of poor dispersion of filler particles. Figure 3.2b presents the surface of film specimen prepared with compatibilizer:filler ratio of 1, the better coverage of filler particles with the matrix and relatively smooth surface with good dispersion indicate an effectiveness of compatibilizer in improvement of the dispersion.

The cryo-fractured surfaces of composite films were analyzed by SEM to study morphological and interfacial characteristics and the micrographs are shown in Figure 3.3. The brittle-fractured surface of uncompatibilized composite film (Figure 3.3a) shows wide gap between the fiber and matrix with significant presence MMT clusters, which decreases film homogeneity [7]. Further, these features provide an evidence of decrease in

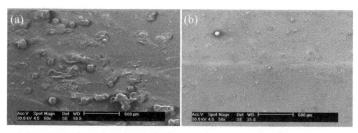

FIGURE 3.2 SEM micrographs of the surface of composite films with varying compatibilizer:filler ratio (a) 0:8 and (b) 1:1.

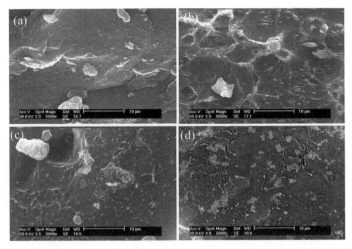

FIGURE 3.3 SEM micrographs of the fractured surface of composite films with varying compatibilizer:filler ratio (a) 0, (b) 0.25, (c) 0.5 and (d) 1.

strength. The better coverage of the filler particles can be confirmed with a compatibilizer:filler ratio of 1 (Figure 3.3d), where a uniform dispersion and sufficient filler/matrix adhesion can be observed, which is not the case for compatibilizer:filler ratio of 0.25 and 0.5. The composite films having compatibilizer:filler ratio 0.25 and 0.5 (Figures 3.3b and 3.3c) show the presence of extensive pull out indicating that the amount of compatibilizer is not sufficient for the fillers to have strong interaction with the matrix. This improvement in dispersion with the incorporation of grafted compatibilizer in the development of natural fiber/inorganic filler filled polyolefin composites has also been reported by other researchers [17, 26].

3.3.3 TENSILE PROPERTIES

The tensile testing was carried out to investigate the effect compatibilizer:filler ratio on the tensile modulus, tensile strength, and elongation at break of RH/MMT hybrid filler filled LDPE based nanocomposite films. Tensile properties were calculated using stress-strain curves and the representative stress strain curves are shown in Figure 3.4.

Figure 3.5 shows the evolution of tensile strength and tensile modulus as a function of compatibilizer:filler ratio. Tensile modulus and tensile

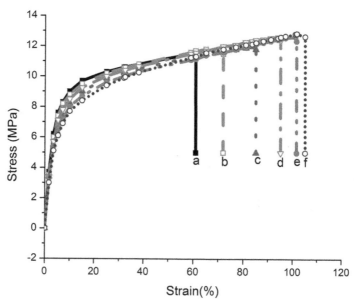

FIGURE 3.4   Stress strain curves of the RH/MMT filler filled and MAPE compatibilized LDPE nanocomposite films having compatibilizer:filler ratio of (a) 0, (b) 0.25, (c) 0.5, (d) 0.75, (e) 1 and (f) 1.5.

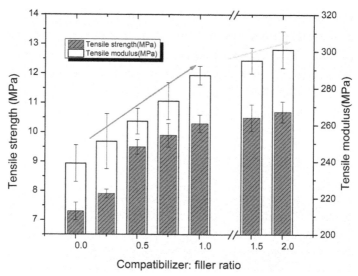

FIGURE 3.5   Tensile strength and tensile modulus of RH/MMT filler filled and MAPE compatibilizer LDPE nanocomposite films.

strength for the uncompatibilized composite film is 239 and 7.3 MPa, respectively. The enhanced strength and modulus with the incorporation of compatibilizer was expected owing to improved interactions between components in the composite blown films. Adding compatibilizer to the uncompatibilized system resulted in significant increase in the tensile modulus (247 to 301 MPa), tensile strength (7.3 to 10.7 MPa). As can be seen, these tensile values for lower compatibilizer:filler ratios are not as higher as with the higher ratios. This indicates that these lower amounts of compatibilizer are not enough to delaminate and/or uniformly disperse the filler. The improvement continued to as the compatibilizer:filler ratio increased; and the best results are obtained for compatibilizer:filler ratio of 1 and higher. However, the higher ratios do not lead to significant increase in the modulus. The enhanced properties are due to improved bonding strength between filler and the polymer matrix; delamination and uniform distribution of the clay platelets and better stress transfer from the matrix to the filler [24]. Further, the results imply that the properties could be enhanced significantly by adapting appropriate compatibilizer:filler ratio. Hemmasi, Khademi-Eslam [12] have also reported an increasing trend in tensile modulus and tensile strength values with increasing the compatibilizer:filler ratio. Further, the variations in these tensile properties are also in agreement with XRD observations.

Regarding elongation at break, the addition of MAPE was found beneficial in enhancing the ductility of RH/MMT filled LDPE films; as it increases from 61% to 109% with increasing compatibilizer:filler ratio from 0 to 2. For the compatibilizer:filler ratio of 0.75 or lower than that, the elongation at break increases abruptly. However, when the ratio is more than 0.75, the increasing trend becomes gradual. Ductility enhancement with increasing graft co-polymers has also been reported by other researchers [12, 31].

3.3.4   BARRIER PROPERTIES

Plastics and their composites are permeable to gases at different degrees and this permeability depend on the internal architecture of the polymer such as its crystallinity, crystalline: amorphous ratio, thermal and mechanical treatment, degree of cross linking, glass transition temperature, molecular mass and so on [32]. The knowledge of permeation behavior of these gases is important for the selection of correct packaging materials for prolonging shelf life. The concept of permeability is normally associated with the quantitative evaluation of the barrier properties of a plastic so, in general, a plastic that is a good barrier has a low permeability. In the present study, oxygen (O_2) and carbon dioxide (CO_2) permeance through the nanocomposite films was evaluated by their volume passing through the films in unit time according to the procedure described in the experimental section. Table 2 presents a summary of the barrier properties of RH/MMT filled LDPE composite films for all the studied compatibilizer:filler ratios.

As the table shows, the MAPE addition to the RH/MMT filler filled LDPE films resulted in a decrease in the O_2 and CO_2 permeability through the film and both the gases followed a similar trend with small variations. Increase in the gas barrier of polymeric materials with the incorporation of nanoplatelets has also been reported by other researchers [13, 14]. This improvement is due to the delamination and distribution of impermeable clay platelets that act as obstacles in the path of the diffusing molecules. These impermeable obstacles force the diffusing molecules to follow a tortuous path, which in turn, retards/reduce the gas molecule diffusion. These findings are consistent with our SEM and XRD observations where RH particles are better dispersed and clay platelets are delaminated with the addition of the compatibilizer. The decrease in permeability also confirms that the impermeable clay platelets are well dispersed as the extent of dispersion of clay platelets plays a major role in creating a tortuous path for the permeating molecules. It is worth mentioning that CO_2 is more permeable than O_2, although its molecular diameter is bigger than that of O_2, suggesting that the permeability behavior is independent of the size of permeating molecule (32). The higher permeability of CO_2 can be attributed to its higher solubility in the cellulosic filler filled film. Further, the barrier increases with the increase of compatibilizer:filler ratio and reaches the maximum value when the ratio is 1. Beyond this limit, the barrier started deteriorating. This decrease in barrier with higher amounts of MAPE may be because of large side groups of MAPEs. These large side groups may affect the structure of LDPE crystal units, increasing the space between polymer chains and leading to more channels for gas transportation [29].

Another important property result that is observed is a wider range of $CO_2:O_2$ permeability ratio (2.47 to 2.68). The determination of $CO_2:O_2$ permeability ratio is necessary to select the packaging films according to the requirement of foods to be stored. Higher ratio will allow less accumulation of CO_2 and vice versa.

3.3.5 THERMAL PROPERTIES

Investigation of thermal properties of a material is important for their processing and final applications and differential scanning calorimetry (DSC) is one of the convenient methods for analyzing first order transitions like melting and crystallization. The RH/MMT filler filled LDPE composite film samples with varying compatibilizer:filler ratios were subjected to DSC analysis and the thermal properties that were obtained from this DSC analysis including melting temperature (T_m), crystallization temperature (T_c) and crystallinity are summarized in Table 3.1. Melting and crystallization temperatures are the peak temperatures of heating and cooling thermograms respectively and degree of crystallinity (X_c) was obtained by the comparison of the heat of fusion with a 100% crystalline PE heat of fusion according to the procedure described in the Experimental section.

Melting and crystallization peaks were observed at 106.9 and 93.5°C for LDPE and 120.3 and 103.2°C for MAPE. Adding RH/MMT filler into

TABLE 3.1 Barrier and Thermal Properties of RH/MMT Filler Filled LDPE Composite Films With Varying Compatibilizer:filler Ratios.

Compatibilizer to filler ratio	Permeability (mL-mm/ m²/24hr/atm)		DSC measurements		
	O_2	CO_2	T_m °C	T_c °C	X_c %
0	1297	3476	108.5	94.2	17.5
0.25	1105	2821	107.8	93.7	17.9
0.5	747	1843	108.3	94.9	18.5
0.75	395	1015	108.9	95.3	19.3
1.0	341	855	108.8	95.1	21.2
1.5	403	995	109.4	94.6	21.1
2.0	475	1253	108.1	94.1	22.3

the LDPE matrix showed no significant difference with the neat matrix as shown in Table 3.1. Further, adding MAPE did not produce any significant difference and for all the studied compatibilizer:filler ratios, the melting peaks were observed at about 107–109°C. Also the crystallization of all samples proceeded in a similar way, with the main peak around 93–95°C. This behavior indicates no significant nucleation activity of MMT in the LDPE based system [30]. In contrast to melting and crystallization behavior, increasing compatibilizer:filler ratio resulted in gradual increase in the degree of crystallinity. This increase could originate from two causes: first, due to the higher crystallinity of MAPE; and second, because of the enhanced interfacial adhesion between filler and matrix material [33]. It is assumed that the crystalline regions of a polymer are essentially impermeable to gases. Thus, the improvement in crystallinity should result in lower permeability. Barrier properties (in the previous section) are also consistent with these findings.

3.4   CONCLUSIONS

MAPE compatibilized RH/MMT hybrid filler filled LDPE nanocomposite blown film was successfully fabricated and the effect of compatibilizer:filler ratio on the morphological, mechanical, oxygen barrier and thermal properties was examined. XRD analysis revealed that the MMT platelets can be delaminated significantly with the appropriate compatibilizer:filler ratio to realize large filler aspect ratio. For instance, the composite films containing compatibilizer:filler ratio 1 increased the interlayer spacing of MMT by about 36% compared with neat MMT. SEM micrographs showed that the interfacial interactions and film homogeneity were improved with the incorporation of the compatibilizer. Tensile and barrier properties improved significantly with the addition of compatibilizer and the optimum ratio for these composites is in the range of 0.75:1 to 1:1. The significant increase in these tensile and barrier properties also confirm that the MMT platelets are well dispersed in the hybrid system. Regarding DSC results, no drastic change was observed in the melting and crystallization behavior of the hybrid system with different compatibilizer:filler ratios.

KEYWORDS

- barrier properties
- compatibilizer to filler ratio
- extrusion blow film
- hybrid filler
- tensile strength

REFERENCES

1. Zhou, X., Shao, X., Shu, J.-J., Liu, M.-M., Liu, H.-L., Feng, X.-H., Liu, F. Thermally stable ionic liquid-based sol–gel coating for ultrasonic extraction–solid-phase microextraction–gas chromatography determination of phthalate esters in agricultural plastic films. *Talanta.* 2012, 89, 129–135.
2. Marsh, K. Bugusu, B. Food Packaging—Roles, Materials, and Environmental Issues. *Journal of Food Science.* 2007, 72, R39–R55.
3. Kyaw, B., Champakalakshmi, R., Sakharkar, M., Lim, C., Sakharkar, K. Biodegradation of Low Density Polythene (LDPE) by Pseudomonas Species. *Indian J. Microbiol.* 2012, 52, 411–419.
4. Bodirlau, R., Teaca, C.-A., Spiridon, I. Influence of natural fillers on the properties of starch-based biocomposite films. *Compos. Part B-Eng.* 2013, 44, 575–583.
5. Sarabi, M. T., Behravesh, A. H., Shahi, P., Soury, E. Reprocessing of Extruded Wood-Plastic Composites; Mechanical Properties. *J. Biobased. Mater. Bio.* 2012, 6, 221–229.
6. Tajeddin, B., Rahman, R. A., Abdulah, L. C. The effect of polyethylene glycol on the characteristics of kenaf cellulose/low-density polyethylene biocomposites. *Int J Biol Macromol.* 2010, 47, 292–297.
7. George, J., Kumar, R., Jayaprahash, C., Ramakrishna, A., Sabapathy, S. N., Bawa, A. S. Rice bran-filled biodegradable low-density polyethylene films: Development and characterization for packaging applications. *Int J Phys Sci.* 2006, 102, 4514–4522.
8. Yang, H.-S., Kim, H.-J., Park, H.-J., Lee, B.-J., Hwang, T.-S. Water absorption behavior and mechanical properties of lignocellulosic filler–polyolefin bio-composites. *Composite Structures.* 2006, 72, 429–437.
9. Zahedi, M., Pirayesh, H., Khanjanzadeh, H., Tabar, M. M. Organo-modified montmorillonite reinforced walnut shell/polypropylene composites. *Mater Design.* 2013, 51, 803–809.
10. Islam, M. S., Hamdan, S., Talib, Z. A., Ahmed, A. S., Rahman, M. R. Tropical wood polymer nanocomposite (WPNC): The impact of nanoclay on dynamic mechanical thermal properties. *Composite Science and Technology.* 2012, 72, 1995–2001.

11. Faruk, O. Matuana, L. M. Nanoclay reinforced HDPE as a matrix for wood-plastic composites. *Compos Sci Technol.* 2008, 68, 2073–2077.

12. Hemmasi, A. H., Khademi-Eslam, H., Talaiepoor, M., Kord, B., Ghasemi, I. Effect of Nanoclay on the Mechanical and Morphological Properties of Wood Polymer Nanocomposite. *Journal of Reinforced Plastics and Composites.* 2010, 29, 964–971.

13. Stephen, R., Ranganathaiah, C., Varghese, S., Joseph, K., Thomas, S. Gas transport through nano and micro composites of natural rubber (NR) and their blends with carboxylated styrene butadiene rubber (XSBR) latex membranes. *Polymer.* 2006, 47, 858–870.

14. Meera, A., Thomas P, S., Thomas, S. Effect of organoclay on the gas barrier properties of natural rubber nanocomposites. *Polymer Composites.* 2012, 33, 524–531.

15. Majeed, K., Jawaid, M., Hassan, A., Abu Bakar, A., Abdul Khalil, H. P. S., Salema, A. A., Inuwa, I. Potential materials for food packaging from nanoclay/natural fibers filled hybrid composites. *Mater Design.* 2013, 46, 391–410.

16. Rozman, H. D., Tay, G. S., Kumar, R. N., Abusamah, A., Ismail, H., Mohd. Ishak, Z. A. Polypropylene–oil palm empty fruit bunch–glass fiber hybrid composites: a preliminary study on the flexural and tensile properties. *European Polymer Journal.* 2001, 37, 1283–1291.

17. Thwe, M. M. Liao, K. Effects of environmental aging on the mechanical properties of bamboo–glass fiber reinforced polymer matrix hybrid composites. *Compos Part A-Appl S.* 2002, 33, 43–52.

18. Sreekala, M. S., George, J., Kumaran, M. G., Thomas, S. The mechanical performance of hybrid phenol-formaldehyde-based composites reinforced with glass and oil palm fibers. *Composite Science and Technology.* 2002, 62, 339–353.

19. Kord, B. Studies on mechanical characterization and water resistance of glass fiber/thermoplastic polymer bionanocomposites. *J Appl Polym Sci.* 2012, 123, 2391–2396.

20. Sliwa, F., Bounia, N. E., Charrier, F., Marin, G., Malet, F. Mechanical and interfacial properties of wood and bio-based thermoplastic composite. *Compos. Sci. Technol.* 2012.

21. Kabir, M. M., Wang, H., Lau, K. T., Cardona, F. Chemical treatments on plant-based natural fiber reinforced polymer composites: An overview. *Compos. Part B-Eng.* 2012, 43, 2883–2892.

22. Chan, M.-L., Lau, K.-T., Wong, T.-T., Ho, M.-P., Hui, D. Mechanism of reinforcement in a nanoclay/polymer composite. *Compos. Part B-Eng.* 2011, 42, 1708–1712.

23. Ku, H., Wang, H., Pattarachaiyakoop, N., Trada, M. A review on the tensile properties of natural fiber reinforced polymer composites. *Compos. Part B-Eng.* 2011, 42, 856–873.

24. Kaewkuk, S., Sutapun, W., Jarukumjorn, K. Effects of interfacial modification and fiber content on physical properties of sisal fiber/polypropylene composites. *Compos. Part B-Eng.* 2012.

25. Kord, B. Effect of nanoparticles loading on properties of polymeric composite based on Hemp Fiber/Polypropylene. *Journal of Thermoplastic Composite Materials.* 2012, 25, 793–806.

26. Gwon, J. G., Lee, S. Y., Chun, S. J., Doh, G. H., Kim, J. H. Physical and mechanical properties of wood–plastic composites hybridized with inorganic fillers. *J Compos Mater.* 2012, 46, 301–309.

27. Najafi, A., Kord, B., Abdi, A., Ranaee, S. The impact of the nature of nanoclay on physical and mechanical properties of polypropylene/reed flour nanocomposites. *Journal of Thermoplastic Composite Materials.* 2012, 25, 717–727.

28. Spencer, M. W., Hunter, D. L., Knesek, B. W., Paul, D. R. Morphology and properties of polypropylene nanocomposites based on a silanized organoclay. *Polymer.* 2011, 52, 5369–5377.

29. Zhong, Y., Janes, D., Zheng, Y., Hetzer, M., De Kee, D. Mechanical and oxygen barrier properties of organoclay-polyethylene nanocomposite films. *Polym Eng Sci.* 2007, 47, 1101–1107.

30. Ali Dadfar, S. M., Alemzadeh, I., Reza Dadfar, S. M., Vosoughi, M. Studies on the oxygen barrier and mechanical properties of low density polyethylene/organoclay nanocomposite films in the presence of ethylene vinyl acetate copolymer as a new type of compatibilizer. *Mater Design.* 2011, 32, 1806–1813.

31. Tjong, S. C., Bao, S. P. Fracture toughness of high density polyethylene/SEBS-g-MA/montmorillonite nanocomposites. *Compos. Sci. Technol.* 2007, 67, 314–323.

32. Siracusa, V., Blanco, I., Romani, S., Tylewicz, U., Dalla Rosa, M. Gas Permeability and Thermal Behavior of Polypropylene Films Used for Packaging Minimally Processed Fresh-Cut Potatoes: A Case Study. *J Food Sci.* 2012, 77, E264-E272.

33. Zaman, H. U., Hun, P. D., Khan, R. A., Yoon, K.-B. Polypropylene/clay nanocomposites: Effect of compatibilizers on the morphology, mechanical properties and crystallization behaviors. *Journal of Thermoplastic Composite Materials.* 2012.

CHAPTER 4

A DETAILED REVIEW ON MODELING OF CNT/GREEN POLYMER COMPOSITES

A. K. HAGHI

University of Guilan, Rasht, Iran

CONTENTS

4.1 INTRODUCTION

The interest manifested, in recent decades, in green composites, is explained by the wide range of their potential applications in many fields. Special attention has been paid to carbon nanotubes, which exhibit superior electrical and mechanical properties that can improve the properties of green polymer composites. The current review focuses on the progress made in carbon nanotube/green polymer composites. In this chapter, the modeling of mechanical properties of CNT/polymer nano-composites is reviewed. The chapter starts with the structural and intrinsic mechanical properties of CNTs. Then we introduce some computational methods that have been applied to polymer nano-composites, covering from molecular scale (e.g., molecular dynamics, Monte Carlo), microscale (e.g., Brownian dynamics, dissipative particle dynamics, lattice Boltzmann, time-dependent Ginzburg–Landau method, dynamic density functional theory method) to mesoscale and macroscale (e.g., micromechanics, equivalent-continuum and self-similar approaches, finite element method). Hence, the knowledge and understanding of the nature and mechanics of length and orientation of nano-tube and load transfer between nano-tube and polymer is critical for manufacturing of enhanced carbon nano-tube-polymer composites and will enable in tailoring of the interface for specific applications or superior mechanical properties. So, in this review a state of these parameters in mechanics of carbon nano-tube polymer composites will be discussed along with some directions for future research in this field.

Carbon nanotubes were first observed by Iijima, almost two decades ago [1], and since then, extensive work has been carried out to characterize their properties [2–4]. A wide range of characteristic parameters has been reported for carbon nanotube nanocomposites. There are contradictory reports that show the influence of carbon nanotubes on a particular property (e.g., Young's modulus) to be improving, indifferent or even deteriorating [5]. However, from the experimental point of view, it is a great challenge to characterize the structure and to manipulate the fabrication of polymer nanocomposites. The development of such materials is still largely empirical and a finer degree of control of their properties cannot be achieved so far. Therefore, computer modeling and simulation will play an ever increasing role in predicting and designing material properties, and guiding such experimental work as synthesis and characterization, For polymer nanocomposites, computer modeling and simulation are especially useful in the hierarchical characteristics of the structure and dynamics of polymer nanocomposites ranging from molecular scale, microscale to mesoscale and macroscale, in particular, the molecular structures and dynamics at the interface between nanoparticles and polymer matrix. The purpose of this review is to discuss the application of modeling and simulation techniques to polymer nanocomposites. This includes a broad subject covering methodologies at various length and time scales and many aspects of polymer nanocomposites. We organize the review as follows. In Section 4.1 we will discuss about Carbon nanotubes (CNTs) and nano composite properties. In Section 4.2, we introduce briefly the computational methods used so far for the systems of polymer nanocomposites which can be roughly divided into three types: molecular scale methods (e.g., molecular dynamics (MD), Monte Carlo (MC)), microscale methods (e.g., Brownian dynamics (BD), dissipative particle dynamics (DPD), lattice Boltzmann (LB), time dependent Ginzburg–Lanau method, dynamic density functional theory (DFT) method), and mesoscale and macroscale methods (e.g., micromechanics, equivalent-continuum and self-similar approaches, finite element method (FEM)) [6]. Many researchers used this method for determine the mechanical properties of nanocomposite that in Section 4.3 will be discussed. In Section 4.4 modeling of interfacial load transfer between CNT and polymer in nanocomposite will be introduced and finally we conclude the review by emphasizing the current challenges and future research directions.

4.2   CARBON NANOTUBES (CNTS) AND NANO COMPOSITE PROPERTIES

4.2.1   INTRODUCTION TO CNTS

CNTs are one dimensional carbon materials with aspect ratio greater than, 1000. They are cylinders composed of rolled-up graphite planes with diameters in nanometer scale [7–10]. The cylindrical nanotube usually has at least one end capped with a hemisphere of fullerene structure. Depending on the process for CNT fabrication, there are two types of CNTs [8–11]: single-walled CNTs (SWCNTs) and multiwalled CNTs (MWCNTs). SWCNTs consist of a single graphene layer rolled up into a seamless cylinder whereas MWCNTs consist of two or more concentric cylindrical shells of graphene sheets coaxially arranged around a central hollow core with van der Waals forces between adjacent layers. According to the rolling angle of the graphene sheet, CNTs have three chirality's: armchair, zigzag and chiral one. The tube chirality is defined by the chiral vector, Ch = na$_1$ + ma$_2$ (Figure 4.1), where the integers (n, m) are the number of steps along the unit vectors (a1 and a2) of the hexagonal lattice [9, 10]. Using this (n, m) naming scheme, the three types of orientation of the carbon atoms around the nanotube circumference are specified. If n = m, the nanotubes are called "armchair." If m = 0, the nanotubes are called "zigzag." Otherwise, they are called "chiral." The chirality of nanotubes has significant impact on their transport properties, particularly the electronic properties. For a given (n, m) nanotube, if (2n + m) is a multiple of 3, then the nanotube is metallic, otherwise the nanotube is a semiconductor. Each MWCNT contains a multi-layer of graphene, and each layer can have different chirality's, so the prediction of its physical properties is more complicated than that of SWCNT. figure 4.1 shows the CNT with different chirality's.

4.2.2 CLASSIFICATION OF CNT/POLYMER NANOCOMPOSITES

Polymer composites, consisting of additives and polymer matrices, including thermoplastics, thermosets and elastomers, are considered to be an important group of relatively inexpensive materials for many engineering applications. Two or more materials are combined to produce composites that possess properties that are unique and cannot be obtained each material acting alone. For example, high modulus carbon fibers or silica particles are added into a polymer to produce reinforced polymer composites that

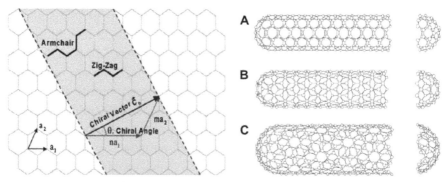

FIGURE 4.1 Schematic diagram showing how a hexagonal sheet of graphene is rolled to form a CNT with different chirality's (A: armchair; B: zigzag; C: chiral).

exhibit significantly enhanced mechanical properties including strength, modulus and fracture toughness. However, there are some bottlenecks in optimizing the properties of polymer composites by employing traditional micron-scale fillers. The conventional filler content in polymer composites is generally in the range of 10–70 wt.%, which in turn results in a composite with a high density and high material cost. In addition, the modulus and strength of composites are often traded for high fracture toughness [12]. Unlike traditional polymer composites containing micron-scale fillers, the incorporation of nanoscale CNTs into a polymer system results in very short distance between the fillers, thus the properties of composites can be largely modified even at an extremely low content of filler. For example, the electrical conductivity of CNT/epoxy nanocomposites can be enhanced several orders of magnitude with less than 0.5 wt.% of CNTs [13]. As described previously, CNTs are among the strongest and stiffest fibers ever known. These excellent mechanical properties combined with other physical properties of CNTs exemplify huge potential applications of CNT/polymer nanocomposites. Ongoing experimental works in this area have shown some exciting results, although the much-anticipated commercial success has yet to be realized in the years ahead. In addition, CNT/polymer nanocomposites are one of the most studied systems because of the fact that polymer matrix can be easily fabricated without damaging CNTs based on conventional manufacturing techniques, a potential advantage of reduced cost for mass production of nanocomposites in the future. Following the first report on the preparation of a CNT/polymer nanocomposite in 1994 [14], many research efforts have been made to understand their structure–property relationship and find useful applications in different fields, and these efforts have become more pronounced after the realization of CNT fabrication in industrial scale with lower costs in the beginning of the twenty-first century [15]. According to the specific application, CNT/polymer nanocomposites can be classified as structural or functional composites [16]. For the structural composites, the unique mechanical properties of CNTs, such as the high modulus, tensile strength and strain to fracture, are explored to obtain structural materials with much improved mechanical properties. As for CNT/polymer functional composites, many other unique properties of CNTs, such as electrical, thermal, optical and damping properties along with their excellent mechanical properties, are used to develop multi-functional composites for applications in the fields of heat resistance, chemical sensing, electrical and thermal management, photoemission, electromagnetic absorbing and energy storage performances, etc.

4.3   MODELING AND SIMULATION TECHNIQUES

4.3.1   MOLECULAR SCALE METHODS

The modeling and simulation methods at molecular level usually employ atoms, molecules or their clusters as the basic units considered. The most popular methods include molecular mechanics (MM), MD and MC simulation. Modeling of polymer nanocomposites at this scale is predominantly directed toward the thermodynamics and kinetics of the formation, molecular structure and interactions. The diagram in Figure 4.1 describes the equation of motion for each method and the typical properties predicted from each of them [17–22]. We introduce here the two widely used molecular scale methods: MD and MC.

4.3.1.1   Molecular Dynamics

MD is a computer simulation technique that allows one to predict the time evolution of a system of interacting particles (e.g., atoms, molecules, granules, etc.) and estimate the relevant physical properties [23, 24]. Specifically, it generates such information as atomic positions, velocities and forces from which the macroscopic properties (e.g., pressure, energy, heat capacities) can be derived by means of statistical mechanics. MD simulation usually consists of three constituents: (i) a set of initial conditions (e.g., initial positions and velocities of all particles in the system); (ii) the interaction potentials to represent the forces among all the particles; (iii) the evolution of the system in time by solving a set of classical Newtonian equations of motion for all particles in the system. The equation of motion is generally given by

$$\overrightarrow{F}_1(t) = m_i \frac{d^2 \overline{r_1}}{dt^2} \qquad (1)$$

where is the force acting on the i-th atom or particle at time t which is obtained as the negative gradient of the interaction potential U, m_i is the atomic mass and $\vec{r_i}$ the atomic position. A physical simulation involves the proper selection of interaction potentials, numerical integration, periodic boundary conditions, and the controls of pressure and temperature to mimic-physically meaningful thermodynamic ensembles. The interaction potentials together with their parameters, that is, the so-called force field, describe in detail how the particles in a system interact with each other, that is, how the potential energy of a system depends on the particle coordinates. Such a force field may be obtained by quantum method (e.g., ab initio), empirical method (e.g., Lennard–Jones, Mores, Born-Mayer) or quantum-empirical method (e.g., embedded atom model, glue model, bond order potential). The criteria for selecting a force field include the accuracy, transferability and computational speed. A typical interaction potential U may consist of a number of bonded and nonbonded interaction terms:

$$U(\vec{r_1},\vec{r_2},\vec{r_3},...,\vec{r_n}) = \sum_{i_{bond}}^{N_{bond}} U_{bond}(i_{bond},\vec{r_a},\vec{r_b}) + \sum_{i_{angle}}^{N_{angle}} U_{angle}(i_{angle},\vec{r_a},\vec{r_b},\vec{r_c})$$

$$+\sum_{i_{torsion}}^{N_{torsion}} U_{torsion}(i_{torsion},\vec{r_a},\vec{r_b},\vec{r_c},\vec{r_d}) + \sum_{i_{inversion}}^{N_{inversion}} U_{inversion}(i_{inversion},\vec{r_a},\vec{r_b},\vec{r_c},\vec{r_d})$$

$$+\sum_{i=1}^{N-1}\sum_{j>i}^{N} U_{vdw}(i,j,\vec{r_a},\vec{r_b}) + \sum_{i=1}^{n-1}\sum_{j>i}^{N} U_{electrostatic}(i,j,\vec{r_a},\vec{r_b}) \qquad (2)$$

The first four terms represent bonded interactions, that is, bond stretching Ubond, bond-angle bend Uangle and dihedral angle torsion Utorsion and inversion interaction Uinversion, while the last two terms are non-bonded interactions, that is, van der Waals energy Uvdw and electrostatic energy Uelectrostatic. In the equation, $\vec{r_a},\vec{r_b},\vec{r_c},\vec{r_d}$ are the positions of the atoms or particles specifically involved in a given interaction; $N_{bond}, N_{angle}, N_{torsion}$ and $N_{inversion}$ stand for the total numbers of these respective interactions in the simulated system; $i_{bond}, i_{angle}, i_{torsion}$ and $i_{inversion}$ uniquely specify an individual interaction of each type; i and j in the van der Waals and electrostatic terms indicate the atoms involved in the interaction. There are many algorithms for integrating the equation of motion using finite difference methods. The algorithms of varlet, velocity varlet, leap-frog and Beeman, are commonly used in MD simulations [23]. All algorithms assume that the atomic position $\vec{r}$, velocities $\vec{v}$ and accelerations $\vec{a}$, can be approximated by a Taylor series expansion:

$$\vec{r}(t+\delta t) = \vec{r}(t) + \vec{v}(t)\delta t + \frac{1}{2}\vec{a}(t)\delta^2 t + ... \tag{3}$$

$$\vec{v}(t+\delta t) = \vec{v}(t)\delta t + \frac{1}{2}\vec{b}(t)\delta^2 t + ... \tag{4}$$

$$\vec{a}(t+\delta t) = \vec{a}(t) + \vec{b}(t)\delta t + ... \tag{5}$$

Generally speaking, a good integration algorithm should conserve the total energy and momentum and be time-reversible. It should also be easy to implement and computationally efficient, and permit a relatively long time step. The verlet algorithm is probably the most widely used method. It uses the positions $\vec{r}(t)$ and accelerations $\vec{a}(t)$ at time t, and the positions $\vec{r}(t-\delta t)$ from the previous step (t − δ) to calculate the new positions $\vec{r}(t+\delta t)$ at (t+δt), we have:

$$\vec{r}(t+\delta t) = \vec{r}(t) + \vec{v}(t)\delta t + \frac{1}{2}\vec{a}(t)\delta t^2 + ... \tag{6}$$

$$\vec{r}(t-\delta t) = \vec{r}(t) - \vec{v}(t)\delta t + \frac{1}{2}\vec{a}(t)\delta t^2 + ... \tag{7}$$

$$\vec{r}(t+\delta t) = 2\vec{r}(t)\delta t - \vec{r}(t-\delta t) + \vec{a}(t)\delta t^2 + ... \tag{8}$$

The velocities at time t and $t + \dfrac{1}{2\delta t}$ can be respectively estimated

$$\vec{v}(t) = \left[\vec{r}(t+\delta t) - \vec{r}(t-\delta t)\right] / 2\delta t \tag{9}$$

$$\vec{v}(t+1/2\delta t) = \left[\vec{r}(t+\delta t) - \vec{r}(t-\delta t)\right] / \delta t \tag{10}$$

MD simulations can be performed in many different ensembles, such as grand canonical (μVT), microcanonical (NVE), canonical (NVT) and isothermal–isobaric (NPT). The constant temperature and pressure can be controlled by adding an appropriate thermostat (e.g., Berendsen, Nose, Nose–Hoover and Nose–Poincare) and barostat (e.g., Andersen, Hoover and Berendsen), respectively. Applying MD into polymer composites allows us to investigate into the effects of fillers on polymer structure and dynamics in the vicinity of polymer–filler interface and also to probe the effects of polymer–filler interactions on the materials properties.

4.3.1.2 Monte Carlo

MC technique, also called Metropolis method [24], is a stochastic method that uses random numbers to generate a sample population of the system from which one can calculate the properties of interest. A MC simulation usually consists of three typical steps. In the first step, the physical problem under investigation is translated into an analogous probabilistic or statistical model. In the second step, the probabilistic model is solved by a numerical stochastic sampling experiment. In the third step, the obtained data are analyzed by using statistical methods. MC provides only the information on equilibrium properties (e.g., free energy, phase equilibrium), different from MD, which gives non-equilibrium as well as equilibrium properties. In a NVT ensemble with N atoms, one hypothesizes a new configuration by arbitrarily or systematically moving one atom from position i→j. Due to such atomic movement, one can compute the change in the system Hamiltonian ΔH:

$$\Delta H = H(j) - H(i) \tag{11}$$

where H(i) and H(j) are the Hamiltonian associated with the original and new configuration, respectively.

This new configuration is then evaluated according to the following rules. If $\Delta H < 0$, then the atomic movement would bring the system to a state of lower energy. Hence, the movement is immediately accepted and the displaced atom remains in its new position. If $\Delta H \geq 0$, the move is accepted only with a certain probability $Pi \rightarrow j$ which is given by

$$Pi \rightarrow j \propto \exp\left(-\frac{\Delta H}{K_B T}\right) \tag{12}$$

where K_B is the Boltzmann constant. According to Metropolis et al. [25] one can generate a random number ζ between 0 and 1 and determine the new configuration according to the following rule:

$$\xi \leq \exp\left(-\frac{\Delta H}{K_B T}\right); \text{ the move is accepted;} \qquad (13)$$

$$\xi \rangle \exp\left(-\frac{\Delta H}{K_B T}\right); \text{ the move is not accepted.} \qquad (14)$$

If the new configuration is rejected, one counts the original position as a new one and repeats the process by using other arbitrarily chosen atoms. In a μVT ensemble, one hypothesizes a new configuration j by arbitrarily choosing one atom and proposing that it can be exchanged by an atom of a different kind. This procedure affects the chemical composition of the system. Also, the move is accepted with a certain probability. However, one computes the energy change ΔU associated with the change in composition. The new configuration is examined according to the following rules. If $\Delta U <0$, the move of compositional change is accepted. However, if $\Delta U \geq 0$, the move is accepted with a certain probability which is given by

$$Pi \rightarrow j \propto \exp(-\frac{\Delta U}{K_B T}) \qquad (15)$$

where ΔU is the change in the sum of the mixing energy and the chemical potential of the mixture. If the new configuration is rejected one counts the original configuration as a new one and repeats the process by using some other arbitrarily or systematically chosen atoms. In polymer nanocomposites, MC methods have been used to investigate the molecular structure at nanoparticle surface and evaluate the effects of various factors.

4.3.2 MICROSCALE METHODS

The modeling and simulation at microscale aim to bridge molecular methods and continuum methods and avoid their shortcomings. Specifically, in nanoparticle–polymer systems, the study of structural evolution (i.e., dynamics of phase separation) involves the description of bulk flow (i.e., hydrodynamic behavior) and the interactions between nanoparticle and polymer components. Note that hydrodynamic behavior is relatively straightforward to handle by continuum methods but is very difficult and expensive to treat by atomistic methods. In contrast, the interactions between components can be examined at an atomistic level but are usually not straightforward to incorporate at the continuum level. Therefore, various simulation methods have been evaluated and extended to study the microscopic structure and phase separation of these polymer nanocomposites, including BD, DPD, LB, time-dependent Ginsburg–Landau (TDGL) theory, and dynamic DFT. In these methods, a polymer system is usually treated with a field description or microscopic particles that incorporate molecular details implicitly. Therefore, they are able to simulate the phenomena on length and time scales currently inaccessible by the classical MD methods.

4.3.2.1   Brownian Dynamics

BD simulation is similar to MD simulations [26]. However; it introduces a few new approximations that allow one to perform simulations on the microsecond timescale whereas MD simulation is known up to a few nanoseconds. In BD the explicit description of solvent molecules used in MD is replaced with an implicit continuum solvent description. Besides, the internal motions of molecules are typically ignored, allowing a much larger time step than that of MD. Therefore, BD is particularly useful for systems where there is a large gap of time scale governing the motion of different components. For example, in polymer–solvent mixture, a short time-step is required to resolve the fast motion of the solvent molecules, whereas the evolution of the slower modes of the system requires a larger time step. However, if the detailed motion of the solvent molecules is concerned, they may be removed from the simulation and their effects on the polymer are represented by dissipative ($-\gamma P$) and random ($\sigma\,\zeta(t)$) force terms. Thus, the forces in the governing Eq. (16) is replaced by a Langevin equation,

$$F_i(t) = \sum_{i \ne j} F_{ij}^{\ c} - \gamma P_i + \sigma \zeta_i(t) \qquad (16)$$

where F_{ij}^c is the conservative force of particle j acting on particle i, γ and σ are constants depending on the system, Pi the momentum of particle i, and $\zeta(t)$ a Gaussian random noise term. One consequence of this approximation of the fast degrees of freedom by fluctuating forces is that the energy and momentum are no longer conserved, which implies that the macroscopic behavior of the system will not be hydrodynamic. In addition, the effect of one solute molecule on another through the flow of solvent molecules is neglected. Thus, BD can only reproduce the diffusion properties but not the hydrodynamic flow properties since the simulation does not obey the Navier–Stokes equations.

4.3.2.2   Dissipative Particle Dynamics

DPD was originally developed by Hoogerbrugge and Koelman [27]. It can simulate both Newtonian and non-Newtonian fluids, including polymer melts and blends, on microscopic length and time scales. Like MD and BD, DPD is a particle-based method. However, its basic unit is not a single atom or molecule but a molecular assembly (i.e., a particle). DPD particles are defined by their mass M_i, position r_i and momentum P_i. The interaction force between two DPD particles i and j can be described by a sum of conservative F_{ij}^C, dissipative F_{ij}^D and random forces F_{ij}^R [28–30]:

$$F_{ij} = F_{ij}^{\ C} + F_{ij}^{\ D} + F^R_{\ ij} \qquad (17)$$

While the interaction potentials in MD are high-order polynomials of the distance between two particles, in DPD the potentials are softened so as to approximate the effective potential at microscopic length scales. The form of the conservative force in particular is chosen to decrease linearly with increasing r_{ij}. Beyond a certain cut-off separation r_c, the weight functions and thus the forces are all zero. Because the forces are pair wise and momentum is conserved, the macroscopic behavior directly incorporates Navier–Stokes hydrodynamics. However, energy is not conserved because of the presence of the dissipative and random force terms, which are similar to those of BD, but incorporate the effects of Brownian motion on larger length scales. DPD has several advantages over MD, for example, the hydrodynamic behavior is observed with far fewer particles than required in a MD simulation because of its larger particle size. Besides, its force forms allow larger time steps to be taken than those in MD.

4.3.2.3   Lattice Boltzmann (LB)

LB [31] is another microscale method that is suited for the efficient treatment of polymer solution dynamics. It has recently been used to investigate the phase separation of binary fluids in the presence of solid particles. The LB method is originated from lattice gas automaton which is constructed as a simplified, fictitious molecular dynamic in which space, time and particle velocities are all discrete. A typical lattice gas automaton consists of a regular lattice with particles residing on the nodes. The main feature of the LB method is to replace the particle occupation variables (Boolean variables), by single-particle distribution functions (real variables) and neglect individual particle motion and particle–particle correlations in the kinetic equation. There are several ways to obtain the LB equation from either the discrete velocity model or the Boltzmann kinetic equation, and to derive the macroscopic Navier–Stokes equations from the LB equation. An important advantage of the LB method is that microscopic physical interactions of the fluid particles can be conveniently incorporated into the numerical model. Compared with the Navier–Stokes equations, the LB method can handle the interactions among fluid particles and reproduce the microscale mechanism of hydrodynamic behavior. Therefore it belongs to the MD in nature and bridges the gap between the molecular level and macroscopic level. However, its main disadvantage is that it is typically not guaranteed to be numerically stable and may lead to physically unreasonable results, for instance, in the case of high forcing rate or high interparticle interaction strength.

4.3.2.4   Time-Dependent Ginzburg–Landau Method

TDGL is a microscale method for simulating the structural evolution of phase-separation in polymer blends and block copolymers. It is based on the Cahn–Hilliard–Cook (CHC) nonlinear diffusion equation for a binary blend and falls under the more general phase-field and reaction-diffusion models [32–34]. In the TDGL method, a free-energy function is minimized to simulate a temperature quench from the miscible region of the phase diagram to the immiscible region. Thus, the resulting time-dependent structural evolution of the polymer blend can be investigated by solving the TDGL/CHC equation for the time dependence of the local blend concentration. Glotzer and co-workers have discussed and applied this method to polymer blends and particle-filled polymer systems [35]. This model reproduces the growth kinetics of the TDGL model, demonstrating that such quantities are insensitive to the precise form of the double-well potential of the bulk free-energy term. The TDGL and CDM methods have recently been used to investigate the phase-separation of polymer nanocomposites and polymer blends in the presence of nanoparticles [36–40].

4.3.2.5   Dynamic DFT Method

Dynamic DFT method is usually used to model the dynamic behavior of polymer systems and has been implemented in the software package Mesodyn TM from Accelrys [41]. The DFT models the behavior of polymer fluids by combining Gaussian mean-field statistics with a TDGL model for the time evolution of conserved order parameters. However, in contrast to traditional phenomenological free-energy expansion methods employed in the TDGL approach, the free energy is not truncated at a certain level, and instead retains the full polymer path integral numerically. At the expense of a more challenging computation, this allows detailed information about a specific polymer system beyond simply the Flory–Huggins parameter and mobilities to be included in the simulation. In addition, viscoelasticity, which is not included in TDGL approaches, is included at the level of the Gaussian chains. A similar DFT approach has been developed by Doi and co-workers [42, 43] and forms the basis for their new software tool Simulation Utilities for Soft and Hard Interfaces (SUSHI), one of a suite of molecular and mesoscale-modeling tools (called OCTA) developed for the simulation of polymer materials [44]. The essence of dynamic DFT method is that the instantaneous unique conformation distribution can be obtained from the off-equilibrium density profile by coupling a fictitious external potential to the Hamiltonian.

Once such distribution is known, the free energy is then calculated by standard statistical thermodynamics. The driving force for diffusion is obtained from the spatial gradient of the first functional derivative of the free energy with respect to the density. Here, we describe briefly the equations for both polymer and particle in the diblock polymer–particle composites [38].

4.3.3   MESOSCALE AND MACROSCALE METHODS

Despite the importance of understanding the molecular structure and nature of materials, their behavior can be homogenized with respect to different aspects which can be at different scales. Typically, the observed macroscopic behavior is usually explained by ignoring the discrete atomic and molecular structure and assuming that the material is continuously distributed throughout its volume. The continuum material is thus assumed to have an average density and can be subjected to body forces such as gravity and surface forces. Generally speaking, the macroscale methods (or called continuum methods hereafter) obey the fundamental laws of: (i) continuity, derived from the conservation of mass; (ii) equilibrium, derived from momentum considerations and Newton's second law; (iii) the moment of momentum principle, based on the model that the time rate of change of angular momentum with respect to an arbitrary point is equal to the resultant moment; (iv) conservation of energy, based on the first law of thermodynamics; and (v) conservation of entropy, based on the second law of thermodynamics. These laws provide the basis for the continuum model and must be coupled with the appropriate constitutive equations and the equations of state to provide all the equations necessary for solving a continuum problem. The continuum method relates the deformation of a continuous medium to the external forces acting on the medium and the resulting internal stress and strain. Computational approaches range from simple closed-form analytical expressions to micromechanics and complex structural mechanics calculations based on beam and shell theory. In this section, we introduce some continuum methods that have been used in polymer nanocomposites, including micromechanics models (e.g., Halpin–Tsai model, Mori–Tanaka model), equivalent-continuum model, self-consistent model and finite element analysis.

4.3.4   MICROMECHANICS

Since the assumption of uniformity in continuum mechanics may not hold at the microscale level, micromechanics methods are used to express the continuum quantities associated with an infinitesimal material element in terms of structure and properties of the micro constituents. Thus, a central theme of micromechanics models is the development of a representative volume element (RVE) to statistically represent the local continuum properties. The RVE is constructed to ensure that the length scale is consistent with the smallest constituent that has a first-order effect on the macroscopic behavior. The RVE is then used in a repeating or periodic nature in the full-scale model. The micromechanics method can account for interfaces between constituents, discontinuities, and coupled mechanical and non-mechanical properties. Our purpose is to review the micromechanics methods used for polymer nanocomposites. Thus, we only discuss here some important concepts of micromechanics as well as the Halpin–Tsai model and Mori–Tanaka model.

4.3.4.1   Basic Concepts

When applied to particle reinforced polymer composites, micromechanics models usually follow such basic assumptions as (i) linear elasticity of fillers and polymer matrix; (ii) the fillers are axisymmetric, identical in shape and size, and can be characterized by parameters such as aspect ratio; (iii) well-bonded filler–polymer interface and the ignorance of interfacial slip, filler–polymer debonding or matrix cracking. The first concept is the linear elasticity, that is, the linear relationship between the total stress and infinitesimal strain tensors for the filler and matrix as expressed by the following constitutive equations:

For filler,

$$\sigma^f = C^f \varepsilon^f \tag{18}$$

For matrix,

$$\sigma^m = C^m \varepsilon^m \tag{19}$$

where C is the stiffness tensor. The second concept is the average stress and strain. Since the pointwise stress field $\sigma(x)$ and the corresponding strain field $\varepsilon(x)$ are usually non-uniform in polymer composites, the volume–average stress $\bar{\sigma}$ and strain $\bar{\varepsilon}$ are then defined over the representative averaging volume V, respectively,

$$\bar{\sigma} = \frac{1}{V}\int \sigma(x)dv \tag{20}$$

$$\bar{\varepsilon} = \frac{1}{V}\int \varepsilon(x)dv \tag{21}$$

Therefore, the average filler and matrix stresses are the averages over the corresponding volumes and, respectively,

$$\bar{\sigma}_f = \frac{1}{V_f}\int \sigma(x)dv \tag{22}$$

$$\bar{\sigma}_m = \frac{1}{V_m}\int \sigma(x)dv \tag{23}$$

The average strains for the fillers and matrix are defined, respectively, as

$$\bar{\varepsilon}_f = \frac{1}{V_f}\int \varepsilon(x)dv \tag{24}$$

$$\bar{\varepsilon}_m = \frac{1}{V_m}\int \varepsilon(x)dv \tag{25}$$

Based on the above definitions, the relationships between the filler and matrix averages and the overall averages can be derived as follows:

$$\bar{\sigma} = \bar{\sigma}_f V_f + \bar{\sigma}_m V_m \tag{26}$$

$$\bar{\varepsilon} = \bar{\varepsilon}_f v_f + \bar{\varepsilon}_m v_m \tag{27}$$

where v_f, v_m are the volume fractions of the fillers and matrix, respectively.

The third concept is the average properties of composites, which are actually the main goal of a micromechanics model. The average stiffness of the composite is the tensor C that maps the uniform strain to the average stress

$$\overline{\sigma} = \overline{\varepsilon}C \tag{28}$$

The average compliance S is defined in the same way:

$$\overline{\varepsilon} = \overline{\sigma}S \tag{29}$$

Another important concept is the strain–concentration and stress–concentration tensors A and B, which are basically the ratios between the average filler strain (or stress) and the corresponding average of the composites.

$$\overline{\varepsilon}_f = \overline{\varepsilon}A \tag{30}$$

$$\overline{\sigma}_f = \overline{\sigma}B \tag{31}$$

Using the above concepts and equations, the average composite stiffness can be obtained from the strain concentration tensor A and the filler and matrix properties:

$$C = C_m + v_f(C_f - C_m)A \tag{32}$$

4.3.4.2 Halpin–Tsai Model

The Halpin–Tsai model is a well-known composite theory to predict the stiffness of unidirectional composites as a functional of aspect ratio. In this model, the longitudinal E11 and transverse E22 engineering moduli are expressed in the following general form:

$$\frac{E}{E_m} = \frac{1 + \zeta\eta v_f}{1 - \eta v_f} \tag{33}$$

where E and E_m represent the Young's modulus of the composite and matrix, respectively, v_f is the volume fraction of filler, and η is given by:

$$\eta = \frac{\dfrac{E}{E_m} - 1}{\dfrac{E_f}{E_m} + \zeta_f} \qquad (34)$$

where E_f represents the Young's modulus of the filler and ζ_f the shape parameter depending on the filler geometry and loading direction. When calculating longitudinal modulus E_{11}, ζ_f is equal to l/t, and when calculating transverse modulus E_{22}, ζ_f is equal to w/t. Here, the parameters of l, w and t are the length, width and thickness of the dispersed fillers, respectively. If $\zeta_f \rightarrow 0$, the Halpin–Tsai theory converges to the inverse rule of mixture (lower bound):

$$\frac{1}{E} = \frac{v_f}{E_f} + \frac{1 - v_f}{E_m} \qquad (35)$$

Conversely, if $\zeta_f \rightarrow \infty$, the theory reduces to the rule of mixtures (upper bound),

$$E = E_f v_f + E_m (1 - v_f) \qquad (36)$$

4.3.4.3 Mori–Tanaka Model

The Mori–Tanaka model is derived based on the principles of Eshelby's inclusion model for predicting an elastic stress field in and around an ellipsoidal filler in an infinite matrix. The complete analytical solutions for longitudinal E_{11} and transverse E_{22} elastic moduli of an isotropic matrix filled with aligned spherical inclusion are [45, 46]:

$$\frac{E_{11}}{E_m} = \frac{A_0}{A_0 + v_f (A_1 + 2v_0 A_2)} \qquad (37)$$

$$\frac{E_{22}}{E_m} = \frac{2A_0}{2A_0 + v_f(-2A_3 + (1-v_0A_4) + (1+v_0)A_5A_0)} \tag{38}$$

where E_m represents the Young's modulus of the matrix, v_f the volume fraction of filler, v_0 the Poisson's ratio of the matrix, parameters, $A_0, A_1,...,A_5$ are functions of the Eshelby's tensor and the properties of the filler and the matrix, including Young's modulus, Poisson's ratio, filler concentration and filler aspect ratio [45].

4.3.4.5   Equivalent-Continuum and Self-Similar Approaches

Numerous micromechanical models have been successfully used to predict the macroscopic behavior of fiber-reinforced composites. However, the direct use of these models for nanotube-reinforced composites is doubtful due to the significant scale difference between nanotube and typical carbon fiber. Recently, two methods have been proposed for modeling the mechanical behavior of single-walled carbon nanotube (SWCN) composites—equivalent-continuum approach and self-similar approach [47]. The equivalent-continuum approach was proposed by Odegard et al. [48]. In this approach, MD was used to model the molecular interactions between SWCN–polymer and a homogeneous equivalent-continuum reinforcing element (e.g., a SWCN surrounded by polymer) was constructed as shown in Figure 4.2. Then, micromechanics are used to determine the effective bulk properties of the equivalent-continuum reinforcing element embedded in a continuous polymer. The equivalent-continuum approach consists of four major steps, as briefly described below. Step 1: MD simulation is used to generate the equilibrium structure of a SWCN–polymer composite and then to establish the RVE of the molecular model and the equivalent-continuum model. Step 2: The potential energies of deformation for the molecular model and effective fiber are derived and equated for identical loading conditions. The bonded and non-bonded interactions within a polymer molecule are quantitatively described by MM. For the SWCN/polymer system, the total potential energy U^m of the molecular model is

$$U^m = \sum U^r(K_r) + \sum U^\theta(K_\theta) + \sum U^{vdw}(K_{vdw}) \tag{39}$$

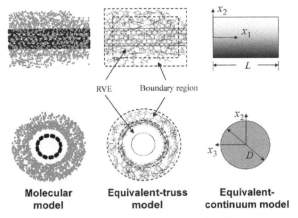

FIGURE 4.2 Equivalent-continuum modeling of effective fiber.

where U^r, U^θ and U^{vdw} are the energies associated with covalent bond stretching, bond-angle bending, and van der Waals interactions, respectively. An equivalent-truss model of the RVE is used as an intermediate step to link the molecular and equivalent-continuum models. Each atom in the molecular model is represented by a pin-joint, and each truss element represents an atomic bonded or non-bonded interaction. The potential energy of the truss model is

$$U^t = \sum U^a(E^a) + \sum U^b(E^b) + \sum U^c(E^c) \qquad (40)$$

where U^a, U^b and U^c are the energies associated with truss elements that represent covalent bond stretching, bond-angle bending, and van der Waals interactions, respectively. The energies of each truss element are a function of the Young's modulus, E.

Step 3: A constitutive equation for the effective fiber is established. Since the values of the elastic stiffness tensor components are not known a priori, a set of loading conditions are chosen such that each component is uniquely determined from

$$U^f = U^t = U^m \qquad (41)$$

$$U^m = \sum U^r(k_r) + \sum U^\Theta(k_\Theta) + \sum U^{vdw}(k_{vdw}) \qquad (42)$$

Step 4: Overall constitutive properties of the dilute and unidirectional SWCN/polymer composite are determined with Mori–Tanaka model with the mechanical properties of the effective fiber and the bulk polymer. The layer of polymer molecules that are near the polymer/nanotube interface (Figure 4.2) is included in the effective fiber, and it is assumed that the matrix polymer surrounding the effective fiber has mechanical properties equal to those of the bulk polymer. The self-similar approach was proposed by Pipes and Hubert [49], which consists of three major steps:

First, a helical array of SWCNs is assembled. This array is termed as the SWCN nanoarray where 91 SWCNs make up the cross-section of the helical nanoarray. Then, the SWCN nanoarrays is surrounded by a polymer matrix and assembled into a second twisted array, termed as the SWCN nanowire Finally, the SWCN nanowires are further impregnated with a polymer matrix and assembled into the final helical array—the SWCN microfiber. The self-similar geometries described in the nanoarray, nanowire and microfiber (Figure 4.3) allow the use of the same mathematical and geometric model for all three geometries [49].

4.3.4.6   Finite Element Method (FEM)

FEM is a general numerical method for obtaining approximate solutions in space to initial-value and boundary-value problems including time-dependent processes. It employs preprocessed mesh generation, which enables the model to fully capture the spatial discontinuities of highly inhomogeneous materials. It also allows complex, nonlinear tensile relationships to be incorporated into the analysis. Thus, it has been widely used in mechanical, biological and geological systems. In FEM, the entire

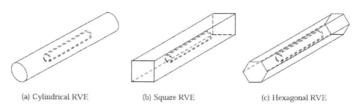

(a) Cylindrical RVE   (b) Square RVE   (c) Hexagonal RVE

FIGURE 4.3   Three nanoscale representative volume elements for the analysis of CNT-based nanocomposites.

domain of interest is spatially discretized into an assembly of simply shaped subdomains (e.g., hexahedra or tetrahedral in three dimensions, and rectangles or triangles in two dimensions) without gaps and without overlaps. The subdomains are interconnected at joints (i.e., nodes). The implementation of FEM includes the important steps shown in Figure 4.4. The energy in FEM is taken from the theory of linear elasticity and thus the input parameters are simply the elastic moduli and the density of the material. Since these parameters are in agreement with the values computed by MD, the simulation is consistent across the scales. More specifically, the total elastic energy in the absence of tractions and body forces within the continuum model is given by [50]:

$$U = U_v + U_k \tag{43}$$

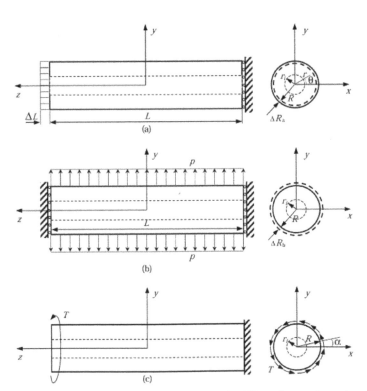

FIGURE 4.4 Three loading cases for the cylindrical RVE used to evaluate the effective material properties of the CNT-based composites. (a) Under axial stretch DL; (b) under lateral uniform load P; (c) under torsional load T.

$$U_k = 1/2 \int dr p(r) \left| \dot{U}_r \right|^2 \tag{44}$$

$$U_v = \frac{1}{2} \int dr \sum_{\mu,\nu,\lambda,\sigma=1}^{3} \varepsilon_{\mu\nu}(r) C_{\mu\nu\lambda\sigma^3\lambda\sigma}(r) \tag{45}$$

where U_v is the Hookian potential energy term which is quadratic in the symmetric strain tensor e, contracted with the elastic constant tensor C. The Greek indices (i.e., m, n, l, s) denote Cartesian directions. The kinetic energy U_k involves the time rate of change of the displacement field $\dot{U}$, and the mass density ρ.

These are fields defined throughout space in the continuum theory. Thus, the total energy of the system is an integral of these quantities over the volume of the sample dυ. The FEM has been incorporated in some commercial software packages and open source codes (e.g., ABAQUS, ANSYS, Palmyra and OOF) and widely used to evaluate the mechanical properties of polymer composites. Some attempts have recently been made to apply the FEM to nanoparticle-reinforced polymer nanocomposites. In order to capture the multiscale material behaviors, efforts are also underway to combine the multiscale models spanning from molecular to macroscopic levels [51, 52].

4.4   MULTI SCALE MODELING OF MECHANICAL PROPERTIES

In Odegard's study [48], a method has been presented for linking atomistic simulations of nano-structured materials to continuum models of the corresponding bulk material. For a polymer composite system reinforced with single-walled carbon nanotubes (SWNT), the method provides the steps whereby the nanotube, the local polymer near the nanotube, and the nanotube/polymer interface can be modeled as an effective continuum fiber by using an equivalent-continuum model. The effective fiber retains the local molecular structure and bonding information, as defined by molecular dynamics, and serves as a means for linking the equivalent-continuum and micromechanics models. The micromechanics method is then available for the prediction of bulk mechanical properties of SWNT/polymer composites as a function of nanotube size, orientation, and volume fraction. The utility of this method was examined by modeling tow composites that both having a interface. The elastic stiffness constants of the composites were determined for both aligned and three-dimensional randomly oriented nanotubes, as a function of nanotube length and volume fraction. They used Mori–Tanaka model [53] for random and oriented fibers position and compare their model with mechanical properties, the interface between fiber and matrix was assumed perfect. Motivated by micrographs showing that embedded nanotubes often exhibit significant curvature within the polymer, Fisher et al. [54] have developed a model combining finite element results and micromechanical methods (Mori-Tanaka) to determine the effective reinforcing modulus of a wavy embedded nanotube with perfect bonding and random fiber orientation assumption. This effective reinforcing modulus (ERM) is then used within a multiphase micromechanics model to predict the effective modulus of a polymer reinforced with a distribution of wavy nanotubes. We found that even slight nanotube curvature significantly reduces the effective reinforcement when compared to straight nanotubes. These results suggest that nanotube waviness may be an additional mechanism limiting the modulus enhancement of nanotube-reinforced polymers. Bradshaw et al. [55] investigated the degree to which the characteristic waviness of nanotubes embedded in polymers can impact the effective stiffness of these materials. A 3D finite element model of a single infinitely long sinusoidal fiber within an infinite matrix is used to numerically compute the dilute strain concentration tensor. A Mori–Tanaka model uses this tensor to predict the effective modulus of the material with aligned or randomly oriented inclusions. This hybrid finite element micromechanical modeling technique is a powerful extension of general micromechanics modeling and can be applied to any composite microstructure containing non-ellipsoidal inclusions. The results demonstrate that nanotube waviness results in a reduction of the effective modulus of the composite relative to straight nanotube reinforcement. The degree of reduction is dependent on the ratio of the sinusoidal wavelength to the nanotube diameter. As this wavelength ratio increases, the effective stiffness of a composite with randomly oriented wavy nanotubes converges to the result obtained with straight nanotube inclusions.

The effective mechanical properties of carbon nanotube-based composites are evaluated by Liu and Chen [56] using a 3-D nanoscale RVE based on 3-D elasticity theory and solved by the finite element method. Formulas to extract the material constants from solutions for the RVE under three loading cases are established using the elasticity. An extended rule of mixtures, which can be used to estimate the Young's modulus in the axial direction of the RVE and to validate the numerical solutions for short CNTs, is also derived using the strength of materials theory. Numerical examples using the FEM to evaluate the effective material constants of a CNT-based composites are presented, which demonstrate that the reinforcing capabilities of the CNTs in a matrix are significant. With only about 2% and 5% volume fractions of the CNTs in a matrix, the stiffness of the composite in the CNT axial direction can increase as many as 0.7 and 9.7 times for the cases of short and long CNT fibers, respectively. These simulation results, which are believed to be the first of its kind for CNT-based composites, are consistent with the experimental results reported in the literature (Schadler et al. [57], Wagner et al. [58]; Qian et al. [59]).

The developed extended rule of mixtures is also found to be quite effective in evaluating the stiffness of the CNT-based composites in the CNT axial direction. Many research issues need to be addressed in the modeling and simulations of CNTs in a matrix material for the development of nanocomposites. Analytical methods and simulation models to extract the mechanical properties of the CNT-based nanocomposites need to be further developed and verified with experimental results. The analytical method and simulation approach developed in this chapter are only a preliminary study. Different type of RVEs, load cases and different solution methods should be investigated. Different interface conditions, other than perfect bonding, need to be investigated using different models to more accurately account for the interactions of the CNTs in a matrix material at the nanoscale. Nanoscale interface cracks can be analyzed using simulations to investigate the failure mechanism in nanomaterials. Interactions among a large number of CNTs in a matrix can be simulated if the computing power is available. Single-walled and multi-walled CNTs as reinforcing fibers in a matrix can be studied by simulations to find out their advantages and disadvantages. Finally, large multiscale simulation models for CNT-based composites, which can link the models at the nano, micro and macro scales, need to be developed, with the help of analytical and experimental work [56]. The three RVEs proposed in Ref. [60] and shown in Figure 4.3 are relatively simple regarding the models and scales and pictures in Figure 4.4 are Three loading cases for the cylindrical RVE. However, this is only the first step toward more sophisticated and large-scale simulations of CNT-based composites. As the computing power and confidence in simulations of CNT-based composites increase, large scale 3-D models containing hundreds or even more CNTs, behaving linearly or nonlinearly, with coatings or of different sizes, distributed evenly or randomly, can be employed to investigate the interactions among the CNTs in a matrix and to evaluate the effective material properties. Other numerical methods can also be attempted for the modeling and simulations of CNT-based composites, which may offer some advantages over the FEM approach. For example, the boundary element method, Liu et al. [60]; Chen and Liu [61], accelerated with the fast multiple techniques, Fu et al. [62]; Nishimura et al. [63], and the mesh free methods (Qian et al. [64]) may enable one to model an RVE with thousands of CNTs in a matrix on a desktop computer. Analysis of the CNT-based composites using the boundary element method is already underway and will be reported subsequently.

The effective mechanical properties of CNT based composites are evaluated using square RVEs based on 3-D elasticity theory and solved by the FEM. Formulas to extract the effective material constants from solutions for the square RVEs under two loading cases are established based on elasticity. Square RVEs with multiple CNTs are also investigated in evaluating the Young's modulus and Poisson's ratios in the transverse plane. Numerical examples using the FEM are presented, which demonstrate that the load-carrying capabilities of the CNTs in a matrix are significant. With the addition of only about 3.6% volume fraction of the CNTs in a matrix, the stiffness of the composite in the CNT axial direction can increase as much as 33% for the case of long CNT fibers [65]. These simulation results are consistent with both the experimental ones reported in the literature [56–59, 66]. It is also found that cylindrical RVEs tend to overestimate the effective Young's moduli due to the fact that they overestimate the volume fractions of the CNTs in a matrix. The square RVEs, although more demanding in modeling and computing, may be the preferred model in future simulations for estimating the effective material constants, especially when multiple CNTs need to be considered. Finally, the rules of mixtures, for both long and short CNT cases, are found to be quite accurate in estimating the effective Young's moduli in the CNT axial direction. This may suggest that 3-D FEM modeling may not be necessary in obtaining the effective material constants in the CNT direction, as in the studies of the conventional fiber reinforced composites. Efforts in comparing the results presented in this chapter using the continuum approach directly with the MD simulations are underway. This is feasible now only for a smaller RVE of one CNT embedded in a matrix. In future research, the MD and continuum approach should be integrated in a multiscale modeling and simulation environment for analyzing the CNT-based composites. More efficient models of the CNTs in a matrix also need to be developed, so that a large number of CNTs, in different shapes and forms (curved or twisted), or randomly distributed in a matrix, can be modeled. The ultimate validation of the simulation results should be done with the nanoscale or microscale experiments on the CNT reinforced composites [64].

Griebel and Hamaekers [67] reviewed the basic tools used in computational nanomechanics and materials, including the relevant underlying principles and concepts. These tools range from subatomic ab initio methods to classical molecular dynamics and multiple-scale approaches. The energetic link between the quantum mechanical and classical systems has been discussed, and limitations of the standing alone molecular dynamics simulations have been shown on a series of illustrative examples. The need for multi-scale simulation methods to tackle nanoscale aspects of material behavior was therefore emphasized; that was followed by a review and classification of the mainstream and emerging multi-scale methods. These simulation methods include the broad areas of quantum mechanics, molecular dynamics and multiple-scale approaches, based on coupling the atomistic and continuum models. They summarize the strengths and limitations of currently available multiple-scale techniques, where the emphasis is made on the latest perspective approaches, such as the bridging scale method, multi-scale boundary conditions, and multi-scale fluidics. Example problems, in which multiple-scale simulation methods yield equivalent results to full atomistic simulations at fractions of the computational cost, were shown. They compare their results with Odegard, et al. [48], the micromechanic method was BEM Halpin-Tsai equation [68] with aligned fiber by perfect bonding.

The solutions of the strain-energy-changes due to a SWNT embedded in an infinite matrix with imperfect fiber bonding are obtained through numerical method by Wan, et al. [69]. A "critical" SWNT fiber length is defined for full load transfer between the SWNT and the matrix, through the evaluation of the strain-energy-changes for different fiber lengths The strain-energy-change is also used to derive the effective longitudinal Young's modulus and effective bulk modulus of the composite, using a dilute solution. The main goal of their research was investigation of strain-energy-change due to inclusion of SWNT using FEM. To achieve full load transfer between the SWNT and the matrix, the length of SWNT fibers should be longer than a 'critical' length if no weak interphase exists between the SWNT and the matrix [69].

A hybrid atomistic/continuum mechanics method is established in the Feng, et al. study [70] the deformation and fracture behaviors of carbon nanotubes (CNTs) in composites. The unit cell containing a CNT embedded in a matrix is divided in three regions, which are simulated by the atomic-potential method, the continuum method based on the modified Cauchy–Born rule, and the classical continuum mechanics, respectively. The effect of CNT interaction is taken into account via the Mori–Tanaka effective field method of micromechanics. This method not only can predict the formation of Stone–Wales (5–7–7–5) defects, but also simulate the subsequent deformation and fracture process of CNTs. It is found that the critical strain of defect nucleation in a CNT is sensitive to its chiral angle but not to its diameter. The critical strain of Stone–Wales defect formation of zigzag CNTs is nearly twice that of armchair CNTs. Due to the constraint effect of matrix, the CNTs embedded in a composite are easier to fracture in comparison with those not embedded. With the increase in the Young's modulus of the matrix, the critical breaking strain of CNTs decreases.

Estimation of effective elastic moduli of nanocomposites was performed by the version of effective field method developed in the framework of quasi-crystalline approximation when the spatial correlations of inclusion location take particular ellipsoidal forms [71]. The independent justified choice of shapes of inclusions and correlation holes provide the formulae of effective moduli, which are symmetric, completely explicit and easily to use. The parametric numerical analyzes revealed the most sensitive parameters influencing the effective moduli which are defined by the axial elastic moduli of nanofibers rather than their transversal moduli as well as by the justified choice of correlation holes, concentration and prescribed random orientation of nanofibers [72].

Li and Chou [73, 74] have reported a multiscale modeling of the compressive behavior of carbon nanotube/polymer composites. The nanotube is modeled at the atomistic scale, and the matrix deformation is analyzed by the continuum finite element method. The nanotube and polymer matrix are assumed to be bonded by van der Waals interactions at the interface. The stress distributions at the nanotube/polymer interface under isostrain and isostress loading conditions have been examined. They have used beam elements for SWCNT using molecular structural mechanics, truss rod for vdW links and cubic elements for matrix. The rule of mixture was used as for comparison in this research. The buckling forces of nanotube/polymer composites for different nanotube lengths and diameters are computed. The results indicate that continuous nanotubes can most effectively enhance the composite buckling resistance.

Anumandla and Gibson [75] describes an approximate, yet comprehensive, closed form micromechanics model for estimating the effective elastic modulus of carbon nanotube-reinforced composites. The model incorporates the typically observed nanotube curvature, the nanotube length, and both 1D and 3D random arrangement of the nanotubes. The analytical results obtained from the closed form micromechanics model for nanoscale representative volume elements and results from an equivalent finite element model for effective reinforcing modulus of the nanotube reveal that the reinforcing modulus is strongly dependent on the waviness, wherein, even a slight change in the nanotube curvature can induce a prominent change in the effective reinforcement provided. The micromechanics model is also seen to produce reasonable agreement with experimental data for the effective tensile modulus of composites reinforced with multi-walled nanotubes (MWNTs) and having different MWNT volume fractions.

Effective elastic properties for carbon nanotube reinforced composites are obtained through a variety of micromechanics techniques [76]. Using the in-plane elastic properties of graphene, the effective properties of carbon nanotubes are calculated using a composite cylinders micromechanics technique as a first step in a two-step process. These effective properties are then used in the self-consistent and Mori–Tanaka methods to obtain effective elastic properties of composites consisting of aligned single or multi-walled carbon nanotubes embedded in a polymer matrix. Effective composite properties from these averaging methods are compared to a direct composite cylinders approach extended from the work of Hashin and Rosen [77] and Christensen and Lo [78]. Comparisons with finite element simulations are also performed. The effects of an interphase layer between the nanotubes and the polymer matrix as result of functionalization is also investigated using a multi-layer composite cylinders approach. Finally, the modeling of the clustering of nanotubes into bundles due to interatomic forces is accomplished herein using a tessellation method in conjunction with a multi-phase Mori–Tanaka technique. In addition to aligned nanotube composites, modeling of the effective elastic properties of randomly dispersed nanotubes into a matrix is performed using the Mori–Tanaka method, and comparisons with experimental data are made.

Selmi, et al. [79] deal with the prediction of the elastic properties of polymer composites reinforced with single walled carbon nanotubes. Our contribution is the investigation of several micromechanical models, while most of the papers on the subject deal with only one approach. They implemented four homogenization schemes, a sequential one and three others based on various extensions of the Mori–Tanaka (M–T) mean-field homogenization model: two-level (M–T/M–T), two-step (M–T/M–T) and two-step (M–T/Voigt). Several composite systems are studied, with various properties of the matrix and the graphene, short or long nanotubes, fully aligned or randomly oriented in 3D or 2D. Validation targets are experimental data or finite element results, either based on a 2D periodic unit cell or a 3D representative volume element. The comparative study showed that there are cases where all micromechanical models give adequate predictions, while for some composite materials and some properties, certain models fail in a rather spectacular fashion. It was found that the two-level (M–T/M–T) homogenization model gives the best predictions in most cases. After the characterization of the discrete nanotube structure using a homogenization method based on energy equivalence, the sequential, the two-step (M–T/M–T), the two-step (M–T/Voigt), the two-level (M–T/M–T) and finite element models were used to predict the elastic properties of SWNT/polymer composites. The data delivered by the micromechanical models are compared against those obtained by finite element analyzes or experiments. For fully aligned, long nanotube polymer composite, it is the sequential and the two-level (M–T/M–T) models, which delivered good predictions. For all composite morphologies (fully aligned, two-dimensional in-plane random orientation, and three-dimensional random orientation), it is the two-level (M–T/M–T) model, which gave good predictions compared to finite element and experimental results in most situations. There are cases where other micromechanical models failed in a spectacular way.

Luo, et al. [80] have used multi-scale homogenization (MH) and FEM for wavy and straight SWCNTs, they have compare their results with Mori-Tanaka et al. [81], and Lauke [82]. Trespass, et al. [83] used 3D elastic beam for C-C bond and, 3D space frame for CNT and progressive fracture model for prediction of elastic modulus, they used rule of mixture for compression of their results. Their assumption was embedded a single SWCNT in polymer with Perfect bonding. The multi-scale modeling, Monte Carlo, FEM and using equivalent continuum method was used by Spanos and Kontsos [84] and compared with Zhu, et al. [85] and Paiva, et al. [86]'s results.

The effective modulus of CNT/PP composites is evaluated using FEA of a 3D RVE which includes the PP matrix, multiple CNTs and CNT/PP interphase and accounts for poor dispersion and non homogeneous distribution of CNTs within the polymer matrix, weak CNT/polymer interactions, CNT agglomerates of various sizes and CNTs orientation and waviness [87]. Currently, there is no other model, theoretical or numerical, that accounts for all these experimentally observed phenomena and captures their individual and combined effect on the effective modulus of nanocomposites. The model is developed using input obtained from experiments and validated against experimental data. CNT reinforced PP composites manufactured by extrusion and injection molding are characterized in terms of tensile modulus, thickness and stiffness of CNT/PP interphase, size of CNT agglomerates and CNT distribution using tensile testing, AFM and SEM, respectively. It is concluded that CNT agglomeration and waviness are the two dominant factors that hinder the great potential of CNTs as polymer reinforcement. The proposed model provides the upper and lower limit of the modulus of the CNT/PP composites and can be used to guide the manufacturing of composites with engineered properties for targeted applications. CNT agglomeration can be avoided by employing processing techniques such as sonication of CNTs, stirring, calendaring, etc., whereas CNT waviness can be eliminated by increasing the injection pressure during molding and mainly by using CNTs with smaller aspect ratio. Increased pressure during molding can also promote the alignment of CNTs along the applied load direction. The 3D modeling capability presented in this study gives an insight on the upper and lower bound of the CNT/PP composites modulus quantitatively by accurately capturing the effect of various processing parameters. It is observed that when all the experimentally observed factors are considered together in the FEA the modulus prediction is in good agreement with the modulus obtained from the experiment. Therefore, it can be concluded that the FEM models proposed in this study by systematically incorporating experimentally observed characteristics can be effectively used for the determination of mechanical properties of nanocomposite materials. Their result is in agreement with the results reported in Ref. [88], The theoretical micromechanical models, shown in Figure 4.5, are used to confirm that our FEM model predictions follow the same trend with the one predicted by the models as expected.

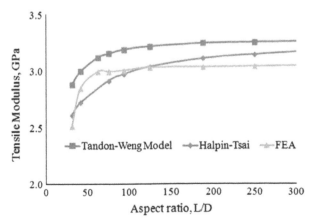

FIGURE 4.5   Effective modulus of 5 wt.% CNT/PP composites: theoretical models vs. FEA.

For reasons of simplicity and in order to minimize the mesh dependency on the results the hollow CNTs are considered as solid cylinders of circular cross-sectional area with an equivalent average diameter, shown in Figure 4.6, calculated by equating the volume of the hollow CNT to the solid one [87].

The micromechanical models used for the comparison were Halpin–Tsai (H–T) [89] and Tandon–Weng (T–W) [90] model and the comparison was performed for 5 wt.% CNT/PP. It was noted that the H–T model results to lower modulus compared to FEA because H–T equation does not account for maximum packing fraction and the arrangement of the reinforcement in the composite. A modified H–T model that account for this has been proposed in the literature [91]. The effect of maximum packing fraction and the arrangement of the reinforcement within the composite becomes less significant at higher aspect ratios [92].

A finite element model of carbon nanotube, inter-phase and its surrounding polymer is constructed to study the tensile behavior of embedded short carbon nanotubes in polymer matrix in presence of vdW interactions in inter-phase region by Shokrieh and Rafiee [93]. The inter-phase is modeled using non-linear spring elements capturing the force-distance curve of vdW interactions. The constructed model is subjected to tensile loading to extract longitudinal Young's modulus. The obtained results of this work have been compared with the results of previous research of the same authors [94] on long embedded carbon nanotube in polymer matrix. It shows that the capped short carbon nanotubes reinforce polymer matrix less efficient than long CNTs.

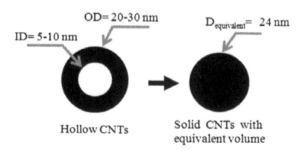

FIGURE 4.6   Schematic of the CNTs considered for the FEA.

Despite the fact that researches have succeeded to grow the length of CNTs up to 4 cm as a world record in US Department of Energy Los Alamos National Laboratory [95] and also there are some evidences on producing CNTs with lengths up to millimeters [96, 97], CNTs are commercially available in different lengths ranging from 100 nm to approximately 30 lm in the market based on employed process of growth [98–101]. Chemists at Rice University have identified a chemical process to cut CNTs into short segments [102]. As a consequent, it can be concluded that the SWCNTs with lengths smaller than, 1000 nm do not contribute significantly in reinforcing polymer matrix. On the other hand, the efficient length of reinforcement for a CNT with (10, 10) index is about 1.2 lm and short CNT with length of 10.8 lm can play the same role as long CNT reflecting the uppermost value reported in our previous research [94]. Finally, it is shown that the direct use of Halpin–Tsai equation to predict the modulus of SWCNT/composites overestimates the results. It is also observed that application of previously developed long equivalent fiber stiffness [94] is a good candidate to be used in Halpin–Tsai equations instead of Young's modulus of CNT. Halpin–Tsai equation is not an appropriate model for smaller lengths, since there is not any reinforcement at all for very small lengths.

Earlier, a nano-mechanical model has been developed by Chowdhury et al. [103] to calculate the tensile modulus and the tensile strength of randomly oriented short carbon nanotubes (CNTs) reinforced nanocomposites, considering the statistical variations of diameter and length of the CNTs. According to this model, the entire composite is divided into several composite segments which contain CNTs of almost the same diameter and length. The tensile modulus and tensile strength of the composite are then calculated by the weighted sum of the corresponding modulus and strength of each composite segment. The existing micro-mechanical approach for modeling the short fiber composites is modified to account for the structure of the CNTs, to calculate the modulus and the strength of each segmented CNT reinforced composites. Multi-walled CNTs with and without inter-tube bridging (see Figure 4.7) have been considered. Statistical variations of the diameter and length of the CNTs are modeled by a normal distribution. Simulation results show that CNTs inter-tube bridging, length and diameter affect the nanocomposites modulus and strength. Simulation results have been compared with the available experimental

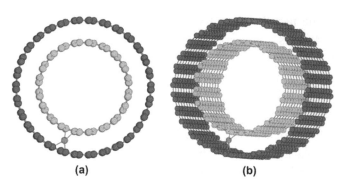

(a)   **(b)**

FIGURE 4.7   Schematic of MWNT with inter-tube bridging. (a) Top view and (b) Oblique view.

results and the comparison concludes that the developed model can be effectively used to predict tensile modulus and tensile strength of CNTs reinforced composites.

The effective elastic properties of carbon nanotube-reinforced polymers have been evaluated by Tserpes and Chanteli [104] as functions of material and geometrical parameters using a homogenized RVE. The RVE consists of the polymer matrix, a multi-walled carbon nanotube (MWCNT) embedded into the matrix and the interface between them. The parameters considered are the nanotube aspect ratio, the nanotube volume fraction as well as the interface stiffness and thickness. For the MWCNT, both isotropic and orthotropic material properties have been considered. Analyzes have been performed by means of a 3D FE model of the RVE. The results indicate a significant effect of nanotube volume fraction. The effect of nanotube aspect ratio appears mainly at low values and diminishes after the value of 20. The interface mostly affects the effective elastic properties at the transverse direction. Having evaluated the effective elastic properties of the MWCNT–polymer at the micro-scale, the RVE has been used to predict the tensile modulus of a polystyrene specimen reinforced by randomly aligned MWCNTs for which experimental data exist in the literature. A very good agreement is obtained between the predicted and experimental tensile moduli of the specimen. The effect of nanotube alignment on the specimen's tensile modulus has been also examined and found to be significant since as misalignment increases the effective tensile modulus decreases radically. The proposed model can be used for the virtual design and optimization of CNT–polymer composites since it has proven capable of assessing the effects of different material and geometrical parameters on the elastic properties of the composite and predicting the tensile modulus of CNT-reinforced polymer specimens.

4.5   MODELING OF THE INTERFACE

4.5.1   INTRODUCTION

The superior mechanical properties of the nanotubes alone do not ensure mechanically superior composites because the composite properties are strongly influenced by the mechanics that govern the nanotube–polymer interface. Typically in composites, the constituents do not dissolve or merge completely and therefore, normally, exhibit an interface between one another, which can be considered as a different material with different mechanical properties. The structural strength characteristics of composites greatly depend on the nature of bonding at the interface, the mechanical load transfer from the matrix (polymer) to the nanotube and the yielding of the interface. As an example, if the composite is subjected to tensile loading and there exists perfect bonding between the nanotube and polymer and/or a strong interface then the load (stress) is transferred to the nanotube; since the tensile strength of the nanotube (or the interface) is very high the composite can withstand high loads. However, if the interface is weak or the bonding is poor, on application of high loading either the interface fails or the load is not transferred to the nanotube and the polymer fails due to their lower tensile strengths. Consider another example of transverse crack propagation. When the crack reaches the interface, it will tend to propagate along the interface, since the interface is relatively weaker (generally) than the nanotube (with respect to resistance to crack propagation). If the interface is weak, the crack will cause the interface to fracture and result in failure of the composite. In this aspect, carbon nanotubes are better than traditional fibers (glass, carbon) due to their ability to inhibit nano and micro cracks. Hence, the knowledge and understanding of the nature and mechanics of load (stress) transfer between the nanotube and polymer and properties of the interface is critical for manufacturing of mechanically enhanced CNT–polymer composites and will enable in tailoring of the interface for specific applications or superior mechanical properties. Broadly, the interfacial mechanics of CNT–polymer composites is appealing from three aspects: mechanics, chemistry, and physics. From a mechanics point of view, the important questions are:

   (i) The relationship between the mechanical properties of individual constituents, that is, nanotube and polymer, and the properties of the interface and the composite overall.

  (ii) The effect of the unique length scale and structure of the nanotube on the property and behavior of the interface.

 (iii) Ability of the mechanics modeling to estimate the properties of the composites for the design process for structural applications.

From a chemistry point of view, the interesting issues are:

   (i) The chemistry of the bonding between polymer and nanotubes, especially the nature of bonding (e.g., covalent or non-covalent and electrostatic).

  (ii) The relationship between the composite processing and fabrication conditions and the resulting chemistry of the interface.

 (iii) The effect of functionalization (treatment of the polymer with special molecular groups like hydroxyl or halogens) on the nature and strength of the bonding at the interface. From the physics point of view, researchers are interested in:

    (a) The CNT–polymer interface serves as a model nano-mechanical or a lower dimensional system (1D) and physicists are interested in the nature of forces dominating at the nano-scale and the effect of surface forces (which are expected to be significant due to the large surface to volume ratio).

    (b) The length scale effects on the interface and the differences between the phenomena of mechanics at the macro (or meso) and the nano-scale.

4.5.2 SOME MODELING METHOD IN INTERFACE MODELING

Computational techniques have extensively been used to study the interfacial mechanics and nature of bonding in CNT–polymer composites. The computational studies can be broadly classified as atomistic simulations and continuum methods. The atomistic simulations are primarily based on molecular dynamic simulations (MD) and density functional theory (DFT) [105, 106–110]. The main focus of these techniques was to understand and study the effect of bonding between the polymer and nanotube (covalent, electrostatic or Van Der Waals forces) and the effect of friction on the interface. The continuum methods extend the continuum theories of micromechanics modeling and fiber-reinforced composites (elaborated in the next section) to CNT–polymer composites [111–114] and explain the behavior of the composite from a mechanics point of view.

On the experimental side, the main types of studies that can be found in literature are as follows:

(i) Researchers have performed experiments on CNT–polymer bulk composites at the macroscale and observed the enhancements in mechanical properties (like elastic modulus and tensile strength) and tried to correlate the experimental results and phenomena with continuum theories like micro-mechanics of composites or Kelly Tyson shear lag model [105, 115–120].

(ii) Raman spectroscopy has been used to study the reinforcement provided by carbon nanotubes to the polymer, by straining the CNT–polymer composite and observing the shifts in Raman peaks [121–125].

(iii) In situ TEM straining has also been used to understand the mechanics, fracture and failure processes of the interface. In these techniques, the CNT–polymer composite (an electron transparent thin specimen) is strained inside a TEM and simultaneously imaged to get real-time and spatially resolved (1 nm) information [110, 126].

4.5.3 NUMERICAL APPROACH

A MD model may serve as a useful guide, but its relevance for a covalent-bonded system of only a few atoms in diameter is far from obvious. Because of this, the phenomenological multiple column models that considers the interlayer radial displacements coupled through the van der Waals forces is used. It should also be mentioned the special features of load transfer, in tension and in compression, in MWNT-epoxy composites studied by Schadler et al. [57] who detected that load transfer in tension was poor in comparison to load transfer in compression, implying that during load transfer to MWNTs, only the outer layers are stressed in tension due to the telescopic inner-wall sliding (reaching at the shear stress 0.5 MPa [127]), whereas all the layers respond in compression. It should be mentioned that NTCMs usually contain not individual, separated SWCNTs, but rather bundles of closest-packed SWCNTs [128], where the twisting of the CNTs produces the radial force component giving the rope structure more stable than wires in parallel. Without strong chemically bonding, load transfer between the CNTs and the polymer matrix mainly comes from weak electrostatic and van der Waals interactions, as well as stress/deformation arising from mismatch in the coefficients of thermal expansion [129]. Numerous researchers [130] have attributed lower than-predicted CNT-polymer composite properties to the availability of only a weak interfacial bonding. So Frankland et al. [106] demonstrated by MD simulation that the shear strength of a polymer/nanotube interface with only van der Waals interactions could be increased by over an order of magnitude at the occurrence of covalent bonding for only 1% of the nanotubes carbon atoms to the polymer matrix. The recent force-field-based molecular-mechanics calculations [131] demonstrated that the binding energies and frictional forces play only a minor role in determining the strength of the interface. The key factor in forming a strong bond at the interface is having a helical conformation of the polymer around the nanotube; polymer wrapping around nanotube improves the polymer-nanotube interfacial strength, although configurationally thermodynamic considerations do not necessarily support these architectures for all polymer chains [132]. Thus, the strength of the interface may result from molecular-level entanglement of the two phases and forced long-range ordering of the polymer. To ensure the robustness of data reduction schemes that are based on continuum mechanics, a careful analysis of continuum approximations used in macromolecular models and possible limitations of these approaches at the nanoscale are additionally required that can be done by the fitting of the results obtained by the use of the proposed phenomenological interface model with the experimental data of measurement of the stress distribution in the vicinity of a nanotube.

Meguid et al. [133] investigated the interfacial properties of carbon nanotube (CNT) reinforced polymer composites by simulating a nanotube pull-out experiment. An atomistic description of the problem was achieved by implementing constitutive relations that are derived solely from interatomic potentials. Specifically, they adopt the Lennard-Jones (LJ) interatomic potential to simulate a non-bonded interface, where only the van der Waals (vdW) interactions between the CNT and surrounding polymer matrix were assumed to exist. The effects of such parameters as the CNT embedded length, the number of vdW interactions, the thickness of the interface, the CNT diameter and the cut-off distance of the LJ potential on the interfacial shear strength (ISS) are investigated and discussed. The problem is formulated for both a generic thermoset polymer and a specific two-component epoxy based on a diglycidyl ether of bisphenol A (DGEBA) and triethylene tetramine (TETA) formulation. The study further illustrated that by accounting for different CNT capping scenarios and polymer morphologies around the embedded end of the CNT, the qualitative correlation between simulation and experimental pull-out profiles can be improved. Only vdW interactions were considered between the atoms in the CNT and the polymer implying a non-bonded system. The vdW interactions were simulated using the LJ potential, while the CNT was described using the Modified Morse potential. The results reveal that the ISS shows a linear dependence on the vdW interaction density and decays significantly with increasing nanotube embedded length. The thickness of the interface was also varied and our results reveal that lower interfacial thicknesses favor higher ISS. When incorporating a 2.5ψ cut-off distance to the LJ potential, the predicted ISS shows an error of approximately 25.7% relative to a solution incorporating an infinite cut-off distance. Increasing the diameter of the CNT was found to increase the peak pull-out force approximately linearly. Finally, an examination of polymeric and CNT capping conditions showed that incorporating an end cap in the simulation yielded high initial pull-out peaks that better correlate with experimental findings. These findings have a direct bearing on the design and fabrication of carbon nanotube reinforced epoxy composites.

Fiber pull-out tests have been well recognized as the standard method for evaluating the interfacial bonding properties of composite materials. The output of these tests is the force required to pullout the nanotube from the surrounding polymer matrix and the corresponding interfacial shear stresses involved. The problem is formulated using a representative volume element (RVE) which consists of the reinforcing CNT, the surrounding polymer matrix, and the CNT/polymer interface as depicted in Figure 4.8(a, b) shows a schematic of the pull-out process, where x is the pullout distance and L is the embedded length of the nanotube. The atomistic-based continuum (ABC) multiscale modeling technique is used to model the RVE. The approach adopted here extends the earlier work of Wernik and Meguid [134].

The new features of the current work relate to the approach adopted in the modeling of the polymer matrix and the investigation of the CNT polymer interfacial properties as appose to the effective mechanical properties of the RVE. The idea behind the ABC technique is to incorporate atomistic interatomic potentials into a continuum framework. In this way, the interatomic potentials introduced in the model capture the underlying atomistic behavior of the different phases considered. Thus, the influence of the nanophase is taken into account via appropriate atomistic constitutive formulations. Consequently, these measures are fundamentally different from those in the classical continuum theory. For the sake of completeness, Wernik and Meguid provided a brief outline of the method detailed in their earlier work [134].

The cumulative effect of the vdW interactions acting on each CNT atom is applied as a resultant force on the respective node, which is then resolved into its three Cartesian components. This process is depicted in Figure 4.9 during each iteration of the pull-out process, the above expression is re-evaluated for each vdW interaction and the cumulative resultant

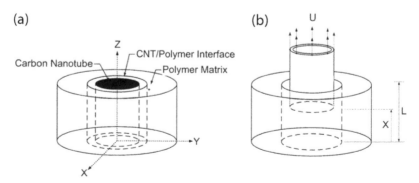

FIGURE 4.8 Schematic depictions of (a) the representative volume element and (b) the pull-out process.

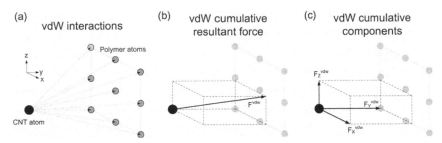

FIGURE 4.9   The process of nodal vdW force application. (a) vdW interactions on an individual CNT atom, (b) the cumulative resultant vdW force, and (c) the cumulative vdW Cartesian components.

force and its three Cartesian components are updated to correspond to the latest pull-out configuration. Figure 4.10 shows a segment of the CNT with the cumulative resultant vdW force vectors as they are applied to the CNT atoms.

Yang et al. [135], investigated the CNT size effect and weakened bonding effect between an embedded CNT and surrounding matrix were characterized using MD simulations. Assuming that the equivalent continuum model of the CNT atomistic structure is a solid cylinder, the transversely isotropic elastic constants of the CNT decreased as the CNT radius increased. Regarding the elastic stiffness of the nanocomposite unit cell, the same CNT size dependency was observed in all independent components, and only the longitudinal Young's modulus showed a positive reinforcing effect whereas other elastic moduli demonstrated negative reinforcing effects as a result of poor load transfer at the interface. To describe the size effect and weakened bonding effect at the interface, a

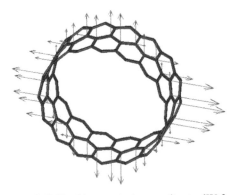

FIGURE 4.10   Segment of CNT with cumulative resultant vdW force vectors.

modified multi-inclusion model was derived using the concepts of an effective CNT and effective matrix. During the scale bridging process incorporating the MD simulation results and modified multi-inclusion model, we found that both the elastic modulus of the CNT and the adsorption layer near the CNT contributed to the size-dependent elastic modulus of the nanocomposites. Using the proposed multiscale bridging model, the elastic modulus for nanocomposites at various volume fractions and CNT sizes could be estimated. Among three major factors (CNT waviness, the dispersion state, and adhesion between the CNT and matrix), the proposed model considered only the weakened bonding effect. However, the present multiscale framework can be easily applied in considering the aforementioned factors and describing the real nanocomposite microstructures. In addition, by considering chemically grafted molecules (covalent or non-covalent bonds) to enhance the interfacial load transfer mechanism in MD simulations, the proposed multiscale approach can offer a deeper understanding of the reinforcing mechanism, and a more practical analytical tool with which to analyze and design functional nanocomposites. The analytical estimation reproduced from the proposed multiscale model can also provide useful information in modeling finite element-based representative volume elements of nanocomposite microstructures for use in multifunctional design.

The effects of the interphase and RVE configuration on the tensile, bending and torsional properties of the suggested nanocomposite were investigated by Ayatollahi et al. [136]. It was found that the stiffness of the nanocomposite could be affected by a strong interphase much more than by a weaker interphase. In addition, the stiffness of the interphase had the maximum effect on the stiffness of the nanocomposite in the bending loading conditions. Furthermore, it was revealed that the ratio of Le/Ln in RVE can dramatically affect the stiffness of the nanocomposite especially in the axial loading conditions.

For carbon nanotubes not well bonded to polymers, Jiang et al. [137] established a cohesive law for carbon nanotube/polymer interfaces. The cohesive law and its properties (e.g., cohesive strength, cohesive energy) are obtained directly from the Lennard–Jones potential from the van der Waals interactions. Such a cohesive law is incorporated in the micromechanics model to study the mechanical behavior of carbon nanotube-reinforced composite materials. Carbon nanotubes indeed improve the mechanical behavior of composite at the small strain. However, such improvement disappears at relatively large strain because the completely debonded nanotubes behave like voids in the matrix and may even weaken the composite. The increase of interface adhesion between carbon nanotubes and polymer matrix may significantly improve the composite behavior at the large strain [138].

Zalamea et al. [139] employed the shear transfer model as well as the shear lag model to explore the stress transfer from the outermost layer to the interior layers in MWCNTs. Basically, the interlayer properties between graphene layers were designated by scaling the parameter of shear transfer efficiency with respect to the perfect bonding. Zalamea et al. pointed out that as the number of layers in MWCNTs increases, the stress transfer efficiency decreases correspondingly. Shen et al. [140] examined load transfer between adjacent walls of DWCNTs using MD simulation, indicating that the tensile loading on the outermost wall of MWCNTs cannot be effectively transferred into the inner walls. However, when chemical bonding between the walls is established, the effectiveness can be dramatically enhanced. It is noted that in the above investigations, the loadings were applied directly on the outermost layers of MWCNTs; the stresses in the inner layers were then calculated either from the continuum mechanics approach [139] or MD simulation [140]. Shokrieh and Rafiee [93, 94] examined the mechanical properties of nanocomposites with capped single-walled carbon nanotubes (SWCNTs) embedded in a polymer matrix. The load transfer efficiency in terms of different CNTs' lengths was the main concern in their examination. By introducing an interphase to represent the vdW interactions between SWCNTs and the surrounding matrix, Shokrieh and Rafiee [93, 94] converted the atomistic SWCNTs into an equivalent continuum fiber in finite element analysis. The idea of an equivalent solid fiber was also proposed by Gao and Li [141] to replace the atomistic structure of capped SWCNTs in the nanocomposites' cylindrical unit cell. The modulus of the equivalent solid was determined based on the atomistic structure of SWCNTs through molecular structure mechanics [142]. Subsequently, the continuum-based shear lag analysis was carried out to evaluate the axial stress distribution in CNTs. In addition, the influence of end caps in SWCNTs on the stress distribution of nanocomposites was also taken into account in their analysis. Tsai and Lu [143] characterized the effects of the layer number, inter-graphic layers interaction, and aspect ratio of MWCNTs on the load transfer efficiency using the conventional shear lag model and finite element analysis. However, in their analysis, the inter-atomistic characteristics of the adjacent graphene layers associated with different degrees of interactions were simplified by a thin interphase with different moduli. The atomistic interaction between the grapheme layers was not taken into account in their modeling of MWCNTs. In light of the forgoing investigations, the equivalent solid of SWCNTs was developed by several researchers and then implemented as reinforcement in continuum-based nanocomposite models. Nevertheless, for MWCNTs, the subjects concerning the development of equivalent continuum solid are seldom explored in the literature. In fact, how to introduce the atomistic characteristics, that is, the interfacial properties of neighboring graphene layers in MWCNTs, into the equivalent continuum solid is a challenging task as the length scales used to describe the physical phenomenon are distinct. Thus, a multi-scale based simulation is required to account for the atomistic attribute of MWCNTs into an equivalent continuum solid. In Lu and Tsai's study [144], the multi-scale approach was used to investigate the load transfer efficiency from surrounding matrix to DWCNTs. The analysis consisted of two stages. First, a cylindrical DWCNTs equivalent continuum was proposed based on MD simulation where the pullout extension on the outer layer was performed in an attempt to characterize the atomistic behaviors between neighboring graphite layers. Subsequently, the cylindrical continuum (denoting the DWCNTs) was embedded in a unit cell of nanocomposites, and the axial stress distribution as well as the load transfer efficiency of the DWCNTs was evaluated from finite element analysis. Both single-walled carbon nanotubes (SWCNTs) and DWCNTs were considered in the simulation and the results were compared with each other.

An equivalent cylindrical solid to represent the atomistic attributes of DWCNTs was proposed in this study. The atomistic interaction of adjacent graphite layers in DWCNTs was characterized using MD simulation based on which a spring element was introduced in the continuum equivalent solid to demonstrate the interfacial properties of DWCNTs. Subsequently, the proposed continuum solid (denotes DWCNTs) was embedded in the matrix to form DWCNTs nanocomposites (continuum model), and the load transfer efficiency within the DWCNTs was determined from FEM analysis. For the demonstration purpose, the DWCNTs with four different lengths were considered in the investigation. Analysis results illustrate that the increment of CNTs' length can effectively improve the load transfer efficiency in the outermost layers, nevertheless, for the inner layers, the enhancement is miniature. On the other hand, when the covalent bonds between the adjacent graphene layers are crafted, the load carrying capacity in the inner layer increases as so does the load transfer efficiency of DWCNTs. As compared to SWCNTs, the DWCNTs still possess the less capacity of load transfer efficiency even though there are covalent bonds generated in the DWCNTs.

4.6   CONCLUDING REMARKS

Many traditional simulation techniques (e.g., MC, MD, BD, LB, Ginzburg–Landau theory, micromechanics and FEM) have been employed, and some novel simulation techniques (e.g., DPD, equivalent-continuum and self-similar approaches) have been developed to study polymer nano-composites. These techniques indeed represent approaches at various time and length scales from molecular scale (e.g., atoms), to microscale (e.g., coarse-grains, particles, monomers) and then to macroscale (e.g., domains), and have shown success to various degrees in addressing many aspects of polymer nanocomposites. The simulation techniques developed thus far have different strengths and weaknesses, depending on the need of research. For example, molecular simulations can be used to investigate molecular interactions and structure on the scale of 0.1–10 nm. The resulting information is very useful to understanding the interaction strength at nanoparticle–polymer interfaces and the molecular origin of mechanical improvement. However, molecular simulations are computationally very demanding, thus not so applicable to the prediction of mesoscopic structure and properties defined on the scale of 0.1–10 mm, for example, the dispersion of nanoparticles in polymer matrix and the morphology of polymer nanocomposites. To explore the morphology on these scales, mesoscopic simulations such as coarse-grained methods, DPD and dynamic mean field theory are more effective. On the other hand, the macroscopic properties of materials are usually studied by the use of mesoscale or macroscale techniques such as micromechanics and FEM. But these techniques may have limitations when applied to polymer nanocomposites because of the difficulty to deal with the interfacial nanoparticle–polymer interaction and the morphology, which are considered crucial to the mechanical improvement of nanoparticle-filled polymer nanocomposites. Therefore, despite the progress over the past years, there are a number of challenges in computer modeling and simulation. In general, these challenges represent the work in two directions. First, there is a need to develop new and improved simulation techniques at individual time and length scales. Secondly, it is important to integrate the developed methods at wider range of time and length scales, spanning from quantum mechanical domain (a few atoms) to molecular domain (many atoms), to mesoscopic domain (many monomers or chains), and finally to macroscopic domain (many domains or structures), to form a useful tool for exploring the structural, dynamic, and mechanical properties, as well as optimizing design and processing control of polymer nanocomposites. The need for the second development is obvious. For example, the morphology is usually determined from the mesoscale techniques whose implementation requires information about the interactions between various components (e.g., nanoparticle–nanoparticle and nanoparticle–polymer) that should be derived from molecular simulations. Developing such a multiscale method is very challenging but indeed represents the future of computer simulation and modeling, not only in polymer nanocomposites but also other fields. New concepts, theories and computational tools should be developed in the future to make truly seamless multiscale modeling a reality. Such development is crucial in order to achieve the longstanding goal of predicting particle–structure property relationships in material design and optimization.

The strength of the interface and the nature of interaction between the polymer and carbon nanotube are the most important factors governing the ability of nanotubes to improve the performance of the composite. Extensive research has been performed on studying and understanding CNT-polymer composites from chemistry, mechanics and physics aspects. However, there exist various issues like processing of composites and experimental challenges, which need to be addressed to gain further insights into the interfacial processes.

APPENDIX A. CARBON NANOTUBES

In, 1960, Bacon of Union Carbide reported observing straight hollow tubes of carbon that appeared as grapheme layers of carbon. In 1970s, Oberlin et al., observed these tubes again by a catalysis-enhanced chemical vapor deposition (CVD) process. In 1985, random events led to the discovery of a new molecule made entirely of carbon, 60 carbons arranged in a soccer ball shape. In fact, what had been discovered was an infinite number of molecules: the fullerenes, C_{60}, C_{70}, C_{84}, etc., every molecule with the characteristic of being a pure carbon cage. These molecules were mostly seen in a spherical shape. However, it is until, 1991 that Iijima of NEC observed a tubular shape in the form of coaxial tubes of graphitic sheets, ranging from two shells to approximately 50. Later this structure was called multiwalled carbon nanotube (MWNT). Two years later, Bethune et al., and Iijima and Ichihashi managed to observe the same tubular structure, but with only a single atomic layer of graphene, which became known as a single-walled carbon nanotube (SWNT).

CNTs have typical diameters in the range of ~1–50 nm and lengths of many microns (even centimeters in special cases). They can consist of one or more concentric graphitic cylinders. In contrast, commercial (PAN and pitch) carbon fibers are typically in the 7–20 µm diameter range, while vapor-grown carbon fibers (VGCFs) have intermediate diameters ranging from a few hundred nanometers up to around a millimeter. The variation in diameter of fibrous graphitic materials is summarized in Figure A1. Crucially, conventional carbon fibers do not have the same potential for structural perfection that can be observed in CNTs. Indeed, there is a general question as to whether the smallest CNTs should be regarded as very small fibers or heavy molecules, since the diameters of the smallest nanotubes are similar to those of common polymer molecules. This ambiguity is characteristic of nanomaterials, and it is not yet clear to what extent conventional fiber composite understanding can be extended to CNT composites (Figure A1).

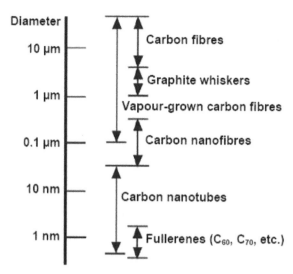

FIGURE A1   Comparison of diameters of various fibrous carbon-based materials.

Properties of Carbon Nanotubes

In, 1991, Japanese researcher Idzhima was studying the sediments formed at the cathode during the spray of graphite in an electric arc. His attention was attracted by the unusual structure of the sediment consisting of microscopic fibers and filaments. Measurements made with an electron microscope showed that the diameter of these filaments does not exceed a few nanometers and a length of one to several microns.

Having managed to cut a thin tube along the longitudinal axis, the researchers found that it consists of one or more layers, each representing a hexagonal grid of graphite, which is based on hexagon with vertices located at the corners of the carbon atoms. In all cases, the distance between the layers is equal to 0.34 nm, that is, the same as that between the layers in crystalline graphite.

Typically, the upper ends of tubes are closed by multilayer hemispherical caps, each layer is composed of hexagons and pentagons, reminiscent of the structure of half a fullerene molecule.

The extended structure consisting of rolled hexagonal grids with carbon atoms at the nodes are called nanotubes.

Lattice structure of diamond and graphite are shown in Figure A2. Graphite crystals are built of planes parallel to each other, in which carbon

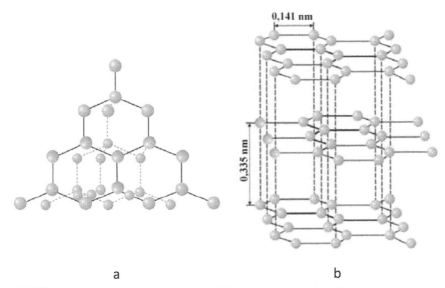

0,141 nm

0,335 nm

a   b

FIGURE A2   The structure of the diamond lattice (a) and graphite (b).

atoms are arranged at the corners of regular hexagons. The distance between adjacent carbon atoms (each side of the hexagon) $d_0 = 0.141 nm$, between adjacent planes – 0.335 nm.

Each intermediate plane is shifted somewhat toward the neighboring planes, as shown in the figure.

The elementary cell of the diamond crystal represents a tetrahedron, with carbon atoms in its center and four vertices. Atoms located at the vertices of a tetrahedron form a center of the new tetrahedron, and thus, are also surrounded by four atoms each, etc. All the carbon atoms in the crystal lattice are located at equal distance (0.154 nm) from each other.

Nanotubes are rolled into a cylinder (hollow tube) graphite plane, which is lined with regular hexagons with carbon atoms at the vertices of a diameter of several nanometers. Nanotubes can consist of one layer of atoms – single-wall nanotubes SWNT and represent a number of "nested" one into another layer pipes – multi-walled nanotubes – MWNT.

Nanostructures can be built not only from individual atoms or single molecules, but the molecular blocks. Such blocks or elements to create nanostructures are graphene, carbon nanotubes and fullerenes.

Graphene

Graphene is a single flat sheet, consisting of carbon atoms linked together and forming a grid, each cell is like a bee's honeycombs (Figure A3). The distance between adjacent carbon atoms in graphene is about 0.14 nm.

Graphite, from which slates of usual pencils are made, is a pile of graphene sheets (Figures A3 and A4). Graphenes in graphite is very poorly connected and can slide relative to each other. So, if you conduct the graphite on paper, then after separating graphene from sheet the graphite remains on paper. This explains why graphite can write.

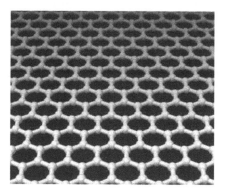

FIGURE A3   Schematic illustration of the graphene. Light balls – the carbon atoms, and the rods between them – the connections that hold the atoms in the graphene sheet.

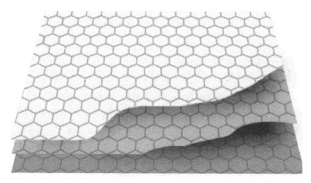

FIGURE A4   Schematic illustration of the three sheets of graphene, which are one above the other in graphite.

Many perspective directions in nanotechnology are associated with carbon nanotubes.

Carbon nanotubes are a carcass structure or a giant molecule consisting only of carbon atoms.

Carbon nanotube is easy to imagine, if we imagine that we fold up one of the molecular layers of graphite – graphene (Figure A5).

The way of folding nanotubes – the angle between the directions of nanotube axis relative to the axis of symmetry of graphene (the folding angle) – largely determines its properties.

The way of folding nanotubes – the angle between the direction of nanotube axis relative to the axis of symmetry of graphene (the folding angle) – largely determines its properties.

Of course, no one produces nanotubes, folding it from a graphite sheet. Nanotubes formed themselves, for example, on the surface of carbon electrodes during arc discharge between them. At discharge, the carbon atoms evaporate from the surface, and connect with each other to form nanotubes of all kinds – single, multi-layered and with different angles of twist (Figure A6).

The diameter of nanotubes is usually about 1 nm and their length is a thousand times more, amounting to about 40 microns. They grow on the cathode in perpendicular direction to surface of the butt. The so-called self-assembly of carbon nanotubes from carbon atoms occurs. Depending on the angle of folding of the nanotube they can have conductivity as

FIGURE A5   Carbon nanotubes.

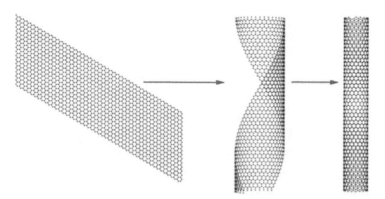

FIGURE A6   One way of imaginary making nanotube (right) from the molecular layer of graphite (left).

high as that of metals, and they can have properties of semiconductors (Figure A7).

Carbon nanotubes are stronger than graphite, although made of the same carbon atoms, because carbon atoms in graphite are located in the sheets. And everyone knows that sheet of paper folded into a tube is much more difficult to bend and break than a regular sheet. That's why carbon nanotubes are strong. Nanotubes can be used as a very strong microscopic rods and filaments, as Young's modulus of single-walled nanotube reaches values of the order of 1–5 TPa, which is much more than steel! Therefore,

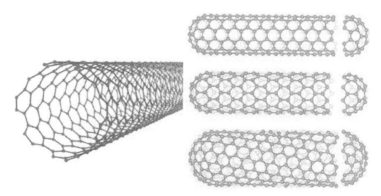

FIGURE A7   Left – schematic representation of a single-layer carbon nanotubes, on the right (top to bottom) – two-ply, straight and spiral nanotubes.

the thread made of nanotubes, the thickness of a human hair is capable to hold down hundreds of kilos of cargo.

It is true that at present the maximum length of nanotubes is usually about a hundred microns – which is certainly too small for everyday use. However, the length of the nanotubes obtained in the laboratory is gradually increasing – now scientists have come close to the millimeter border. So there is every reason to hope that in the near future, scientists will learn how to grow a nanotube length in centimeters and even meters!

Fullerenes

The carbon atoms, evaporated from a heated graphite surface, connecting with each other, can form not only nanotube, but also other molecules, which are closed convex polyhedral, for example, in the form of a sphere or ellipsoid. In these molecules, the carbon atoms are located at the vertices of regular hexagons and pentagons, which make up the surface of a sphere or ellipsoid.

The molecules of the symmetrical and the most studied fullerene consisting of 60 carbon atoms (C_{60}), form a polyhedron consisting of 20 hexagons and 12 pentagons and resembles a soccer ball (Figure A8). The diameter of the fullerene C_{60} is about 1 nm. The image of the fullerene C_{60} many consider as a symbol of nanotechnology.

FIGURE A8   Schematic representation of the fullerene C_{60}.

Classification of Nanotubes

The main classification of nanotubes is conducted by the number of constituent layers.

Single-walled nanotubes (Figure A9) – the simplest form of nanotubes. Most of them have a diameter of about 1 nm in length, which can be many thousands of times more. The structure of the nanotubes can be represented as a "wrap" a hexagonal network of graphite (graphene), which is based on hexagon with vertices located at the corners of the carbon atoms in a seamless cylinder. The upper ends of the tubes are closed by hemispherical caps, each layer is composed of hexa – and pentagons, reminiscent of the structure of half of a fullerene molecule. The distance d between adjacent carbon atoms in the nanotube is approximately equal to $d = 0.15$ nm.

Multi-walled nanotubes consist of several layers of graphene stacked in the shape of the tube. The distance between the layers is equal to 0.34 nm, that is the same as that between the layers in crystalline graphite (Figure A.10).

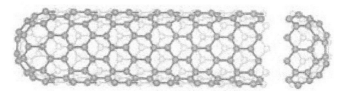

FIGURE A9   Graphical representation of single-walled nanotube.

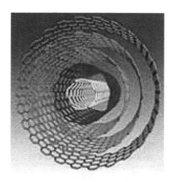

FIGURE A10   Graphic representation of a multiwalled nanotube.

Due to its unique properties (high fastness (63 GPa), superconductivity, capillary, optical, magnetic properties, etc.), carbon nanotubes could find applications in numerous areas:

- additives in polymers;
- catalysts (autoelectronic emission for cathode ray lighting elements, planar panel of displays, gas discharge tubes in telecom networks);
- absorption and screening of electromagnetic waves;
- transformation of energy;
- anodes in lithium batteries;
- keeping of hydrogen;
- composites (filler or coating);
- nanosondes;
- sensors;
- strengthening of composites;
- supercapacitors.

Chirality – a set of two integer positive indices (n, m), which determines how the graphite plane folds and how many elementary cells of graphite at the same time fold to obtain the nanotube.

From the value of parameters (n, m) are distinguished

- direct (achiral) high-symmetry carbon nanotubes
  - armchair $n = m$
  - zigzag $m = 0$ or $n = 0$
- helical (chiral) nanotube

In Figure A11a is shown a schematic image of the atomic structure of graphite plane – graphene, and shown how a nanotube can be obtained from it. The nanotube is fold up with the vector connecting two atoms on a graphite sheet. The cylinder is obtained by folding this sheet so that were combined the beginning and end of the vector. That is, to obtain a carbon nanotube from a graphene sheet, it should turn so that the lattice vector $\overline{R}$ has a circumference of the nanotube in Figure A11b. This vector can be expressed in terms of the basis vectors of the elementary cell graphene sheet $\vec{R} = n\vec{r}_1 + m\vec{r}_2$. Vector $\overline{R}$, which is often referred to simply by a pair of indices (n, m), called the chiral vector. It is assumed that $n > m$. Each pair of numbers (n, m) represents the possible structure of the nanotube.

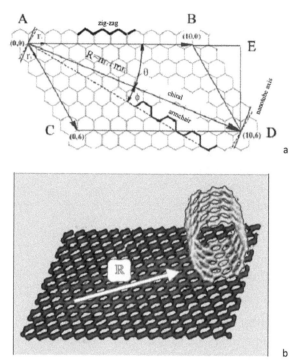

FIGURE A11   Schematic representation of the atomic structure of graphite plane.

In other words the chirality of the nanotubes (n, m) indicates the coordinates of the hexagon, which as a result of folding the plane has to be coincide with a hexagon, located at the beginning of coordinates (Figure A12).

Many of the properties of nanotubes (e.g., zonal structure or space group of symmetry) strongly depend on the value of the chiral vector. Chirality indicates what property has a nanotube – a semiconductor or metallicheskm. For example, a nanotube (10, 10) in the elementary cell contains 40 atoms and is the type of metal, whereas the nanotube (10, 9) has already in 1084 and is a semiconductor.

If the difference $n - m$ is divisible by 3, then these CNTs have metallic properties. Semimetals are all achiral tubes such as "chair." In other cases, the CNTs show semiconducting properties. Just type chair CNTs ($n = m$) are strictly metal (Figure A13).

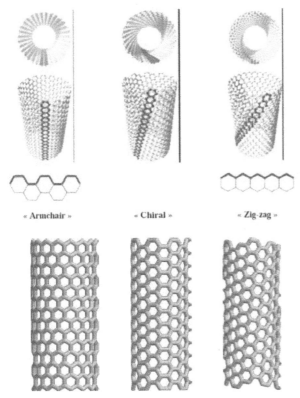

« Armchair »   « Chiral »   « Zig-zag »

FIGURE A12 Single-walled carbon nanotubes of different chirality (in the direction of convolution). Left to right: the zigzag (16, 0), armchair (8, 8) and chiral (10, 6) carbon nanotubes.

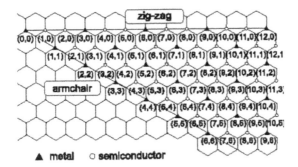

FIGURE A13 The scheme of indices (n, m) of lattice vector $\overline{R}$ tubes having semiconductor and metallic properties.

KEYWORDS

- carbon nanotube
- interfaces
- mechanical properties
- modeling
- multi-scale
- polymer composites
- simulation

REFERENCES

1. Iijima, S. Helical microtubules of graphitic carbon. Nature, 1991, 354(6348), 56.
2. Dresselhaus, M. S., Dresselhaus, G., Eklund, P. C. Science of fullerenes and carbon nanotubes. New York: Academic Press; 1996.
3. Saito, R., Dresselhaus, G., Dresselhaus, M. S. Physical properties of carbon nanotubes. Imperial College Press; 1998.
4. Harris, P. J. F. Carbon nanotubes and related structures: new materials for the 20-first century. Cambridge, United Kingdom: Cambridge University Press; 1999, 279.
5. Wagner, H. D., Vaia, R. A. Nanocomposites: issues at the interface. Mater Today, 2004, 7(11), 38.
6. Zeng, Q. H., Yu, A. B., Lu, G. Q. Multiscale modeling and simulation of polymer nanocomposites, Prog. Polym. Sci. 33 (2008) 191–269.
7. Iijima, S. Helical microtubules of graphitic carbon. Nature, 1991, 354, 56–8.
8. Bethune, D. S., Klang, C. H., De Vries, M. S., et al. Cobalt-catalyzed growth of carbon nanotubes with single-atomic-layer walls. Nature, 1993, 363, 605–7.
9. Dresselhaus, M. S., Dresselhaus, G., Saito, R. Physics of carbon nanotubes. Carbon, 1995, 33, 883–91.
10. Thostenson, E. T., Ren, Z. F., Chou, T. W. Advances in the science and technology of CNTs and their composites: a review. Compos Sci Technol, 2001, 61, 1899–912.
11. Yakobson, B. I., Avouris, P. Mechanical properties of carbon nanotubes. Top Appl Phys, 2001, 80, 287–327.
12. Ajayan, P. M., Schadler, L. S., Braun, P. V. Nanocomposite science and technology. Weinheim: Wiley-VCH; 2003. p. 77–80.
13. Li, J., Ma, P. C., Chow, W. S., To, C. K., Tang, B. Z., Kim, J. K., Correlations between percolation threshold, dispersion state and aspect ratio of carbon nanotube, Adv Funct Mater, 2007, 17, 3207–3215.

14. Ajayan, P. M., Stephan, O., Colliex, C., Trauth, D. Aligned carbon nanotube arrays formed by cutting a polymer resin-nanotube composite. Science, 1994, 265, 1212–4.
15. Summary of Searching Results. http://www.scopus.com (accessed January, 2010).
16. Du, J. H., Bai, J., Cheng, H. M. The present status and key problems of carbon nanotube based polymer composites. Express Polym Lett, 2007, 1, 253–73.
17. Lee, J. Y., Baljon, A. R. C., Loring, R. F., Panagiotopoulos, A. Z. Simulation of polymer melt intercalation in layered nanocomposites. J Chem Phys, 1998, 109, 10321–30.
18. Smith, G. D., Bedrov, D., Li, L. W., Byutner, O. A molecular dynamics simulation study of the viscoelastic properties of polymer nanocomposites. J Chem Phys, 2002, 117, 9478–89.
19. Smith, J. S., Bedrov, D., Smith, G. D. A molecular dynamics simulation study of nanoparticle interactions in a model polymer–nanoparticle composite. Compos Sci Technol, 2003, 63, 1599–605.
20. Zeng, Q. H., Yu, A. B., Lu, G. Q., Standish, R. K. Molecular dynamics simulation of organic-inorganic nanocomposites: layering behavior and interlayer structure of organoclays. Chem Mater, 2003, 15, 4732–8.
21. Vacatello, M. Predicting the molecular arrangements in polymer-based nanocomposites. Macromol Theory Simul, 2003, 12, 86–91.
22. Zeng, Q. H., Yu, A. B., Lu, G. Q. Interfacial interactions and structure of polyurethane intercalated nanocomposite. Nanotechnology, 2005, 16, 2757–63.
23. Allen, M. P., Tildesley, D. J. Computer simulation of liquids. Oxford: Clarendon Press; 1989.
24. Frenkel, D., Smit, B. Understanding molecular simulation: from algorithms to applications. 2nd ed. San Diego: Academic Press; 2002.
25. Metropolis, N., Rosenbluth, A. W., Marshall, N., Rosenbluth, M. N., Teller, A. T. Equation of state calculations by fast computing machines. J Chem Phys, 1953, 21, 1087–92.
26. Carmesin, I., Kremer, K. The bond fluctuation method: a new effective algorithm for the dynamics of polymers in all spatial dimensions. Macromolecules, 1988, 21, 2819–23.
27. Hoogerbrugge, P. J., Koelman JMVA. Simulating microscopic hydrodynamic phenomena with dissipative particle dynamics. Europhys Lett, 1992, 19, 155–60.
28. Gibson, J. B., Chen, K., Chynoweth, S. Simulation of particle adsorption onto a polymer-coated surface using the dissipative particle dynamics method. J Colloid Interface Sci, 1998, 206, 464–74.
29. Dzwinel, V., Yuen, D. A. A two-level, discrete particle approach for large-scale simulation of colloidal aggregates. Int J Mod Phys C 2000, 11, 1037–61.
30. Dzwinel, W., Yuen, D. A. A two-level, discrete-particle approach for simulating ordered colloidal structures. J Colloid Interface Sci, 2000, 225, 179–90.
31. Chen, S., Doolen, G. D. Lattice Boltzmann method for fluid flows. Annu Rev Fluid Mech, 1998, 30, 329 64.
32. Cahn, J. W. On spinodal decomposition. Acta Metall, 1961, 9, 795–801.
33. Cahn, J. W., Hilliard, J. E. Spinodal decomposition: a reprise. Acta Metall, 1971, 19, 151–61.
34. Cahn, J. W. Free energy of a nonuniform system. II. Thermodynamic basis. J Chem Phys, 1959, 30, 1121–4.

35. Lee, B. P., Douglas, J. F., Glotzer, S. C. Filler-induced composition waves in phase-separating polymer blends. Phys Rev E 1999, 60, 5812–22.
36. Ginzburg, V. V., Qiu, F., Paniconi, M., Peng, G. W., Jasnow, D., Balazs, A. C. Simulation of hard particles in a phase separating binary mixture. Phys Rev Lett, 1999, 82, 4026–9.
37. Qiu, F., Ginzburg, V. V., Paniconi, M., Peng, G. W., Jasnow, D., Balazs, A. C. Phase separation under shear of binary mixtures containing hard particles. Langmuir, 1999, 15, 4952–6.
38. Ginzburg, V. V., Gibbons, C., Qiu, F., Peng, G. W., Balazs, A. C. Modeling the dynamic behavior of diblock copolymer/ particle composites. Macromolecules, 2000, 33, 6140–7.
39. Ginzburg, V. V., Qiu, F., Balazs, A. C. Three-dimensional simulations of diblock copolymer/particle composites. Polymer, 2002, 43, 461–6.
40. He, G., Balazs, A. C. Modeling the dynamic behavior of mixtures of diblock copolymers and dipolar nanoparticles. J Comput Theor Nanosci, 2005, 2, 99–107.
41. Altevogt, P., Ever, O. A., Fraaije JGEM, Maurits, N. M., van Vlimmeren, B. A. C. The MesoDyn project: software for mesoscale chemical engineering. J Mol Struct, 1999, 463, 139–43.
42. Kawakatsu, T., Doi, M., Hasegawa, A. Dynamic density functional approach to phase separation dynamics of polymer systems. Int J Mod Phys C 1999, 10, 1531–40.
43. Morita, H., Kawakatsu, T., Doi, M. Dynamic density functional study on the structure of thin polymer blend films with a free surface. Macromolecules, 2001, 34, 8777–83.
44. Doi, M. OCTA-a free and open platform and softwares of multiscale simulation for soft materials /http://octa.jp/S. 2002.
45. Tandon, G. P., Weng, G. J. The effect of aspect ratio of inclusions on the elastic properties of unidirectionally aligned composites. Polym Compos, 1984, 5, 327–33.
46. Fornes, T. D., Paul, D. R. Modeling properties of nylon 6/clay nanocomposites using composite theories. Polymer, 2003, 44, 4993–5013.
47. Odegard, G. M., Pipes, R. B., Hubert, P. Comparison of two models of SWCN polymer composites. Compos Sci Technol, 2004, 64, 1011–1120.
48. Odegard, G. M., Gates, T. S., Wise, K. E., Park, C., Siochi, E. J. Constitutive modeling of nanotube-reinforced polymer composites. Compos Sci Technol, 2003, 63, 1671–1687.
49. Pipes, R. B., Hubert, P. Helical carbon nanotube arrays: mechanical properties. Compos Sci Technol, 2002, 62, 419–428.
50. Rudd, R. E., Broughton, J. Q. Concurrent coupling of length scales in solid state systems. Phys Stat Sol B 2000, 217, 251–291.
51. Starr, F. W., Glotzer, S. C. Simulations of filled polymers on multiple length scales, In: Nakatani, A. I., Hjelm, R. P., Gerspacher, M., Krishnamoorti, R., editors. Filled and nanocomposite polymer materials, Materials research symposium proceedings. Warrendale: Materials Research Society, 2001. p. KK4.1.1–KK4.1.13.
52. Glotzer, S. C., Starr, F. W. Towards multiscale simulations of filled and nanofilled polymers, In: Cummings, P. T., Westmoreland, P. R., Carnahan, B., editors. Foundations of molecular modeling and simulation: Proceedings of the 1st international conference on molecular modeling and simulation. Keystone: American Institute of Chemical Engineers, 2001. p. 44–53.

53. Mori, T., Tanaka, K. Average stress in matrix and average elastic energy of materials with misfitting inclusions. Acta Metallurgica, 1973, 21, 571–575.

54. Fisher, F. T., Bradshaw, R. D., Brinson, L. C. Fiber waviness in nanotube-reinforced polymer composites—I: modulus predictions using effective nanotube properties, Comp Sci and Tech, 2003, 63, 1689–1703.

55. Fisher, F. T., Bradshaw, R. D., Brinson LC Fiber waviness in nanotube-reinforced polymer composites—II: modeling via numerical approximation of the dilute strain concentration tensor. Comp Sci and Tech, 2003, 63, 1705–1722.

56. Liu, Y. J., Chen, X. L. Evaluations of the effective material properties of carbon nanotube-based composites using a nanoscale representative volume element, Mech of Mat, 2003, 35, 69–81.

57. Schadler, L. S., Giannaris, S. C., Ajayan, P. M., 1998. Load transfer in carbon nanotube epoxy composites. Applied Physics Letters 73 (26), 3842–3844.

58. Wagner, H. D., Lourie, O., Feldman, Y., Tenne, R., 1998. Stress-induced fragmentation of multiwall carbon nanotubes in a polymer matrix. Applied Physics Letters 72 (2), 188–190.

59. Qian, D., Dickey, E. C., Andrews, R., Rantell, T., 2000. Load transfer and deformation mechanisms in carbon nanotube polystyrene composites. Applied Physics Letters 76 (20), 2868–2870.

60. Liu, Y. J., Chen, X. L., 2002. Modeling and analysis of carbon nanotube-based composites using the FEM and, B. E. M. Submitted to CMES: Computer Modeling in Engineering and Science.

61. Liu, Y. J., Xu, N., Luo, J. F., 2000. Modeling of inter phases in fiber-reinforced composites under transverse loading using the boundary element method. Journal of Applied Mechanics. 67 (1), 41–49.

62. Fu, Y., Klimkowski, K. J., Rodin, G. J., Berger, E., et al., 1998. A fast solution method for three-dimensional many-particle problems of linear elasticity. International Journal for Numerical Methods in Engineering 42, 1215–1229.

63. Nishimura, N., Yoshida, K.-i., Kobayashi, S., 1999. A fast multipole boundary integral equation method for crack problems in 3D. Engineering Analysis with Boundary Elements 23, 97–105.

64. Qian, D., Liu, W. K., Ruoff, R. S., 2001. Mechanics of C60 in nanotubes. The Journal of Physical Chemistry B 105 (44), 10753–10758.

65. Chen, X. L., Liu, Y. J. Square representative volume elements for evaluating the effective material properties of carbon nanotube-based composites, Comput Mater Sci, 2004, 29, 1–11.

66. C. Bower, R. Rosen, L. Jin, J. Han, O. Zhou, Deformation of carbon nanotubes in nanotube–polymer composites, Applied Physics Letters 74 (1999) 3317–3319.

67. Gibson, R. F. Principles of composite material mechanics, CRC Press, 2nd edition. 2007, 97–134.

68. Wan, H., Delale, F., Shen, L. Effect of CNT length and CNT-matrix interphase in carbon nanotube (CNT) reinforced composites, Mech Res Commun, 2005, 32, 481– 489.

69. Shi, D., Feng, X., Jiang, H., Huang, Y. Y., Hwang, K. Multiscale analysis of fracture of carbon nanotubes embedded in composites, Int J of Fract, 2005, 134, 369–386.

70. Buryachenko, V. A., Roy, A. Effective elastic moduli of nanocomposites with prescribed random orientation of nanofibers, Comp: Part B 2005, 36(5), 405–416.

71. Buryachenko, V. A., Roy, A., Lafdi, K., Andeson, K. L., Chellapilla, S. Multi-scale mechanics of nanocomposites including interface: experimental and numerical investigation, Comp Sci and Tech, 2005, 65, 2435–246.

73. Li, C., Chou, T. W. Multiscale modeling of carbon nanotube reinforced polymer composites, J of Nanosci Nanotechnol, 2003, 3, 423–430.

73. Li, C., Chou, T. W., Multiscale modeling of compressive behavior of carbon nanotube/polymer composites, Comp Sci and Tech, 2006, 66, 2409–2414.

74. Anumandla, V., Gibson, R. F. A comprehensive closed form micromechanics model for estimating the elastic modulus of nanotube-reinforced composites, Com: Part A2006, 37, 2178–2185.

75. Seidel, G. D., Lagoudas, D. C. Micromechanical analysis of the effective elastic properties of carbon nanotube reinforced composites, Mech of Mater, 2006, 38, 884–907.

76. Z. Hashin and, B. Rosen. The elastic moduli of fiber-reinforced materials. 1964, Journal of Applied Mechanics 31, 223–232.

77. R. Christensen and, K. Lo. Solutions for effective shear properties in three phase sphere and cylinder models. Journal of the Mechanics and Physics of Solids, 1979, 27, 315–330.

78. Selmi, A., Friebel, C., Doghri, I., Hassis, H. Prediction of the elastic properties of single walled carbon nanotube reinforced polymers: A comparative study of several micromechanical models, Comp Sci and Tech, 2007, 67, 2071–2084.

79. Luo, D., Wang, W. X., Takao, Y. Effects of the distribution and geometry of carbon nanotubes on the macroscopic stiffness and microscopic stresses of nanocomposites, Comp Sci and Tech, 2007, 67, 2947–2958.

80. Fu, S. Y., Yue, C. Y., Hu, X., Mai, Y. W. On the elastic transfer and longitudinal modulus of unidirectional multi-short-fiber composites, Compos Sci Technol, 2000, 60, 3001–3013.

81. Lauke, B. Theoretical considerations on deformation and toughness of short-fiber reinforced polymers, J Polym Eng, 1992, 11, 103–154.

82. Tserpes, K. I., Panikos, P., Labeas, G., Panterlakis SpG. Multi-scale modeling of tensile behavior of carbon nanotube-reinforced composites, Theoret and Appl Fract Mech, 2008, 49, 51–60.

83. Spanos, P. D., Kontsos, A. A multiscale Monte Carlo finite element method for determining mechanical properties of polymer nanocomposites, Prob Eng Mech, 2008, doi:10.1016/j.probengmech.2007.09.002.

84. Paiva, M. C., Zhou, B., Fernando, K. A. S., Lin, Y., Kennedy, J. M., Sun Y-P. Mechanical and morphological characterization of polymer-carbon nanocomposites from functionalized carbon nanotubes. Carbon, 2004, 42, 2849–54.

85. J. Zhu, H. Peng, F. Rodriguez-Macias, J. Margrave, V. Khabashesku, A. Imam, K. Lozano, E. Barrera, "Reinforcing epoxy polymer composites through covalent integration of functionalized nanotubes," (2004) Advanced Functional Materials 14 (7), 643–648.

86. Md. A. Bhuiyan a, Raghuram V Pucha a, Johnny Worthy b, Mehdi Karevan a, Kyriaki Kalaitzidou, Defining the lower and upper limit of the effective modulus of CNT/polypropylene composites through integration of modeling and experiments. Composite Structures 95 (2013) 80–87.

87. Papanikos, P., Nikolopoulos, D. D., Tserpes, K. I. Equivalent beams for carbon nanotubes. Comput Mater Sci, 2008, 43(2), 345–52.
88. Affdl, J. C. H., Kardos, J. L. The Halpin–Tsai equations: a review. Polym Eng Sci, 1976, 16(5), 344–52.
89. Tandon, G. P., Weng, G. J. The effect of aspect ratio of inclusions on the elastic properties of unidirectionally aligned composites. Polym Composite, 1984, 5(4), 327–33.
90. Nielsen, L. E. Mechanical properties of polymers and composites, vol. 2. New York: Marcel Dekker; 1974.
91. Tucker III CL, Liang, E. Stiffness predictions for unidirectional short-fiber composites: review and evaluation. Compos Sci Technol, 1999, 59(5), 655–71.
92. Shokrieh, M. M., Rafiee, R., Investigation of nanotube length effect on the reinforcement efficiency in carbon nanotube based composites Composite Structures 92 (2010) 2415–2420.
93. Shokrieh, M. M., Rafiee, R. On the tensile behavior of an embedded carbon nanotube in polymer matrix with non-bonded interphase region. J Compos Struct, 2009, 22, 23–5.
94. Press Release. US Consulate. World-record-length carbon nanotube grown at US Laboratory. Mumbai-India; September 15, 2004.
95. Evans, J. Length matters for carbon nanotubes: long carbon nanotubes hold promise for new composite materials. Chemistry World News, 2004. <http:// www/rsc.org/ chemistryworld/news>.
96. Pan, Z., Xie, S. S., Chang, B., Wang, C. Very long carbon nanotubes. Nature, 1998, 394, 631–2.
97. http://www.carbonsolution.com
98. http://www.fibermax.eu/shop/
99. http://www.nanoamor.com
100. www.thomas-swan.co.uk
101. Rice University's chemical 'Scissors' yield short carbon nanotubes. New process yields nanotubes small enough to migrate through cells. Science Daily July, 2003. <http://www.sciencedaily.com/releases/2003/07/030723083644.htm>.
102. Chowdhury, S. C., (Gama) Haque, B. Z., Okabe, T., Gillespie Jr. J. W., Modeling the effect of statistical variations in length and diameter of randomly oriented CNTs on the properties of CNT reinforced nanocomposites, Composites: Part B 43 (2012) 1756–1762.
103. Tserpes, K. I., Chanteli, A. Parametric numerical evaluation of the effective elastic properties of carbon nanotube-reinforced polymers, Composite Structures 99 (2013) 366–374.
104. Gou, J., Minaie, B., Wang, B., Liang, Z., Zhang, C. Computational and experimental study of interfacial bonding of single-walled nanotube reinforced composites. Comput Mater Sci, 2004, 31, 225–36.
105. Frankland, S. J. V., Harik, V. M. Analysis of carbon nanotube pull-out from a polymer matrix. Surf Sci, 2003, 525, 103–8.
106. Natarajan, U., Misra, S., Mattice, W. L. Atomistic simulation of a polymer-polymer interface: interfacial energy and work of adhesion. Comput Theor Polym Sci, 1998, 8, 323–9.
107. Lordi, V., Yao, N. Molecular mechanics of binding in carbon-nanotube polymer composites. J Mater Res, 2000, 15, 2770–9.

108. Wong, M., Paramsothy, M., Xu, X. J., Ren, Y., Li, S., Liao, K. Physical interactions at carbon nanotube-polymer interface. Polymer, 2003, 44, 7757–64.
109. Qian, D., Liu, W. K., Ruoff, R. S. Load transfer mechanism in carbon nanotube ropes. Compos Sci Technol, 2003, 63, 1561–9.
110. Liu, Y. J., Chen, X. L. Continuum models of carbon nanotube-based composites using the boundary element method. J Boundary Elem, 2003, 1, 316–35.
111. Chen, X. L., Liu, Y. J. Square representative volume elements for evaluating the effective material properties of carbon nanotube-based composites. Comput Mater Sci, 2004, 29, 1–11.
112. Chen, X. L., Liu, Y. J. Evaluations of the effective material properties of carbon nanotube-based composites using a nanoscale representative volume element. Mech Mater, 2003, 35, 69–81.
113. Qian, D., Dickey, E. C., Andrews, R., Rantell, T. Load transfer and deformation mechanisms in carbon nanotube polystyrene composites. Appl Phys Lett, 2000, 76, 2868.
114. Thostenson, E. T., Chou T-W. Aligned multi-walled carbon nanotube-reinforced composites: processing and mechanical characterization. J Phys D Appl Phys, 2002, 35, 77–80.
115. Bower, C., Rosen, R., Jin, L., Han, J., Zhou, O. Deformation of carbon nanotubes in nanotube-polymer composites. Appl Phys Lett, 1999, 74, 3317–9.
116. Cooper, C. A., Cohen, S. R., Barber, A. H., Wagner, H. D. Detachment of nanotubes from a polymer matrix. Appl Phys Lett, 2002, 81, 3873–5.
117. Qian, D., Dickey, E. C. In-situ transmission electron microscopy studies of polymer-carbon nanotube composite deformation. J Microsc, 2001, 204, 39–45.
118. Schadler, L. S., Giannaris, S. C., Ajayan, P. M. Load transfer in carbon nanotube epoxy composites. Appl Phys Lett, 1998, 73, 3842.
119. Wagner, H. D. Nanotube-polymer adhesion: a mechanics approach. Chem Phys Lett, 2002, 361, 57–61.
120. Ajayan, P. M., Schadler, L. S., Giannaris, C., Rubio, A. Single-walled carbon nano-tube-polymer composites: strength and weakness. Adv Mater, 2000, 12, 750–3.
121. Cooper, C. A., Young, R. J., Halsall, M. Investigation into the deformation of carbon nanotubes and their composites through the use of Raman spectroscopy. Compos Part A Appl Sci Manuf, 2001, 32, 401–11.
122. Hadjiev, V. G., Iliev, M. N., Arepalli, S., Nikolaev, P., Files, B. S. Raman scattering test of single-wall carbon nanotube composites. Appl Phys Lett, 2001, 78, 3193.
123. Paipetis, A., Galiotis, C., Liu, Y. C., Nairn, J. A. Stress transfer from the matrix to the fiber in a fragmentation test: Raman experiments and analytical modeling. J Compos Mater, 1999, 33, 377–99.
124. Valentini, L., Biagiotti, J., Kenny, J. M., Lopez Manchado, M. A. Physical and mechanical behavior of single-walled carbon nanotube/polypropylene/ethylene-pro-pylene-diene rubber nanocomposites. J Appl Polym Sci, 2003, 89, 2657–63.
125. Qian, D., Wagner, G. J., Liu, W. K., Yu M-F, Ruoff, R. S. Mechanics of carbon nano-tubes. Appl Mech Rev, 2002, 55, 495–532.
126. Yu M-F, Yakobson, B. I., Ruo, R. S. Controlled sliding and pullout of nested shells in individual multiwalled nanotubes. J Phys Chem B 2000, 104, 8764–7.

127. Qian, D., Liu, W. K., Ruoff, R. S. Load transfer mechanism in carbon nanotube ropes. Compos Sci Technol, 2003, 63, 1561–9.

128. Liao, K., Li, S. Interfacial characteristics of a carbon nanotube polystyrene composite system. Appl Phys Lett, 2001, 79, 4225–7.

129. Andrews, R., Weisenberger, M. C. Carbon nanotube polymer composites. Curr Opin Solid State Mater Sci, 2004, 8, 31–7.

130. Lordi, V., Yao, N. Molecular mechanics of binding in carbon nanotube- polystyrene composite system. J Mater Res, 2000, 5, 2770–9.

131. Wagner, H. D., Vaia, R. A. Nanocomposites: issue at the interface. Mater Today, 2004, 7, 38–42.

132. Wernik, J. M., Cornwell-Mott, B. J., Meguid, S. A. Determination of the interfacial properties of carbon nanotube reinforced polymer composites using atomistic-based continuum model, Inte. Jour. of Sol. and Struct. 2012, 49, 1852–1863.

133. Wernik, J. M., Meguid, S. A., 2011. Multiscale modeling of the nonlinear response of nano-reinforced polymers. Acta Mech. 217, 1–16.

134. Yang, S., W. Kyoung, S.Yu., Han, D. S., Cho, M. Multiscale modeling of size-dependent elastic properties of carbon nanotube/polymer nanocomposites with interfacial imperfections Polymer 2012, 53, 623–633.

135. Ayatollahi, M. R., Shadlou, S., Shokrieh, M. M. Multiscale modeling for mechanical properties of carbon nanotube reinforced nanocomposites subjected to different types of loading. Composite Structures 2011, 93, 2250–2259.

136. Jiang, L. Y., Huang, Y., Jiang, H., Ravichandran, G., Gao, H., Hwang, K. C., et al. A cohesive law for carbon nanotube/polymer interfaces based on the van der Waals force. J Mech Phys Solids, 2006, 54, 2436–52.

137. Tan, H., Jiang, L. Y., Huang, Y., Liu, B., Hwang, K. C. The effect of van der Waals-based interface cohesive law on carbon nanotube-reinforced composite materials, Composites Science and Technology 2007, 67, 2941–2946.

138. Zalamea, L., Kim, H., Pipes, R. B. Stress transfer in multi-walled carbon nanotubes. Compos Sci Technol, 2007, 67(15–16), 3425–33.

139. 140.Shen, G. A., Namilae, S., Chandra, N. Load transfer issues in the tensile and compressive behavior of multiwall carbon nanotubes. Mater Sci Eng A 2006, 429(1–2), 66–73.

140. Gao, X. L., Li, K. A shear-lag model for carbon nanotube-reinforced polymer composites. Int J Solids Struct, 2005, 42(5–6), 1649–67.

141. Li, C., Chou, T. W. A structural mechanics approach for the analysis of carbon nanotubes. Int J Solids Struct, 2003, 40(10), 2487–99.

142. Tsai, J. L., Lu, T. C. Investigating the load transfer efficiency in carbon nanotubes reinforced nanocomposites. Compos Struct, 2009, 90(2), 172–9.

143. Lu, T. C., Tsai, J. L. Characterizing load transfer efficiency in double-walled carbon nanotubes using multiscale finite element modeling, Composites: Part B 2013, 44, 394–402.

CHAPTER 5

MOVING TOWARDS GREENER CATIONIC POLYMERIZATIONS

STEWART P. LEWIS,[1] DEIVASAGAYAM DAKSHINAMOORTHY,[2] and ROBERT T. MATHERS[2]

[1]*Innovative Science Corp. (www.innovscience.com), Salem, VA 24153, USA; Visiting Scientist, Department of Chemistry, Pennsylvania State University, New Kensington, PA 15068, USA*

[2]*Department of Chemistry, Pennsylvania State University, New Kensington, PA 15068, USA*

CONTENTS

ABSTRACT

This mini-review discusses what constitutes an ideal green cationic polymerization process and recaps progress that has been made towards developing such systems. Emphasis is given to both petroleum and naturally derived monomers but discussion is limited to olefins. Although this chapter attempts to point out advances that have been made in the cationic polymerization of such monomers, due to brevity it cannot be exhaustively comprehensive.

5.1   THE IDEAL GREEN CATIONIC POLYMERIZATION

An ideal cationic polymerization can be visualized as a pyramid (Figure 5.1) where *sustainable components* form its base. Ideally, this translates to monomers and solvents (if required) that are derived solely from renewable resources that have no food value and are non-invasive species. The next course in the pyramid is *energy and waste reduction*. In a truly green system, both monomer and polymer are produced in high yields with no by-products/waste and without the need for excessive heating or cooling so synthesis is neutral in terms of both materials and energy consumption. Moreover, if production requires solvents and/or catalysts, they should be readily recyclable. Closer to the top of the pyramid are the *toxicological* and *environmental characteristics* of the overall process. For a system to be "green, " all

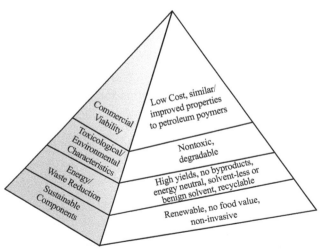

FIGURE 5.1   Green cationic polymerizations master pyramid.

reagents involved must have low toxicity and be biodegradable or chemically degradable (e.g., via hydrolysis). *Commercial viability* forms the capstone of the pyramid and contains a number of factors that must be addressed.

- Is the chemistry easily implementable or does it require specialized reaction conditions (e.g., anhydrous, anaerobic, etc.)?
- Can it be accomplished by a limited number of process steps within a useful time frame and in a continual manner using reactors of simplified design so as to reduce plant footprint and equipment investment while maintaining a distinct cost advantage compared to traditional polymerizations?
- Are the materials identical to existing products (i.e., same monomer structure but plant-derived) or if new monomer types are involved are the polymers superior or inferior to their petroleum counterparts?

5.2   THE PITFALLS OF CATIONIC POLYMERIZATIONS

From an environmental standpoint cationic polymerizations [1–4] leave much to be desired. Some of the main limitations of this method include:

1. *Energy Consumption*: Cyrogenic temperatures are required for the synthesis of high molecular weight (MW) polymers. Reduced reaction temperature is the main method for suppression of chain transfer (CT) processes, which are responsible for reduction of polymer chain length [1–4]. Since the activation energy for CT ($\sim 6.5 \pm 1$ kcal·mol^{-1}) exceeds that of propagation (~ 0.2–2.0 kcal·mol^{-1}), lower polymerization temperatures have a greater effect on reducing the former process without adversely affecting chain growth [3]. Use of counter anions of low nucleophilicity helps to further reduce chain transfer by forcing it to occur by the higher activation energy "spontaneous" process as opposed to the more facile "counter anion-assisted" route and, as a result, reduces the amount of cooling necessary to reach a given MW.

2. *Toxic Solvents*: Due to the highly reactive nature of carbocations, solvent basicity must not exceed that of the monomer being polymerized; otherwise, solvent will readily scavenge active species [2]. Only in the instance of highly reactive monomers that contain basic functionalities {e.g. vinyl ethers (VEs) and N-vinyl carbazole} can

polymerization be effected in solvents common to many organic reactions (e.g., ethers, ketones) [2, 5–7]. In the case of most olefins, solvents are restricted to aliphatic or aromatic compounds and/ or their nitro (i.e., NO_2) or halogen bearing derivatives. Careful use of the latter group of solvents (i.e., aromatics) must be made since they can also engender side reactions such as Friedel-Crafts alkylation [2]. Furthermore, solvent choice can have a dramatic impact on the outcome of polymerization. Depending on the initiator system, solvent polarity may be of paramount importance in determining if polymerization will occur. Typically, polymerization is more facile in solvents of higher dielectric strength, and the majority of cationic polymerization processes rely on the use of toxic halogenated alkanes (e.g., MeCl) that are damaging to the environment [1–4, 8–10].

3. *Recyclability and Sensitivity of Reaction Components*: The majority of Lewis acid (LA) coinitiators that make up initiator systems for cationic polymerizations are hydrolytically unstable and homogeneous [11]. As such, they are difficult to remove from the reaction mixture without undergoing degradation that precludes their reuse. This translates to increased materials consumption and waste generation. Due to their reactive nature, safe handling of these materials requires specialized equipment which increases facility related costs and plant footprint. Carbocations will also react with basic impurities commonly present in monomers and solvents (e.g., water). As a result, most cationic polymerizations must be conducted under anhydrous conditions, which can be costly and difficult to implement. In some instances (e.g., processes involving radical ionic species), oxygen also has a detrimental effect on cationic polymerization [12–26].

5.3   SUSTAINABLE COMPONENTS

5.3.1   RENEWABLE MONOMERS

Renewable olefinic monomers, some of which are depicted in Chart 1, can be grouped into four distinct subsets [27].

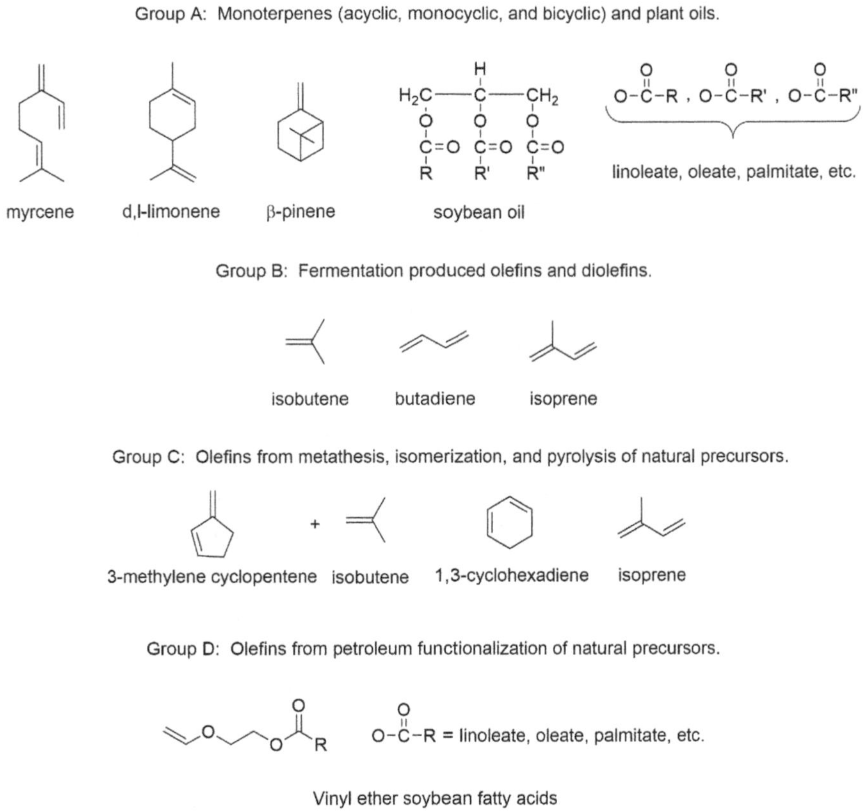

Group A: Monoterpenes (acyclic, monocyclic, and bicyclic) and plant oils.

myrcene   d,l-limonene   β-pinene   soybean oil   linoleate, oleate, palmitate, etc.

Group B: Fermentation produced olefins and diolefins.

isobutene   butadiene   isoprene

Group C: Olefins from metathesis, isomerization, and pyrolysis of natural precursors.

3-methylene cyclopentene   isobutene   1,3-cyclohexadiene   isoprene

Group D: Olefins from petroleum functionalization of natural precursors.

O-C-R = linoleate, oleate, palmitate, etc.

Vinyl ether soybean fatty acids

CHART 1   Renewable monomers for cationic polymerizations.

1. Polymerizable monomers produced by plants such as terpenes and unsaturated plant oils (Chart 1, Group A) are one of the most desired groups. Of these, terpenes are the most important class from a commercial standpoint [28]. These polymers find widespread use as tackifying agents in adhesives. The cationic polymerization of plant oils (e.g., soybean oil) is known but the resultant materials do not possess useful physical properties and therefore they are typically copolymerized with petroleum-derived monomers (e.g., divinylbenzene, dicyclopentadiene) [29–35].

2. Another important group of olefins are those produced by fermentation processes. Isobutene (IB), butadiene (BD) and isoprene (IP) are three important monomers that can be produced by these

methodologies (Chart 1, Group B). In some instances (i.e., IP) olefin is produced directly as fermentation off-gas by genetically engineered bacteria [36]. Other fermentation routes involve the synthesis of precursor alcohols (e.g., ethanol, isobutanol, 2,3-butanediol, and isoamyl alcohol), which undergo subsequent reactions (e.g., dehydration) to form olefins or diolefins [37–44]. During WWII, Union Carbide produced an average of 54% of all BD used in the synthetic rubber program from ethanol using a modified version of the aldol based process originally described by Ostromislensky [39]. Although fermentation processes currently cannot produce enough of these materials to meet current demand they do hold promise as they are cost-competitive with petroleum [45]. Moreover, even though edible plant materials (e.g., corn) are currently used as feedstocks, progress is being made in the area of renewable plant materials (e.g., cellulosics) that do not burden food production [36, 45]. One disadvantage of fermentation is the generation of CO_2; however, CO_{2s} equestration technologies are currently under consideration and may eventually negate this drawback [45]. As progress in this field continues to accelerate, fermentation produced olefins may supplant a large portion of monomers once derived from petroleum.

3.   A third subset consists of natural olefins that require self-metathesis, isomerization, or pyrolysis to impart increased reactivity towards polymerization. The synthesis of 1,4-cyclohexadiene (1,4-CHD) by ring closing metathesis of soybean oil and its subsequent isomerization to the more reactive 1,3-cyclohexadiene [46], production of isoprene by pyrolysis of d, l-limonene [47], myrcene [48], and natural rubber [49], and the combined production of isobutene and 3-methylene cyclopentene by metathesis of myrcene [50] are illustrative of this subset (Chart 1, Group C). The main detraction of this group of monomers is that in some instances catalysts based on expensive metals (e.g., Pd, Ru) are required to effect the necessary transformation.

4.   Renewable feedstocks that require additional materials input for the generation of a polymerizableolefinic group make up the fourth and least desirable subset due requisite use of petroleum derived functionalizing agents (Chart 1, Group D). A recent example of

this class of monomers is vinyl ether soybean fatty acids (VESFAs) [51, 52]. These materials are produced by base catalyzed transesterification of soybean oil with VEs bearing a reactive hydroxyl on the ethereal substituent (e.g., ethylene glycol vinyl ether) and are readily polymerized by cationic techniques.

5.3.2   RENEWABLE SOLVENTS

Despite the fact that solvents typically make of up a significant portion of cationic polymerizations from a materials standpoint, very little work has been conducted in the area of systems that use renewable reaction media. Currently, research is being conducted on the use of hydrogenated terpenes as solvents for the copolymerization of isobutene and 3-methylene cyclopentene as derived from metathesis of myrcene [53, 54]. Such work is most likely the first truly sustainable cationic polymerization in regards to components as even polymerization of terpenes is generally conducted in petroleum derived solvents (e.g., xylenes) [28]. Likewise, exploratory work on the homopolymerization of other naturally derived monomers (e.g., terpenes and CHD) in hydrogenated terpenes is underway [54]. These polymerizations constitute the first examples of systems that operate by a cationic mechanism where both monomers and solvents are renewable.

5.4   REDUCED ENERGY CONSUMPTION, ELIMINATION OF TOXIC SOLVENTS AND WASTE REDUCTION

The development of systems that operate in non-halogenated solvents at temperatures closer to ambient has been a long-standing goal in the cationic polymerization field. This is especially true in regards to the copolymerization of isobutene with isoprene to make butyl rubber; industrial production is conducted as a slurry in MeCl using $AlCl_3$ at low temperatures ($\sim -100°C$) [55–61]. The majority of improvements that have been made in terms of reduced energy consumption, omission of chlorinated solvents, and elimination of waste have come from developments in the chemistry of initiator systems. The bulk of these stems from research conducted in the area of IB polymerization and have been previously covered in great detail [62]. This information is presented below in a condensed manner along with new material.

5.4.1 HOMOGENEOUS INITIATOR SYSTEMS

Many strong LAs (e.g., $AlCl_3$) do not function efficiently in media of low polarity. For example, polymerization of isobutene (IB) using $AlCl_3$ in *n*-pentane only gives rise to low yields of low MW polyisobutene (PIB) even at reaction temperatures as low as −78°C [63]. One reason for this appears to be that these coinitiators and species derived from them (e.g., counter anions) are essentially insoluble in aliphatic solvents and this not only precludes effective initiation but also hampers propagation. Some of the earliest attempts at conducting cationic polymerization in aliphatic solvents involved application of Lewis acids that are soluble in nonpolar media (e.g., $AlBr_3$ and $EtAlCl_2$) [64–74]. Not only does this strategy allow for polymerization in the absence of toxic solvents but these systems have improved MW-T profiles and are capable of making much higher MW butyl at a given temperature compared to systems based on $AlCl_3$ (Figure 5.2). Another benefit of copolymerizing IB with IP in solution

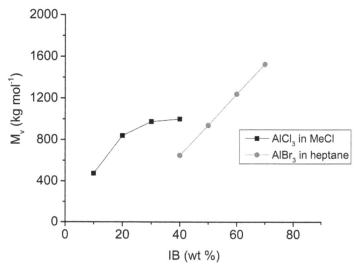

FIGURE 5.2 Comparison of butyl rubber made by slurry versus solution polymerization at −95°C [65].

versus slurry is the ability to produce copolymers with high IP content [66, 72–74]. Although initiation by self-ionization [64, 75, 76] is a possibility with these LAs (Scheme 1), under industrial conditions polymerization by protonic initiation whereby adventitious moisture or hydrohalogen impurities act as a proton source (Scheme 2) is more likely.

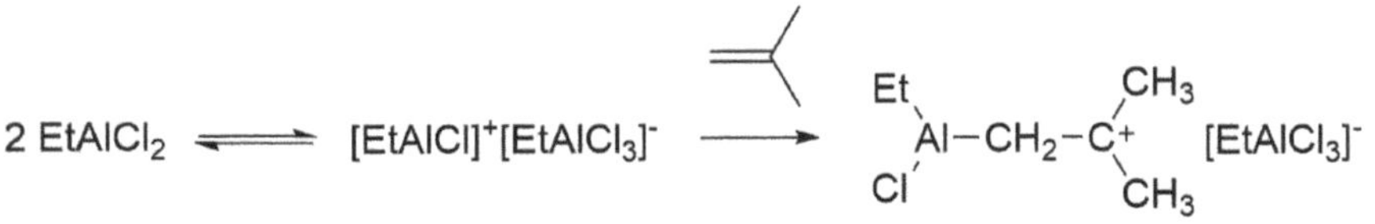

$$2\ EtAlCl_2 \ \rightleftharpoons\ [EtAlCl]^+[EtAlCl_3]^- \ \longrightarrow \ \underset{Cl}{\overset{Et}{Al}}-CH_2-\underset{CH_3}{\overset{CH_3}{C^+}} \ [EtAlCl_3]^-$$

SCHEME 1 Initiation by self-ionization of the coinitiator.

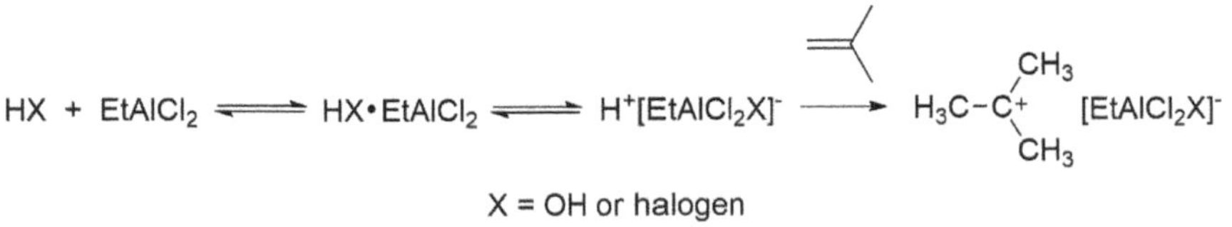

$$HX\ +\ EtAlCl_2 \ \rightleftharpoons\ HX\cdot EtAlCl_2 \ \rightleftharpoons\ H^+[EtAlCl_2X]^- \ \longrightarrow \ H_3C-\underset{CH_3}{\overset{CH_3}{C^+}} \ [EtAlCl_2X]^-$$

X = OH or halogen

SCHEME 2 Protonic initiation by impurities bearing active hydrogen.

Mixtures of hydrocarbon soluble LAs can be more effective for solution polymerization of IB than when used separately (Chart 2, Group A) [77–79]. A related strategy can be employed to effect solubilization of LAs that typically do not operate in low polarity media with concomitant enhancement in polymerization activity. This involves addition of a second LA that is hydrocarbon soluble but not necessarily active for polymerization by itself (Chart 2, Group B) [80]. Both approaches form the basis for a large number of initiator systems consisting of mixtures of LAs (Chart 2, Group C) [81–88]. In many instances, both LAs are hydrocarbon soluble; however, in all variants one is weaker and typically incapable of inducing polymerization by itself, yet its inclusion boosts the activity of the stronger LA.

Mixed LA systems may operate by ionization of the weaker LA to form ionic species which then go on to initiate polymerization either directly (Scheme 3, reaction 1) or indirectly (Scheme 3, reaction 2) in conjunction with adventitious moisture. The literature is replete with numerous examples of the ionization of a weaker LA by a stronger one to form ionic species [89–93];

Group A

$$EtAlCl_2 + Et_2AlCl.^{refs\ 77-79}$$

Group B

Metal alkyls, alkoxides, or oxy-alkoxides {e.g. $Al(OEt)_3$} + metal halides (e.g. $AlCl_3$).[80]

Group C

$$Et_3Al + SnCl_4.^{ref\ 87}$$

$VOCl_3$ or CrO_2Cl_2 or $SnCl_4$ + Et_3Al or Et_2AlCl.[88]

Group 8, 9, or 10 metal alkoxide {e.g. $Fe(OBu)_3$} + BF_3.[83-84]

Metal amide {e.g. $Al(NEt_2)_3$} + BF_3.[82]

Alkoxy aluminum or titanium halides {e.g. Cl_3TiOBu} + BF_3.[85]

Mixed metal oxide-metal alkoxides {e.g. $Zn[OAl(OEt)_2]_2$} + BF_3.[81]

Alkoxy aluminum halides (e.g. Cl_2AlOCH_3) or alkoxy alkyl aluminum halides {e.g. $EtAl(OEt)Cl$} + metal halide (e.g. $TiCl_4$).[86]

CHART 2   Initiator systems based on mixtures of Lewis acids.

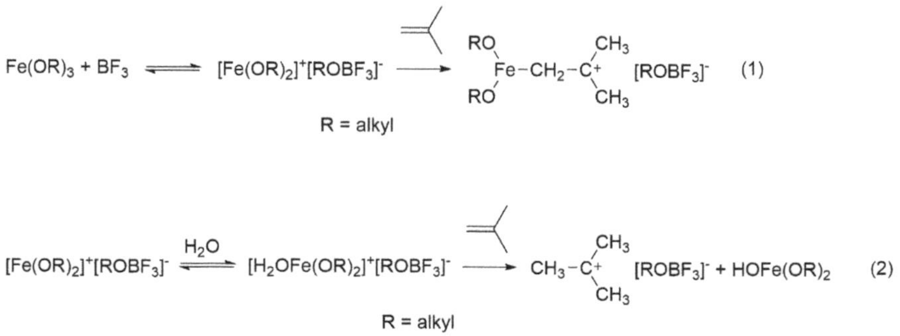

$$Fe(OR)_3 + BF_3 \rightleftharpoons [Fe(OR)_2]^+[ROBF_3]^- \longrightarrow \underset{RO}{\overset{RO}{\diagup}}Fe-CH_2-\overset{CH_3}{\underset{CH_3}{\diagdown}}C^+ \quad [ROBF_3]^- \quad (1)$$

R = alkyl

$$[Fe(OR)_2]^+[ROBF_3]^- \underset{H_2O}{\rightleftharpoons} [H_2OFe(OR)_2]^+[ROBF_3]^- \longrightarrow CH_3-\overset{CH_3}{\underset{CH_3}{\diagdown}}C^+ \quad [ROBF_3]^- + HOFe(OR)_2 \quad (2)$$

R = alkyl

SCHEME 3   Plausible mechanisms of initiation involving ionization of mixed Lewis acids.

however, evidence for direct initiation by these systems has been limited [75, 76, 89]. Another plausible mode of initiation is the in-situ formation of highly reactive LAs from precursors of lower activity (Scheme 4) that then induce polymerization in conjunction with adventitious moisture. Polymerization could also be induced by strong Brønsted acids (BAs) formed from the interaction of both LAs with adventitious moisture (Scheme 5). Many mixed LAs operate in aliphatic solvents or in neat monomer and have MW-T profiles

$$Et_3Al + SnCl_4 \rightleftharpoons Et_2AlCl + EtSnCl_3 \rightleftharpoons EtAlCl_2 + Et_2SnCl_2$$

SCHEME 4   In-situ formation of highly reactive Lewis acids.

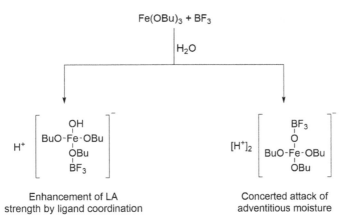

SCHEME 5   Generation of strong Brønsted acids by interaction of mixed Lewis acids and moisture.

(Figure 5.3) that compare favorably or are even superior (Figure 5.4) to those for systems based on perfluoroarylated Lewis acids (PFLAs).

Dialkylaluminum halides form the basis of a larger number of chemically distinct initiator systems capable of producing relatively high MW IB polymers at elevated temperatures (Chart 3) [12, 22, 94–117]. Most of these systems only operate in chlorinated solvents limiting their utility from a "green" standpoint. Of the different initiators useful with dialkyaluminum halides, halogens and electron acceptors (EAs) provide some of the best MW-T profiles and yields (Figure 5.3) mirroring results obtained using PFLAs (Figure 5.4). From a chemistry standpoint, polymerization generally results from addition of monomer to an electrophile generated

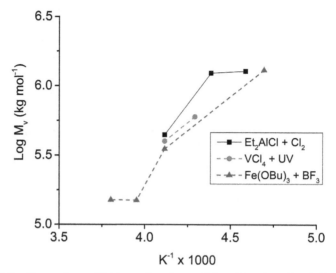

FIGURE 5.3 Comparison of PIBs made using traditional Lewis acid coinitiators. Solid line = MeCl, dashed line = hexane or neat IB [13, 83–84, 116].

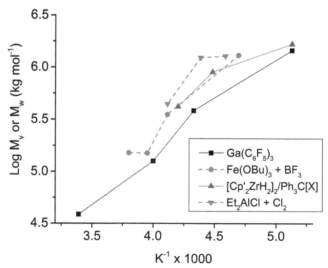

FIGURE 5.4 PIBs made using traditional Lewis acid coinitiators versus those prepared with weakly coordinating anions. Cp' = trimethylsilylcyclopentadienyl, X = $[B(C_6F_5)_4]^-$, solid line = $\overline{M}_W$, dashed line = $\overline{M}_V$. All systems conducted in neat IB or hexane with the exception of $Et_2AlCl + Cl_2$ which was conducted in MeCl [83–84, 116, 124, 149].

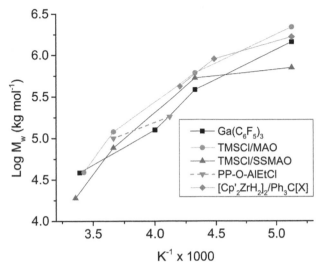

FIGURE 5.5 PIBs made using initiator systems containing weakly coordinating anions. TMSCl = chlorotrimethylsilane, SSMAO = silica supported MAO, Cp' = trimethylsilylcyclopentadienyl, X = [B(C_6F_5)$_4$]$^-$, solid line = neat IB, dashed line = CH_2Cl_2 [124, 149, 180–181, 190].

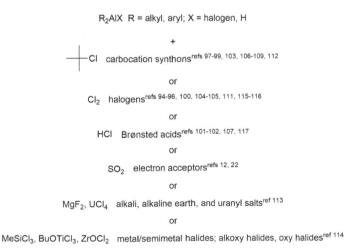

R_2AlX R = alkyl, aryl; X = halogen, H

+

⊢Cl   carbocation synthons[refs 97-99, 103, 106-109, 112]

or

Cl_2 halogens[refs 94-96, 100, 104-105, 111, 115-116]

or

HCl   Brønsted acids[refs 101-102, 107, 117]

or

SO_2 electron acceptors[refs 12, 22]

or

MgF_2, UCl_4 alkali, alkaline earth, and uranyl salts[ref 113]

or

$MeSiCl_3$, $BuOTiCl_3$, $ZrOCl_2$ metal/semimetal halides; alkoxy halides, oxy halides[ref 114]

CHART 3   Dialkylaluminum halide/hydride based initiator systems.

by ionization of the initiator component (Scheme 6). In a few instances (e.g., EAs) initiation is a much more complicated process-giving rise to species such as radical carbocation/radical anion pairs (Scheme 7).

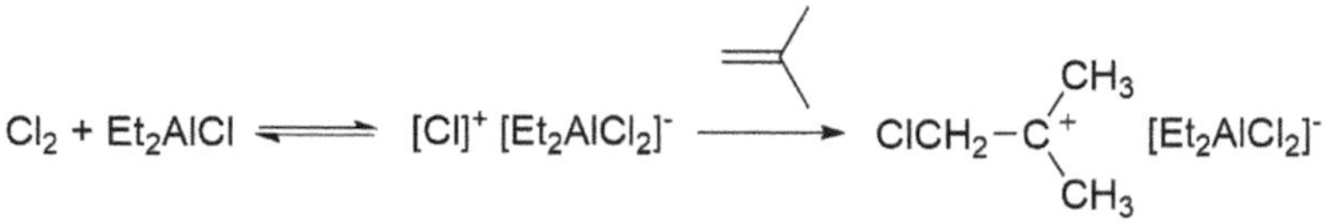

SCHEME 6 Generation of an initiating electrophile via ionization of initiator by dialkylaluminum halide.

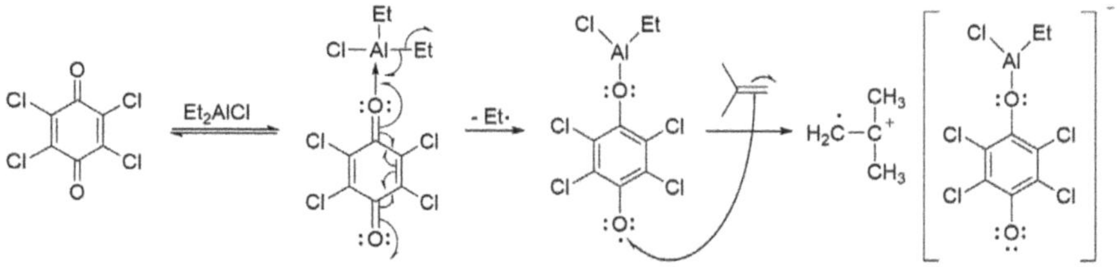

SCHEME 7 Initiation by charge transfer complexes derived from dialkylaluminum halides in conjunction with electron acceptors.

Groups 4 and 5 LA halides have been shown to produce high MW IB polymers in solution at relatively high temperatures (Chart 4) [13–21, 23–26, 118, 119]. Polymerization can be initiated in aliphatic solvents either by electron donors (e.g., naphthalene), alkali metal alkyl compounds, light, or even by thermal energy when conducted in polar media. The majority of these systems operate by a mechanism involving electron transfer from a charge transfer complex (CTC) to produce radical ions (Scheme 8) and are therefore sensitive to oxygen in addition to moisture. The MW-T profiles for these systems are quite favorable and polymerization behavior can be further enhanced by the addition of various additives, which may work as internal desiccants (e.g., CaH_2).

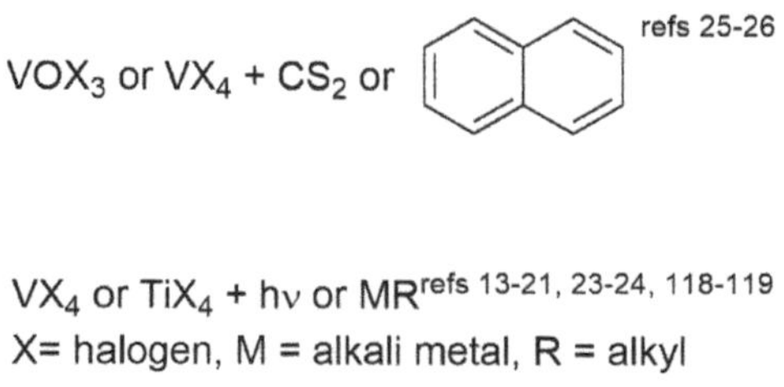

VOX_3 or VX_4 + CS_2 or   refs 25-26

VX_4 or TiX_4 + hv or MR[refs 13-21, 23-24, 118-119]
X= halogen, M = alkali metal, R = alkyl

CHART 4   Initiator systems based on Group 4 and 5 metal halides.

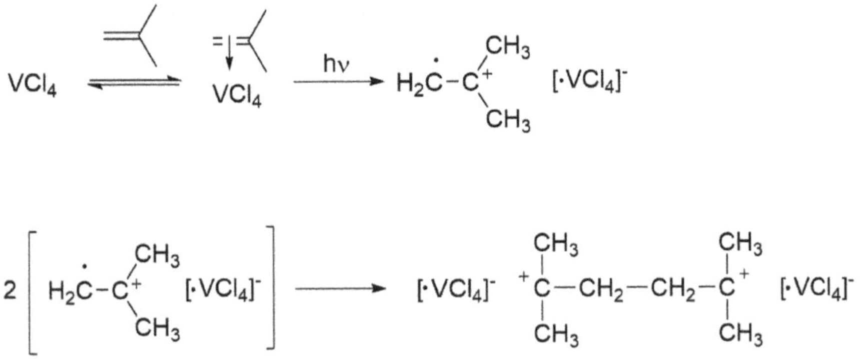

SCHEME 8 Initiation by electron transfer from a charge transfer complex.

More recently, emphasis has been placed on the development of initiator systems based on PFLAs. These systems can be categorized as follows.

1. Those in which a PFLA and adventitious moisture constitute the initiator system (Chart 5) [120–134]. For the majority of these systems, stopping experiments involving sterically hindered pyridines (SHPs) have demonstrated protonic initiation (Scheme 9). Of the PFLAs

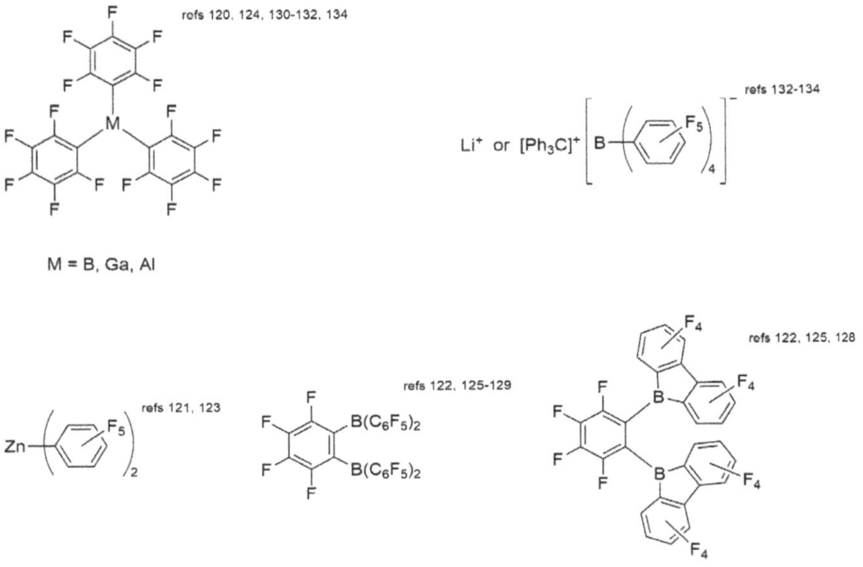

CHART 5 Perfluoroarylated Lewis acid based systems that initiate polymerization protonically.

known to induce protonic initiation, only tris(pentafluorophenyl) gallium, aluminum and the chelating PFLAs $\{1,2\text{-}C_6F_4[B(C_6F_5)_2]_2$ and $1,2\text{-}C_6F_4[B(C_{12}F_8)_2]_2\}$ function effectively in nonpolar media and are even active in aqueous media.

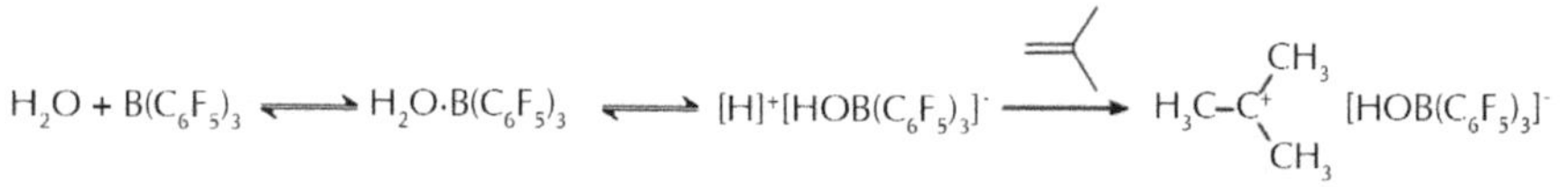

SCHEME 9 Protonic initiation by tris(pentafluorophenyl)boron in conjunction with moisture.

2. Systems in which PFLA coinitiators are used in conjunction with carbocation synthons or silicenium ion precursors (Chart 6)

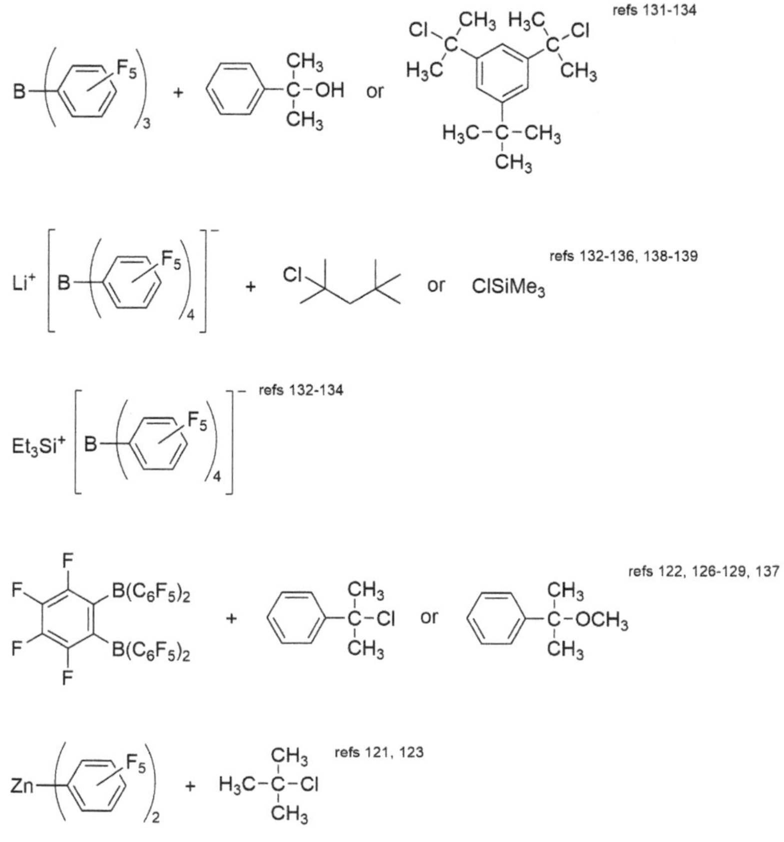

CHART 6 Initiator systems based on perfluoroarylated Lewis acids and carbocation/silicenium ion precursors.

[121–123, 126–129, 131–139]. Modeling studies involving sterically encumbered monomers (e.g., 2,4,4-trimethyl-1-pentene),
stable trityl salts, variable temperature NMR spectroscopy, and
SHP stopping experiments have demonstrated initiation by species
produced from ionization of the carbocation synthon (Scheme 10).
Of this type of system, only those based on $1,2\text{-}C_6F_4[B(C_6F_5)_2]_2$
are known to function in the complete absence of halogenated solvents for the polymerization of IB. It should be mentioned that
tris(pentafluorophenyl)boron in conjunction with hydroxyl substituted carbocation synthons (e.g., cumyl alcohol) polymerizes ST
and IP to low MWs under aqueous conditions where initiation is
purportedly by ionization of the carbocation synthon and not protonic in manner [140, 141]. *The lead author of this chapter was
unable to replicate the results reported for aqueous polymerization
of ST with B(C6F5)3 and this system is still under investigation*

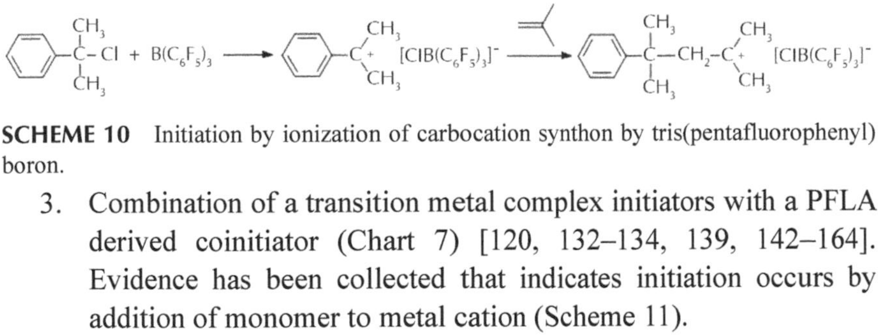

SCHEME 10 Initiation by ionization of carbocation synthon by tris(pentafluorophenyl)
boron.

3. Combination of a transition metal complex initiators with a PFLA
   derived coinitiator (Chart 7) [120, 132–134, 139, 142–164].
   Evidence has been collected that indicates initiation occurs by
   addition of monomer to metal cation (Scheme 11).

Most PFLA based systems offer MW-T profiles (Figures 5.4 and 5.5)
approaching that yielded by γ-radiation [165–169] (i.e., the theoretical
limit) and yet many of these systems are capable of functioning in aliphatic
solvents and in several instances in aqueous media. The main detractions
of PFLA systems are:

1. The expense of PFLAs and transition metal complex initiators.
   In the majority of instances these materials are not recoverable
   although recent work in the area of supported PFLAs hold promise
   for recyclability [170].
2. Most PFLA based systems, especially those that make use
   of carbocation, silicenium, or transition metal ion precursors are overly sensitive to trace moisture [133, 144, 147, 149].

refs 132-134, 139, 147, 155, 161

Cp$_2$MMe$_2$ or Cp'$_2$ZrMe$_2$ or Cp*$_2$ZrMe$_2$ or Cp$_n${MeC$_6$H$_4$C(Me)$_2$Cp}$_m$ZrMe$_2$+ B(C$_6$F$_5$)$_3$ or Ph$_3$C$^+$[B(C$_6$F$_5$)$_4$]$^-$

M = Zr, Hf; Cp = cyclopentadienyl; Cp' = trimethylsilylcyclopentadienyl;
Cp* = pentamethylcyclopentadienyl; n + m = 2

ref 149

[Cp'$_2$ZrH$_2$]$_2$ + Ph$_3$C$^+$[B(C$_6$F$_5$)$_4$]$^-$ or Ph$_3$C$^+$[CN{B(C$_6$F$_5$)$_3$}$_2$]$^-$ or Ph$_3$C$^+$[H$_2$N{B(C$_6$F$_5$)$_3$}$_2$]$^-$

Cp' = trimethylsilylcyclopentadienyl

refs 144, 146, 151, 154, 156

Cp$_2$AlMe or Cp"$_2$AlMe or Cp*$_2$AlMe + B(C$_6$F$_5$)$_3$ or Ph$_3$C$^+$[Al{OC(CF$_3$)$_3$}$_4$]$^-$

Cp" = tetramethylcyclopentadienyl; Cp* = pentamethylcyclopentadienyl

ref 162

[Cp'$_2$Y(μ–Me)]$_2$ + B(C$_6$F$_5$)$_3$ or Ph$_3$C$^+$[B(C$_6$F$_5$)$_4$]$^-$

Cp' = trimethylsilylcyclopentadienyl

refs 120, 142-143, 152-153, 163

Cp*TiMe$_3$ + B(C$_6$F$_5$)$_3$ or Al(C$_6$F$_5$)$_3$ or Ph$_3$C$^+$[B(C$_6$F$_5$)$_4$]$^-$ or n-C$_{18}$H$_{37}$EH $\cdot$B(C$_6$F$_5$)$_3$ or n-C$_{17}$H$_{35}$CO$_2$H$\cdot$ B(C$_6$F$_5$)$_3$

Cp* = pentamethylcyclopentadienyl; E = O, S

ref 158

Ti(CH$_2$Ph)$_4$ + Ph$_3$C$^+$[B(C$_6$F$_5$)$_4$]$^-$

ref 148

MeZr[N(SiMe$_3$)$_2$]$_3$ + B(C$_6$F$_5$)$_3$

refs 145, 150 157, 159-160, 164

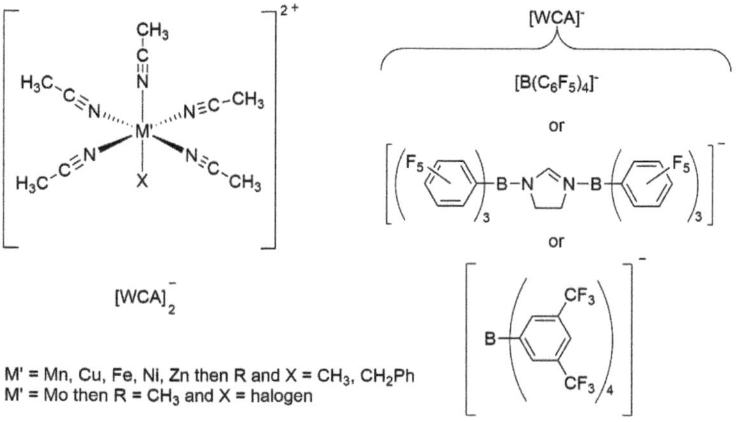

M' = Mn, Cu, Fe, Ni, Zn then R and X = CH$_3$, CH$_2$Ph
M' = Mo then R = CH$_3$ and X = halogen

CHART 7   Perfluoroarylated Lewis acid-transition metal complex initiator systems.

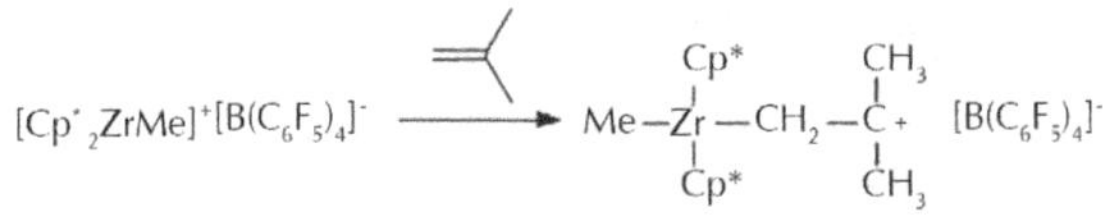

$$[Cp^{*}{}_{2}ZrMe]^{+}[B(C_{6}F_{5})_{4}]^{-} \xrightarrow{\quad =\!\!\!\!\!\!\big\backslash\quad} Me-\overset{\overset{\displaystyle Cp^{*}}{|}}{\underset{\underset{\displaystyle Cp^{*}}{|}}{Zr}}-CH_{2}-\overset{\overset{\displaystyle CH_{3}}{|}}{\underset{\underset{\displaystyle CH_{3}}{|}}{C}}+\quad [B(C_{6}F_{5})_{4}]^{-}$$

SCHEME 11 Initiation by direct addition of monomer to metal cation.

Exceptions are systems based on chelating PFLAs [125–128], tris(pentafluorophenyl) gallium and aluminum [130], and tris(pentafluorophenyl) boron [140, 141, 171].

3. From the data reported, the bulk of PFLA based systems do not appear capable of making high MW PIBs at temperatures greater than −20°C. The inability to polymerize at higher temperatures may in part be due to the instability of the B-C_6F_5 bond to cleavage by C^+ [172]. Onlytris(pentafluorophenyl) gallium and aluminum have been reported capable of producing high MW PIBs in excess of this temperature [124].

One benefit attributed to PFLA based systems is the ability to make butyl rubber of high IP content [121, 123, 135]. Higher IP content could be the result of conducting polymerization under homogeneous conditions and not due to the weakly coordinating nature of the PFLA based anion since even standard LAs (e.g., $AlBr_3$) can produce copolymers with high IP content when used in solution [66, 72–74]. Likewise, it is worthwhile to point out that some PFLA based systems may be capable of producing PIBs of higher MW than originally reported, as these experiments [131, 133, 139, 146, 163] were conducted in dichloromethane, a solvent where the cutoff MW for solubility of PIB is ~ 30 kg·mol^{-1} [173].

Another but lesser explored class of initiator systems containing weakly coordinating anions is those based on alkylaluminoxanes (Chart 8). Initial work focused on aluminoxanes produced from hydrolysis of $EtAlCl_2$ [174] and subsequent modifications [175–177]. Several groups of researchers investigated the use of carbocation synthons in conjunction with toluene solutions of methylaluminoxane (MAO) to produce high MW PIBs in aliphatic solvents at temperatures up to −30°C [178, 179]. Solid but unsupported MAO in neat IB has expanded the number of initiator subsets to include silicon and germanium halides, halogens/interhalogens, boron halides, and Brønsted acids. A drawback to these systems was the upper temperature limit of $-20°C$ [180, 181]. The means by which these initiator systems operate has not been thoroughly investigated. For carbocation synthons, initiation most likely involves ionization of the carbocation precursor (Scheme 12) but for silyl halides the process is more complicated and appears to involve ligand exchange reactions to generate species that initiate polymerization in an unknown manner (Scheme 13) [181]. Regardless, MAO based systems offer MW-T profiles that compared favorably to γ-radiation at a cost lower than PFLAs (Figures 5.4 and 5.5) and still provide for the ability to operate in aliphatic solvents or neat monomer.

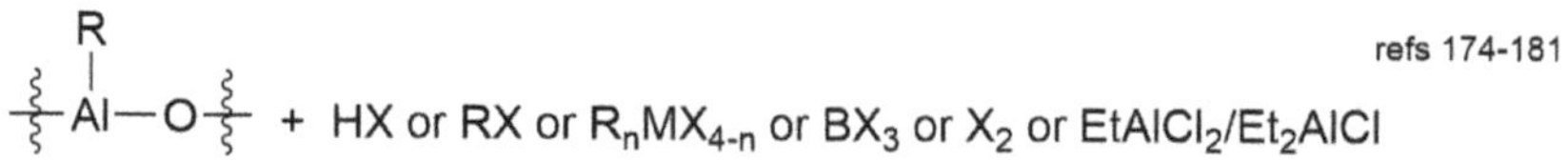

refs 174-181

$$\xi\text{-Al}-\text{O}\xi + \text{HX or RX or } R_nMX_{4-n} \text{ or } BX_3 \text{ or } X_2 \text{ or } EtAlCl_2/Et_2AlCl$$

R = alkyl, aryl; X = halogen; M = Si, Ge

This structure represents a simplified depiction of an alkylaluminoxane.

CHART 8   Alkylaluminoxane based initiator systems.

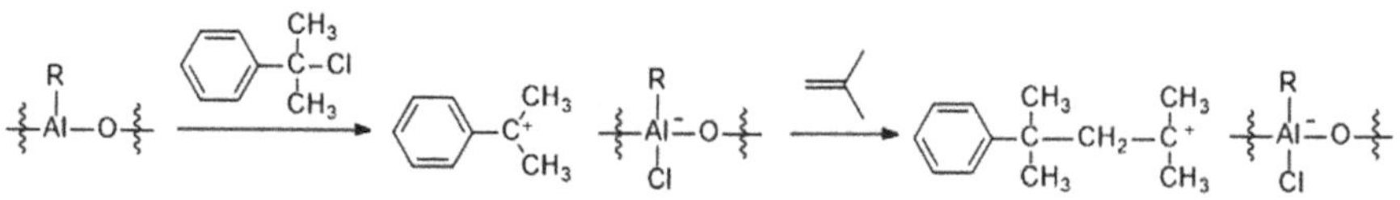

R = alkyl This structure represents a simplified depiction of an alkylaluminoxane

SCHEME 12   Proposed initiation of polymerization by ionization of carbocation synthon by alkylaluminoxane.

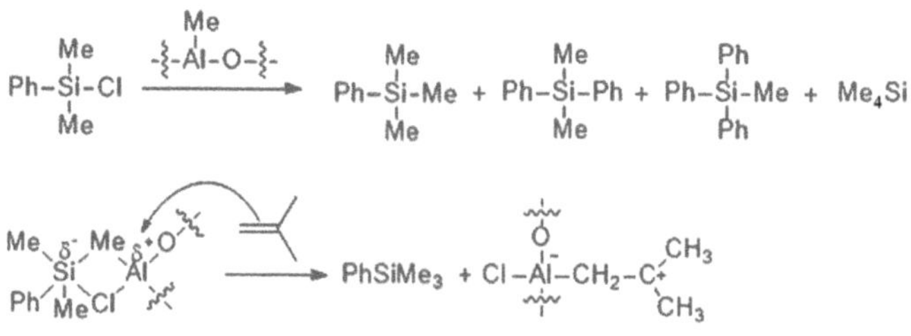

MAO is depicted by a simplified structure.

SCHEME 13   Reaction of methylaluminoxane with silicon halides and plausible mechanism for initiation of polymerization thereby.

5.4.2 HETEROGENEOUS INITIATOR SYSTEMS

Very little progress has been made in the area of heterogeneous initiator systems that allow for waste reduction by preventing contamination of the polymer with LA. The first successful system was a green gel resulting from addition of $TiCl_4$ to a solid produced from reaction of $Al(OsecBu)_3$ with BF_3 [182, 183]. Although it affords good yields of polymer in aliphatic solvents and has a favorable MW-T profile, due to thermally instability, catalyst aging does not produce desirable results. The exact mode of initiation by this system is currently unknown. Around the same time, freshly milled $CdCl_2$ layer structure metal dihalides (e.g., $MgCl_2$) were found to be active coinitiators for polymerization of IB [184, 185]. These metal dihalides are capable of operating in aliphatic solvents at elevated temperatures but successful polymerization requires careful manipulation of water concentration. From its sensitivity to moisture content and the fact that polymer encapsulates the coinitiator, initiation is believed to occur by protonation of monomer by a strong BA formed from exposed metal ions (e.g., Mg^{2+}) and adventitious moisture (Scheme 14). In a similar manner, $Mg(ClO_4)_2$ is capable of producing high MW PIBs at 0°C in bulk IB; however, yields are low ($\sim$ 1–4%) [186].

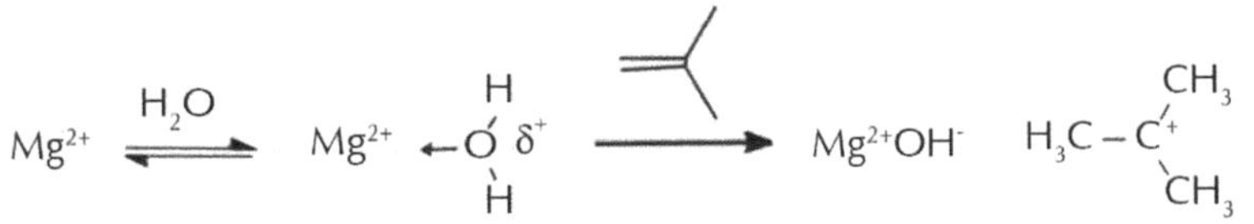

SCHEME 14 Initiation of polymerization by strong heterogeneous Brønsted acids derived from $MgCl_2$ and moisture.

Many years later, two heterogeneous systems based on polymer supported LAs emerged. One class (Chart 9) is based on crosslinked polystyrene (PS) supports where the LA components can be bound as pi complexes (PS-π→LA), n-donor complexes (PS-n→LA), or by covalent bonds (PS-LA) depending on the functional groups present on the support [187, 188]. The first two variants are capable of producing medium MW PIBs at ambient temperatures in nonpolar media. This system is also unique in that the supported LA can be reused multiple times. The other system is based on

EtAlCl$_2$, Et$_2$AlCl or BF$_3$ supported on hydroxyl functionalized polypropylene (PP) or polybutene-1 (Chart 10) [189–193]. For this polymer-supported system, both PP-O-AlCl$_2$ and PP-O-AlEtCl make high MW PIBs at elevated temperatures, but only in chlorinated solvents (Figures 5.4 and 5.5). Most likely, PP supported LAs initiate polymerization protonically in conjunction with adventitious moisture (Scheme 15). Most recently, systems based on supported MAO in conjunction with numerous types of initiators (Chart 11) have emerged and are capable of producing high MW PIBs in the absence of chlorinated solvents at temperatures up to ambient (Figures 5.4 and 5.5) [180, 181]. MAO can be fixed on a number of inorganic supports (e.g., alumina, boric acid) and functions effectively with all classes of initiators

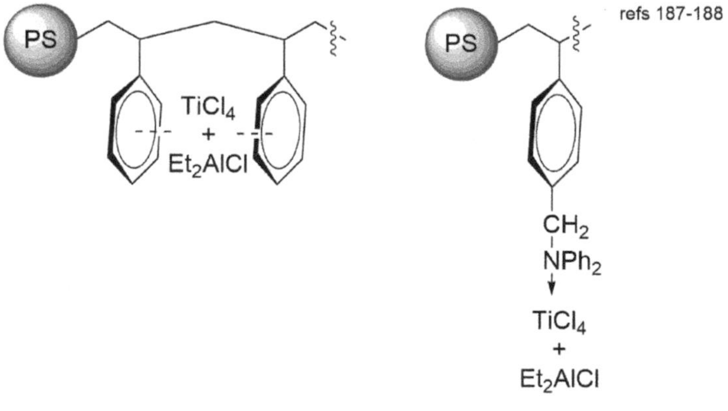

CHART 9 Heterogeneous coinitiators based on Lewis acids supported on crosslinked polystyrene.

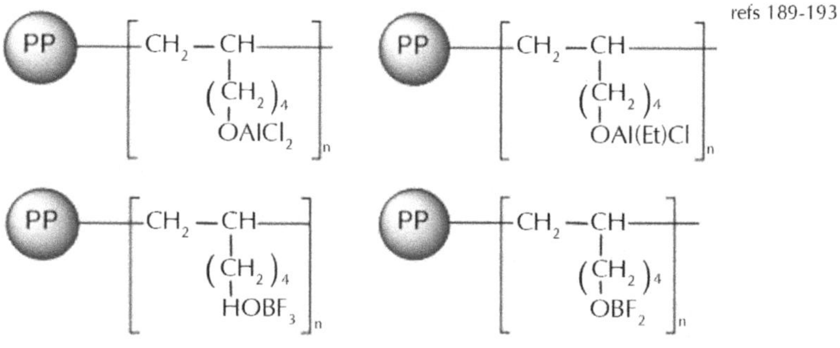

CHART 10 Heterogeneous coinitiators based on Lewis acids supported on crosslinked polypropylene.

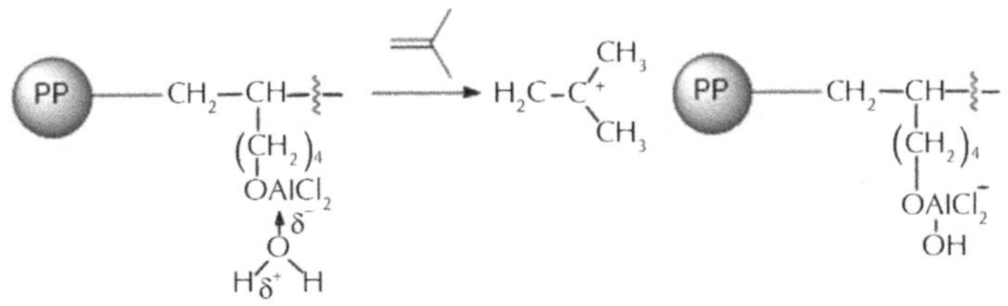

SCHEME 15 Proposed mechanism for protonic initiator by polypropylene supported Lewis acids in combination with moisture.

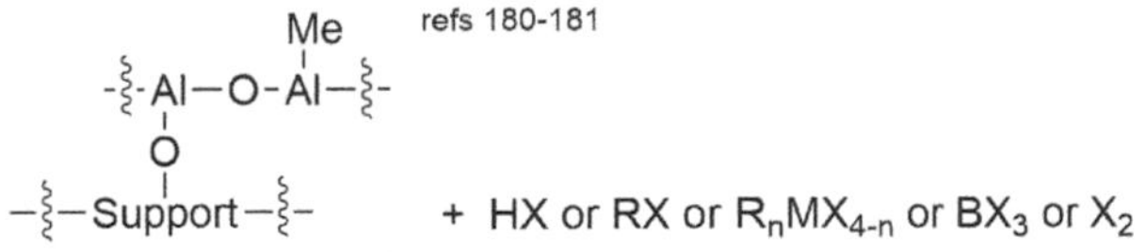

Support = silica, alumina, boric acid; R = alkyl, aryl; X = halogen; M = Si, Ge

Structure depicts simplified representation of MAO.

CHART 11 Heterogeneous alkylaluminoxane based initiator systems.

known to be compatible with MAO in solution. In a somewhat related manner, the use of supported PFLAs has been reported and these materials are currently under investigation for the polymerization of IB [170].

5.5 TOXICOLOGICAL AND ENVIRONMENTAL CHARACTERISTICS

To date, all cationic polymerizations entail the use of materials that are toxic and/or detrimental to the environment. Systems that operate in the absence of chlorinated solvents are a step in the right direction but only solventless or aqueous polymerizations truly alleviate the burden imposed by organic solvents. As suggested by a review of the literature, few such systems exist. Several new aqueous polymerization systems have been developed recently and are in the process of being patented [194]. Likewise, heterogeneous initiator systems are less cumbersome from a green standpoint in that toxic waste streams are reduced, but the majority of them still rely on toxic and/or expensive precursors.

5.6 COMMERCIAL VIABILITY

Many of the systems described in this book chapter are cost-effective in comparison to older technologies based on $AlCl_3$ or its soluble counterparts (e.g., $AlBr_3$, $EtAlCl_2$) and yet provide distinct improvements from a green standpoint. Systems based on mixtures of LAs may be one of the most cost-competitive methods for polymerization at higher temperatures in aliphatic solvents. PFLAs and recently reported heterogeneous systems offer further improvements, but currently their usage is prohibited due to cost associated with manufacture of the initiator system. Most likely, as time progresses, these systems will become more palatable from a cost standpoint. Heterogeneous variants offer further advantages in that they are conducive to continual production methods using fixed-bed technology, which has the potential to improve output and simplify reactor design.

Despite the environmental consequences, little changes have been made to existing commercial procedures involving cationic polymerization even though major advances have been made in cleaning up the chemistry. For example, most butyl rubber (one of the largest commodity polymers by volume produced by the cationic technique) is still made by the "slurry process" based on MeCl [55–61]. MeCl is a toxic solvent which has a global warming potential that exceeds alkanes [195, 196]. Only a few butyl plants are reported to operate in solution [4]. In the slurry process, MeCl acts a solvent for $AlCl_3$ and nonsolvent (i.e., diluent) for butyl and thus this form of suspension polymerization. Its widespread use comes as a further surprise since one of the main detractions of the slurry process (i.e., reactor fouling) is completely avoided by conducting reaction in solution. Fouling reduces productivity in that periodic shutdown and cleaning of reactors is required [55–59, 61]. It furthermore increases plant footprint as multiple reactors are required in order to allow for continual production by rotation of fouled and cleaned reactors. The slurry process is also not conducive to the production of value added variants of butyl such as halobutyl [55–59, 61, 197, 198] or butyl rubber of intermediate unsaturation [66, 72–74]. In the former instance, MeCl must be replaced with an inert solvent (i.e., alkane) prior to conducting halogenation. As for the latter case, solution copolymerization allows for the production of butyl with increased IP incorporation without gel formation which occurs to a considerable extent under slurry conditions when [IP] is raised beyond a few percentage of the total feed. The use of homogeneous LAs also necessitates de-ashing of the product by hydrolytic work up, which generates waste and consumes materials adding to cost.

Most likely, the main barrier to implementing these improvements is the cost of modifying existing plants. Nonetheless, after contacting all producers of IB and terpene based polymers in North America and Europe, the lead author found that Arizona Chemical [199], Chevron-Oronite [200], and Gevo [45] are interested in development of more environmentally acceptable business practices. Hopefully, this mini-review will help prompt the replacement of older systems with cleaner and greener methodologies.

5.7 SUMMARY

Over the past 60–70 years great strides have been made in the field of cationic polymerization in the following areas.

- renewable monomers and solvents.
- reduced energy consumption.
- ability to operate in the absence of chlorinated solvents.
- elimination of post-polymerization de-ashing of initiator components.
- reduction of waste associated with initiator systems.

Many hurdles still need to be surmounted such as reducing the toxicity of all materials involved, increasing the number and quantities of renewable feedstocks, and enhancing the economic viability of certain systems. Such difficulties may be overcome in the future as momentum in the area of green polymer science continues to increase.

ACKNOWLEDGEMENTS

The authors would like to thank Mr. B. Hickory (Sustainable Syntheses LLC) for assistance in the preparation of this review.

KEYWORDS

- commercial viability
- environmental characteristics
- green cationic polymerization
- recyclability
- renewable monomers
- toxicological characteristics

REFERENCES

1. *The Chemistry of Cationic Polymerization*; Plesch, P., Ed., Pergamon Press: Oxford, 1963.
2. Gandini, A., Cheradame, H. *Adv. Polym. Sci.* 1980, *34–35*, 1–284.
3. Kennedy, J. P., Marechal, E. *Carbocationic Polymerization*; John Wiley and Sons: New York, 1982; pp. 1–510.
4. *Cationic Polymerizations*; Matyjaszewski, K., Ed., Marcel Dekker, Inc.: New York, 1996.
5. Bowyer, P. M., Ledwith, A., Sherrington, D. C. *Polymer,* 1975, *16*, 509–520.
6. Ellinger, L. P. *Polymer,* 1964, *5*, 549–560.
7. Tazuke, S., Tjoa, T. B., Okamura, S. *J. Polym. Sci., Part A: Polym. Chem.* 1967, *5*, 1911–1925.
8. Matyjaszewski, K., Lin, C.-H., Pugh, C. *Macromolecules,* 1993, *26*, 2649–2654.
9. Jordan, D. O., Mathieson, A. R. *J. Chem. Soc.* 1952, 2354–2358.
10. Pepper, D. C. *Nature,* 1946, *158*, 789–790.
11. Olah, G. A. *Friedel-Crafts Chemistry*; John Wiley and Sons: New York, 1973.
12. Cesca, S., Priola, A., Ferraris, G., Busetto, C., Bruzzone, M. *J. Polym Sci., Polym. Symp.* 1977, *56*, 159–172.
13. Marek, M. *J. Polym Sci., Polym. Symp.* 1976, *56*, 149–158.
14. Marek, M., Toman, L. *J. Polym. Sci., Part C.: Polym. Symp.* 1973, *42*, 339–343.
15. Marek, M., Toman, L. US Patent, 3998713, 1976.
16. Marek, M., Toman, L. *Makromol. Chemie, Rapid Commun.* 1980, *1*, 161–163.
17. Marek, M., Toman, L. *J. Macromol. Sci., Part A-Chem.* 1981, *A15*, 1533–1543.
18. Marek, M., Toman, L., Pecka, J. US Patent, 3897322, 1975.
19. Marek, M., Toman, L., Pecka, J. US Patent, 3997417, 1976.
20. Marek, M., Toman, L., Pilar, J. *J. Polym. Sci., Part A: Polym. Chem.* 1975, *43*, 1565–1573.
21. Pilar, J., Toman, L., Marek, M. *J. Polym. Sci., Part A: Polym. Chem.* 1976, *14*, 2399–2405.
22. Priola, A., Ferraris, G., Cesca, S. US Patent, 3850895, 1974.

23. Toman, L., Pecka, J., Wichterle, O., Sulc, J. GB Patent, 1349381, 1974.
24. Toman, L., Pilar, J., Spevacek, J., Marek, M. *J. Polym. Sci., Part A: Polym. Chem.* 1978, *16*, 2759–2770.
25. Yamada, M., Shimada, K., Takemura, T. US Patent, 3326879, 1967.
26. Yamada, N., Shimada, K., Hayashi, T. *J. Polym. Sci., Part B: Polym. Lett.* 1966, *4*, 477–480.
27. Mathers, R. T. *J. Polym. Sci., Part A: Polym. Chem.* 2012, *50*, 1–15.
28. Mathers, R. T., Lewis, S. P. *Monoterpenes as Polymerization Solvents and Monomers in Polymer Chemistry*; In *Green Polymerization Methods: Renewable Starting Materials, Catalysis and Waste Reduction;* Mathers, R. T., Meier, M. A. R., Eds., Wiley-VCH: New York, 2011; pp. 91–128.
29. Andjelkovic, D. D., Larock, R. C. *Biomacromolecules,* 2006, *7*, 927–936.
30. Andjelkovic, D. D., Larock, R. C. *Macromol. Mater. Eng.* 2009, *294*, 472–473.
31. Ronda, J. C., Lligadas, G., Galina, M., Cadiz, V. *Reactive and Functional Polymers,* 1013, *73*, 381–395.
32. Sacristan, M., Ronda, J. C., Galina, M., Cadiz, V. *Biomacromolecules,* 2009, *10*, 2678–2685.
33. Sacristan, M., Ronda, J. C., Galina, M., Cadiz, V. *Macromol. Chem. Phys.* 2010, *211*, 801–808.
34. Sacristan, M., Ronda, J. C., Galina, M., Cadiz, V. *Polymer,* 2010, *51*, 6099–6106.
35. Sacristan, M., Ronda, J. C., Galina, M., Cadiz, V. *J. Appl. Polym. Sci.* 2011, *122*, 1649–1658.
36. Whited, G. M., Feher, F. J., Benko, D. A., Cervin, M. A., Chotani, G. K., McAuliffe, J. C., LaDuca, R. J., Ben-Shoshan, E. A., Sanford, K. J. *Ind. Biotech.* 2010, *6*, 152–163.
37. Caventou, E. *Ann.* 1863, *127*, 93.
38. Gruber, P. R., Peters, M. W., Griffith, J. M., Obaidi, Y. A., Manzer, L. E., Taylor, J. D., Henton, D. E. US Patent, 8378160, 2013.
39. Ostromislensky, I. I. *J. Russ. Phys. Chem. Soc.* 1915, *47*, 1472–1506.
40. Peters, M. W., Taylor, J. D., Manzer, L. E., Henton, D. E. US Patent, 2010/0216958 A1, 2010.
41. Schotz, S. P. *Synthetic Rubber*; Benn: London, 1926; pp. 124–125.
42. Toussaint, W. J., Dunn, J. T. US Patent, 2421361, 1947.
43. Toussaint, W. J., Dunn, J. T. *Ind. Eng. Chem.* 1947, *39*, 120–125.
44. Young, C. O. US Patent, 1977750, 1934.
45. Smith, J., Gevo, personal communication, 2013.
46. Mathers, R. T., Shreve, M. J., Meyler, E., Damodaran, K., Iwig, D. F., Kelley, D. J. *Macromol. Rapid Commun.* 2011, *32*, 1338–1342.
47. Staudinger, H., Klever, H. W. *Ber.* 1911, *44*, 2212–2215.
48. Davis, B. L., Goldblatt, L. A., Palkin, S. *Ind. Eng. Chem.* 1946, *38*, 53–57.
49. Boonstra, B. B. S. T., van Amerongen, G. J. *Ind. Eng. Chem.* 1949, *41*, 161.
50. Kobayashi, S., Lu, C., Hoye, T. R., Hillmyer, M. A. *J. Am. Chem. Soc.* 2009, *131*, 7960–7961.
51. Chernykh, A., Alam, S., Jayasooriya, A., Bahr, J., Chisholm, B. J. *Green Chem.* 2013, *15*, 1834–1838.
52. Chisholm, B. J., Alam, S. US Patent, 2012/0316309 A1, 2012.
53. "Recent News, Bio-based Replacements for Butyl Rubber" October, 22 2013, www. innovscience.com.

54. Dakshinamoorthy, D., Mathers, R. T., Lewis, S. P. unpublished research data, 2013.
55. Dias, A. J. *Isobutylene Copolymers*; In *Polymeric Materials Encyclopedia;* Salamone, J. C., Ed., CRC Press: Boca Raton, 1996; Vol. 5, pp. 3484–3492.
56. Duffy, J., Wilson, G. J. *Synthesis of Butyl Rubber by Cationic Polymerization*, 5th ed., In *Ullman's Encyclopedia of Industrial Chemistry;* Elvers, B., Hawkins, S., Russey, W., Schulz, G., Eds., VCH Publishers: Weinheim, 1993; Vol. A23, pp. 288–294.
57. Kennedy, J. P., Kirshenbaum, I. *High Polymers,* 1971, *24,* 691–756.
58. Kresge, E., Wang, H.-C. *Butyl Rubber,* 4th ed., In *Kirk-Othmer Encyclopedia of Chemical Technology;* Kroschwitz, J. I., Howe-Grant, M., Eds., John Wiley and Sons: New York, 1993; Vol. 8, pp. 934–955.
59. Kresge, E. N., Schatz, R. H., Wang, H.-C. *Isobutylene Polymers*, 2nd ed., In *Encyclopedia of Polymer Science and Engineering;* Mark, H. F., Bikales, N. M., Overberger, C. G., Menges, G., Eds., John Wiley and Sons: New York, 1987; Vol. 8, pp. 423–448.
60. Thomas, R. M., Sparks, W. J. US Patent, 2356128, 1944.
61. Webb, R. N., Shaffer, T. D., Tsou, A. H. *Butyl Rubber*, 5th ed., In *Kirk-Othmer Encyclopedia of Chemical Technology;* Seidel, A., Ed., John Wiley and Sons: Hoboken, N.J., 2004; Vol. 4, pp. 433–458.
62. Lewis, S. P., Mathers, R. T. *Advances in Acid Mediated Polymerizations*; In *Renewable Polymers, Synthesis, Technology and Processing;* Vikas, M., Ed., Wiley-VCH: New York, 2011; pp. 69–173.
63. Cesca, S., Priola, A., Ferraris, G. *Macromol. Chem. Phys.* 1972, *156,* 325–328.
64. Chmelir, M., Marek, M., Wichterle, O. *J. Polym. Sci., Part C: Polym. Symp.* 1967, *16,* 833–839.
65. Ernst, J. L., Rose, H. J. US Patent, 2772255, 1956.
66. Ernst, J. L., Rose, H. J. CA Patent, 556596, 1958.
67. Sangalov, Y. A., Nelkenbaum, Y. Y., Ponomarev, O. A., Minsker, K. S. *Vysokomol. Soedin., Ser. A* 1979, *21,* 2267.
68. Scherbakova, N. V., Petrova, V. D., Prokofiev, Y. N., Timofeev, E. G., Lazariants, E. G., G.A., S., Pautov, P. G., Nabilkina, V. A., Dobrovinsky, V. E., Arkhipov, N. B., Sobolev, V. M., Minsker, K. S., Emelyanova, G. V., Vinogradova, A. V., Lebedev, J. V., Krapivina, K. Y., Kolosova, E. V., Vladykin, L. N., Sletova, L. I., Orlova, A. P., Yatsyshina, T. M., Bugrov, V. P., Rodionova, N. N., Tsvetkova, A. G. DE Patent, 104985, 1974.
69. Scherbakova, N. V., Petrova, V. D., Prokofiev, Y. N., Timofeev, E. G., Lazariants, E. G., G.A., S., Pautov, P. G., Nabilkina, V. A., Dobrovinsky, V. E., Arkhipov, N. B., Sobolev, V. M., Minsker, K. S., Emelyanova, G. V., Vinogradova, A. V., Lebedev, J. V., Krapivina, K. Y., Kolosova, E. V., Vladykin, L. N., Sletova, L. I., Orlova, A. P., Yatsyshina, T. M., Bugrov, V. P., Rodionova, N. N., Tsvetkova, A. G. CA Patent, 1019095, 1977.
70. Small, A. B., Ernst, J. L. US Patent, 2830977, 1958.
71. Small, A. B., Ernst, J. L. CA Patent, 605748, 1960.
72. Thaler, W. A., Buckley, D. J., Sr. *Rubber Chem. Technol.* 1976, *4,* 960–966.
73. Thaler, W. A., Buckley, D. J., Sr., Kennedy, J. P. US Patent, 3856763, 1974.
74. Thaler, W. A., Kennedy, J. P., Buckley, D. J., Sr. US Patent, 3808177, 1974.
75. Balogh, L., Wang, L., Faust, R. *Macromolecules,* 1994, *27,* 3453–3458.
76. Faust, R., Balogh, L. US Patent, 5665837, 1997.

77. Kraus, C. A. US Patent, 2220930, 1940.
78. Kraus, C. A. US Patent, 2387517, 1945.
79. Parker, P. T., Hanan, J. A. US Patent, 3361725, 1968.
80. Young, D. W., Kellog, H. B. US Patent, 2440498, 1948.
81. Matsushima, S. US Patent, 3773733, 1973.
82. Matsushima, S., Ueno, K. US Patent, 3766155, 1973.
83. Miyoshi, M., Uemura, S., Tsuchiya, S., Kato, O. US Patent, 3402164, 1968.
84. Nippon Oil Co. Ltd. GB Patent, 1056730, 1967.
85. Nippon Petrochemicals Co. Ltd. G.B. Patent, 1183118, 1970.
86. Priola, A., Cesca, S., Ferraris, G. US Patent, 3850897, 1974.
87. Strohmayer, H. F., Minckler, L. S., Jr., Simko, J. P., Jr., Stogryn, E. L. US Patent, 3066123, 1962.
88. Tanaka, S., Nakamura, A., Kubo, E. US Patent, 3324094, 1967.
89. Marek, M., Chmelir, M. *J. Polym. Sci., Part C.: Polym. Symp.* 1968, *23*, 223–229.
90. Marek, M., Chmelir, M. *J. Polym. Sci., Part C.: Polym. Symp.* 1968, *22*, 177–183.
91. Marek, M., Pecka, J., Halaska, V. *Makromol. Chem., Macromol. Symp.* 1988, *13*, 443–455.
92. Rissoan, G., Randriamahefa, S., Cheradame, H. *Heterogeneous Cationic Polymerization Initiators*; In *Cationic Polymerization;* Faust, R., Shaffer, T. D., Eds., American Chemical Society: Washington, D.C., 1997; pp. 135–150.
93. Webster, M. *Chem. Rev.* 1966, *1*, 87–118.
94. Baccaredda, M., Giusti, P., Priola, A., Cesca, S. GB Patent, 1362295, 1974.
95. Cesca, S., Giusti, P., Magagnini, P. L., Priola, A. *Macromol. Chem. Phys.* 1975, *176*, 2319–2337.
96. Cesca, S., Priola, A., Bruzzone, M., Ferraris, G., Giusti, P. *Macromol. Chem. Phys.* 1975, *176*, 2339–2358.
97. Gasparoni, F., Longiave, C. BE Patent, 605351, 1961.
98. Gasparoni, F., Longiave, C. FR Patent, 1292702, 1962.
99. Gasparoni, F., Longiave, C. US Patent, 3123592, 1964.
100. Giusti, P., Priola, A., Magagnini, P. L., Narducci, P. *Macromol. Chem. Phys.* 1975, *176*, 2303–2317.
101. Kennedy, J. P. US Patent, 3349065, 1967.
102. Kennedy, J. P. *J. Polym Sci., Part A-1* 1968, *6*, 3139–3150.
103. Kennedy, J. P. *J. Macromol. Sci., Part A-Chem.* 1969, A*3*, 885–895.
104. Kennedy, J. P. US Patent, 4029866, 1977.
105. Kennedy, J. P. US Patent, 4081590, 1978.
106. Kennedy, J. P., Baldwin, F. P. US Patent, 3560458, 1971.
107. Kennedy, J. P., Gillham, J. K. *Adv. Polym. Sci.* 1972, *10*, 1–33.
108. Kennedy, J. P., Milliman, G. E. *Adv. Chem. Ser.* 1969, *91*, 287–305.
109. Kennedy, J. P., Rengachary, S. *Adv. Polym. Sci.* 1974, *14*, 1–48.
110. Kennedy, J. P., Sivaram, S. *J. Macromol. Sci., Part A-Chem.* 1973, A*7*, 969–989.
111. Maina, M. D., Cesca, S., Giusti, P., Ferraris, G., Magagnini, P. L. *Macromol. Chem. Phys.* 1977, *178*, 2223–2234.
112. Priola, A., Cesca, S., Ferraris, G. *Macromol. Chem. Phys.* 1972, *160*, 41–57.
113. Priola, A., Cesca, S., Ferraris, G. GB Patent, 1407420, 1975.

114. Priola, A., Cesca, S., Ferraris, G. US Patent, 3965078, 1976.
115. Priola, A., Cesca, S., Ferraris, G., Maina, M. *Macromol. Chem. Phys. 1975, 176,* 2289–2302.
116. Priola, A., Ferraris, G., Maina, M., Giusti, P. *Macromol. Chem. Phys. 1975, 176,* 2271–2288.
117. Tinyakova, E. I., Zhuravleva, T. G., Kurengina, T. M., Kirikova, N. S., Dolgoplosk, B. A. *Dokl. Akad. Nauk, 1962, 144,* 592.
118. Langstein, G., Bohnenpoll, M., Commander, R. US Patent, 14726 A1, 2001.
119. Langstein, G., Bohnenpoll, M., Denninger, U., Obrecht, W., Plesch, P. US Patent, 6015841, 2000.
120. Baird, M. C. US Patent, 5448001, 1995.
121. Bochmann, M., Garratt, S. US Patent, 7041760, 2006.
122. Collins, S., Piers, W. E., Lewis, S. P. US Patent, 7196149, 2007.
123. Garratt, S., Guerrero, A., Hughes, D. L., Bochmann, M. *Angew. Chem., Int. Ed. 2004, 43,* 2166–2169.
124. Hand, N., Mathers, R. T., Damodaran, K., Lewis, S. P., submitted for publication in *Ind. Eng. Chem.,* 2013.
125. Kennedy, J. P., Collins, S., Lewis, S. P. U.S. Patent 7, 202, 317, 2007.
126. Lewis, S. P., PhD Thesis, The Univ. of Akron, Diss. Abstr. Int. 2004, vol. 65, p. 770. Cf: Chem. Abs. 2004, vol. 143, p. 173195., 2004.
127. Lewis, S. P., Henderson, L., Parvez, M. R., Piers, W. E., Collins, S. *J. Am. Chem. Soc.* 2005, *127,* 46–47.
128. Lewis, S. P., Jianfang, C., Collins, S., Sciarone, T. J. J., Henderson, L. D., Fan, C., Parvez, M., Piers, W. E. *Organometallics, 2009, 28,* 249–263.
129. Lewis, S. P., Piers, W. E., Taylor, N., Collins, S. *J. Am. Chem. Soc. 2003, 125,* 14686–14687.
130. Mathers, R. T., Lewis, S. P. *J. Polym. Sci., Part A. Polym. Chem.* 2012, *50,* 1325–1332.
131. Shaffer, T. D., In *Cationic Polymerization; F*aust, R., Shaffer, T. D., Eds., American Chemical Society: Washington, D.C., 1997; pp. 96–105.
132. Shaffer, T. D. US Patent, 6699938, 2004.
133. Shaffer, T. D., Ashbaugh, J. R. *J. Polym. Sci., Part A: Polym. Chem. 1997, 35,* 329.
134. Shaffer, T. D., Dias, A. J., Finkelstein, I. D., Kurtzman, M. B. US Patent, 6291389, 2001.
135. Jacob, S., Pi, Z., Kennedy, J. P. *Polym. Bull. 1998, 41,* 503–510.
136. Jacob, S., Pi, Z., Kennedy, J. P. *Polym. Mater. Sci. Eng. 1999, 80,* 495.
137. Jianfang, C., Lewis, S. P., Kennedy, J. P., Collins, S. *Macromolecules, 2007, 40,* 7421–7424.
138. Pi, Z., Kennedy, J. P. *Cationic Polymerizations at Elevated Temperatures by Novel Initiating Systems Having Weakly Coordinating Counteranions. 1. High Molecular Weight Polyisobutylenes,* Nato Sci. Ser., Ser. E. ed., In I*onic Polymerizations and Related Processes; P*uskas, J. E., Ed., Kluwer: Dordrecht, Neth., 1999; Vol. 359, pp. 1–12.
139. Shaffer, T. D., Ashbaugh, J. R. P*olym. Prepr., Am. Chem. Soc. Div. Polym. Chem. 1996, 37,* 339–340.
140. Kostjuk, S. V., Ganachaud, F. M*acromolecules, 2006, 39,* 3110–3113.
141. Kostjuk, S. V., Ouardad, S., Peruch, F., Deffieux., A., Absalon, C., Puskas, J. E., Ganachaud, F. *Macromolecules, 2011, 44,* 1372–1384.

142. Baird, M. C. *Chem. Rev.* 2000, *100*, 1471–1478.
143. Barsan, F., Karan, A. R., Parent, M. A., Baird, M. C. *Macromolecules*, 1998, *31*, 8439–8447.
144. Bochmann, M., Dawson, D. M. *Angew. Chem., Int. Ed.* 1996, *35*, 2226–2228.
145. Bohnenpoll, M., Ismeier, J., Nuyken, O., Vierle, M., Schon, D. K., Kuhn, F. US Patent, 7291758, 2007.
146. Burns, C. T., Shapiro, P. J., Budzelaar, P. H. M., Willett, R., Vij, A. *Organometallics*, 2000, *19*, 3361–3367.
147. Carr, A. G., Dawson, D. M., Bochmann, M. *Macromolecules*, 1998, *31*, 2035–2040.
148. Carr, A. G., Dawson, D. M., Bochmann, M. *Macromol. Rapid Commun.* 1998, *19*, 205–207.
149. Garratt, S., Carr, A. G., Langstein, G., Bochmann, M. *Macromolecules*, 2003, *36*, 4276–4287.
150. Hijazi, A. K., Yeong, H. Y., Zhang, Y., Herdtweck, E., Nuyken, O., Kühn, F. E. *Macromol. Rapid Commun.* 2007, *28*, 670–675.
151. Huber, M., Kurek, A., Krossing, I., Mulhaupt, R., Schnockel, H. *Z. Anorg. Allg. Chem.* 2009, *635*, 1787–1793.
152. Kumar, K. R., Hall, C., Penciu, A., Drewitt, M. J., Mcinenly, P. J., Baird, M. C. *J. Polym. Sci., Part A: Polym. Chem.* 2002, *40*, 3302–3311.
153. Kumar, K. R., Penciu, A., Drewitt, M. J., Baird, M. C. *J. Org. Chem.* 2004, *689*, 2900–2904.
154. Langstein, G., Bochmann, M., Dawson, D. M. US Patent, 5703182, 1997.
155. Langstein, G., Bochmann, M., Dawson, D. M., Carr, A. G., Commander, R. DE Patent, 19836663 A1, 2000.
156. Lee, S.-J., Shapiro, P. J., Twamley, B. *Organometallics*, 2006, *25*, 5582–5586.
157. Li, Y., Voon, L. T., Yeong, H. Y., Hijazi, A. K., Radhakrishnan, N., Köhler, K., Voit, B., Nuyken, O., Kühn, F. E. *Chem. Eur. J.* 2008, *14*, 7997–8003.
158. Lin, M., Baird, M. C. *J. Org. Chem.* 2001, *619*, 62–73.
159. Nuyken, O., Vierle, M., Kuhn, F. E., Zhang, Y. *Macromol. Symp.* 2006, *236*, 69–77.
160. Radhakrishnan, N., Hijazi, A. K., Komber, H., Voit, B., Zschoche, S., Kühn, F. E., Nuyken, O., Walter, M., Hanefeld, P. J. *Polym. Sci., Part A* 2007, *45*, 5636–5648.
161. Sabmannshausen, J. *Dalton Trans.* 2009, 9026–9032.
162. Song, X., Thornton-Pett, M., Bochmann, M. *Organometallics*, 1998, *17*, 1004–1006.
163. Tse, C. J. W., Kumar, K. R., Drewitt, M. J., Baird, M. C. *Macromol. Chem. Phys.* 2004, *205*, 1439–1444.
164. Vierle, M., Zhang, Y., Herdtweck, E., Bohnenpoll, M., Nuyken, O., Kuhn, F. E. *Angew. Chem., Int. Ed.* 2003, *42*, 1307–1310.
165. Davison, W. H. T., Pinner, S. H., Worrall, R. *Chem. Ind.* 1957, 1274.
166. Davison, W. H. T., Pinner, S. H., Worrall, R. *Proc. Roy. Soc. (London), A* 1959, *252*, 187–196.
167. Kennedy, J. P., Shinkawa, A., Williams, F. *J. Polym Sci., Part A-1* 1971, *9*, 1551–1561.
168. Williams, F., Shinkawa, A., Kennedy, J. P. *J. Polym. Sci., Part C.: Polym. Symp.* 1976, *56*, 421–430.
169. Worrall, R., Pinner, S. H. *J. Polym. Sci.* 1959, *34*, 229–242.
170. Lewis, S. P. US Patent, 8283427, 2012.
171. Kostjuk, S. V., Radchenko, A. V., Ganachaud, F. *Macromolecules*, 2007, *40*, 482–490.
172. Reed, C. A. *Chem. Commun.* 2005, 1669–1677.

173. Biddulph, R. H., Plesch, P. H., Rutherford, P. P. *J. Chem. Soc. 1965*, 275–294.
174. Belov, G. P. *Macromol. Symp. 1995*, *97*, 63–78.
175. Gronowski, A. CA Patent, 2252295, 2000.
176. Gronowski, A. US Patent, 6403747, 2002.
177. Gronowski, A. US Patent, 6630553, 2003.
178. Langstein, G., Freitag, D., Lanzendörfer, M., Weiss, K. US Patent, 5668232, 1997.
179. Lisovskii, A., Nelkenbaum, E., Volkis, V., Semiat, R., Eisen, M. S. *Inorg. Chim. Acta*, 2002, *334*, 243–252.
180. Lewis, S. US Patent, 2010/0273964, 2010.
181. Liu, Q., Mathers, R. T., Damodaran, K., Godugu, B., Lewis, S. P. *Green Materials*, 2013, 1, 161–175.
182. Imanishi, Y., Yamamoto, R., Higashimura, T., Kennedy, J. P., Okamura, S. J. *Macromol. Sci., Part A-Chem. 1967*, A*1*, 877–890.
183. Wichterle, O., Marek, M., Trekoval, I. J. *Polym. Sci. 1961*, *53*, 281–287.
184. Addecott, K. S. B., Mayor, L., Turton, C. N. *Eur. Polym. J. 1967*, *3*, 601–617.
185. Turton, C. N. GB Patent, 1091083, 1967.
186. Collomb, J., Morin, B., Gandini, A., Cheradame, H. *Eur. Polym. J. 1980*, *16*, 1135–1144.
187. Ran, R. J. *Polym. Sci., Part A: Polym. Chem. 1993*, *31*, 1561–1569.
188. Ran, R. *Support Catalysts (Lewis Acid and Ziegler-Natta); In Polymeric Materials Encyclopedia; S*alamone, J. C., Ed., CRC Press: Boca Raton, FL, 1996; Vol. 10, pp. 8063–8078.
189. Chung, T.-C. *Supported Lewis Acid Catalysts (on Polypropylene; Recoverable and Resuable); In Polymeric Materials Encyclopedia; S*alamone, J. C., Ed., CRC Press: Boca Raton, 1996; Vol. 10, pp. 8093–8100.
190. Chung, T.-C., Chen, F. J., Stanat, J. E., Kumar, A. US Patent, 5288677 1994.
191. Chung, T.-C., Chen, F. J., Stanat, J. E., Kumar, A. US Patent, 5409873, 1995.
192. Chung, T.-C., Kumar, A. *Polym. Bull. 1992*, *28*, 123–128.
193. Chung, T.-C., Kumar, A., Rhubright, D. *Polym. Bull. 1993*, *30*, 385–391.
194. Lewis, S. P. unpublished research data, 2013.
195. US Environmental Protection Agency, Methyl Chloride (Chloromethane), http://www.epa.gov/ttn/atw/hlthef/methylch.html, 2000.
196. Forster, P., Ramaswamy, V., Artaxo, P., Berntsen, T., Betts, R., Fahey, D. W., Haywood, J., Lean, J., Lowe, D. C., Myhre, G., Nganga, J., Prinn, R., Raga, G., Schulz, M., Dorland, R. V. *Changes in Atmospheric Constituents and in Radiative Forcing In Climate Change; In The Physical Science Basis. Contribution of Working Group I to the Fourth Assessment Report of the Intergovernmental Panel on Climate Change; S*olomon, S., Qin, D., Manning, M., Chen, Z., Marquis, M., Averyt, K. B., M. Tignor, Miller, H. L., Eds., Cambridge University Press: Cambridge, 2007; pp. 129–234.
197. Duffy, J., Wilson, G. J. *Halobutyl Rubber* 5th ed., In *Ullman's Encyclopedia of Industrial Chemistry; E*lvers, B., Hawkins, S., Russey, W., Schulz, G., Eds., VCH Publishers: Weinheim, 1993; Vol. A23, pp. 314–318.
198. Paul, H.-I., Feller, R., Lovegrove, J. G. A., Gronowski, A., Jupke, A., Hecker, M., Kirchhoff, J., Bellinghausen, R. WO Patent, 2010.
199. Deshpande, A., Williams, P., Arizona Chemical, personal communication, 2013.
200. Morgan, D., Chevron-Oronite, personal communication, 2013.

CHAPTER 6

LAYER-BY-LAYER ASSEMBLY OF BIOPOLYMERS ONTO SOFT AND POROUS GELS

ANA M. DIEZ-PASCUAL[1,2] and PETER S. SHUTTLEWORTH[1]

[1] Institute of Polymer Science and Technology, ICTP-CSIC, Juan de la Cierva 3, 28006 Madrid, Spain

[2] Analytical Chemistry, Physical Chemistry and Chemical Engineering Department, Faculty of Biology, Environmental Sciences and Chemistry, Alcalá University, 28871 Alcalá de Henares, Madrid, Spain

CONTENTS

ABSTRACT

This chapter deals with the layer-by-layer (LbL) assembly of polyelectrolyte multilayers of polypeptides (poly L-lysine/poly L-glutamic acid) and polysaccharides (chitosan/dextran sulfate) onto soft and porous thermo-responsive poly (N-isopropylacrylamide-*co*-methacrylic acid) (P(NiPAM-co-MAA)) nanogels. The formation of the LbL structures and characterization of hydrodynamic radius, electrophoretic mobility, bilayer thickness and stability over time of the systems will be comparatively discussed. Further, the assembly of biopolymers onto similar microgels will be analyzed. The results presented demonstrate that the structure and properties of biocompatible multilayer films can be finely tuned by confinement onto soft and porous gels, which provides new perspectives for biomedical applications, particularly storage and controlled release of biomolecules.

6.1 INTRODUCTION

In 1992, Decher et al. [1, 2] reported a new approach to build controlled multilayer polymer film architectures based on the layer-by-layer (LbL) assembly of oppositely charged polyelectrolytes (PE) onto a charged surface. Initially this work focused on the assembly of PE on to flat and rigid substrates, analyzing parameters such as the PE nature, conformation and charge density, pH, ionic strength, as well as the influence of solution temperature on multilayer formation [3–5]. The LbL technique was then extended to 3D hard colloidal particles; Donath et al. [6] and Caruso et al. [7, 8] undertook intensive studies on the assembly of PE on core particles, which became the precursor route for the fabrication of hollow capsules. Subsequently, these authors also demonstrated that the LbL assembly was feasible on hard and porous templates like mesoporous silica [9] or $CaCO_3$ [10]. By coating the interior pores of the particles followed by dissolution of the core, interconnecting networks made of PE complexes were formed. This approach can be applied to encapsulate high therapeutic loads due to the high surface area of the porous particles, and is applicable to a wide range of materials of different sizes, ranging from proteins to low molecular weight drug compounds.

A more recent approach is to carry out LbL assembly onto soft and porous templates at the nanometer scale, such as poly(N-isopropylacrylamide) (PNiPAM) thermoresponsive nanogels (NGs). Smart NGs are stimuli-responsive 3D soft and porous polymeric networks [11], which can either be neutral or charged that swell or contract in response to external stimuli such as temperature, pH, ionic strength, light and electromagnetic fields, etc. They are characterized by an abrupt and reversible volume phase transition (VPT) from a swollen to a collapsed state upon heating. Below the VPT, the formation of H-bonds between the amide groups of PNiPAM and the water molecules are thermodynamically favorable. Nevertheless, increasing the temperature above the lower critical solution temperature (LCST) [12] hydrophobic interactions dominate and provoke the precipitation of the polymer. NGs respond faster to changes of external stimuli than their macroscopic counterparts (gels), and exhibit a higher surface area to volume ratio. This makes them highly suitable for a number of applications ranging from sensors to dye or drug encapsulation and delivery, as well as for the immobilization of enzymes or proteins within their interior or outer surface [13, 14]. In addition, the NG properties can be tailored and their stability improved by depositing a PE shell, which can govern the transport of substances into and out of the resulting core-shell ensemble. With the proper choice of PE pairs, selective permeability can be achieved as well as sustained release of a variety of substances. The porosity of the NG plays a key role in LbL assembly since the adsorbing PE layers can not only interdigitate among themselves, like when dealing with hard and rigid templates, but also penetrate into the gel, conferring novel surface properties. The extent of interpenetration is conditioned by the mesh size of the polymeric network (related to the degree of crosslinking), and also by the PE molecular weight and degree of branching [15]. The larger the pore or mesh size the easier for the PE to move within the NG. On the other hand, a high molecular weight or a highly branched PE is expected to have a lower degree of interpenetration with the NG. The PE adsorption process is principally electrostatically driven, although secondary cooperative interactions such as hydrophobic–hydrophobic and H-bonds are also important, especially with uncharged gels. By modifying the NG chemical structure or copolymerization with another monomer the VPT can be varied. In the case of the copolymers poly(N-isopropylacrylamide-co-methacrylic acid) and P(NiPAM-co-MAA), the main monomer, NiPAM, lends thermosensitivity to the NG while the co-monomer, methacrylic acid, enables pH tuneability, thus modifying its swelling behavior.

It has been demonstrate that by simply modifying either the total number of layers deposited or the conditions of the assembly process, the deposited films can range from nanometers to microns in thickness [4]. In fact, variation in parameters such as the PE concentration [5], ionic strength of the dipping and rinsing solutions [16], or the time allowed for polymer adsorption [17], all greatly influence the film properties. Polyelectrolytes can be either synthetic or natural and may be further classified as being strong or weak depending whether they are fully or partially charged. Typically, multilayers have been prepared from strong PEs like poly(diallyldimethylammonium chloride) (PDACMAC) or poly(sodium 4-styrenesulfonate) (PSS) that remain fully charged over the entire pH range, with the primary means to control the adsorbed layer thickness being the addition of salt [15]. However, this approach is limited to a certain extent by the poor solubility of high molecular weight PEs in solutions of high ionic strength. To overcome this limitation and make the LbL process more versatile, weak PEs such us poly(allylamine hydrochloride) (PAH) and poly(acrylic acid) (PAA) can be used [18]. Thus, it is possible to systematically vary the linear charge density of these PEs by simply adjusting the pH of the dipping solutions. When a weak PE is incorporated into a multilayer system, its degree of ionization may change considerably from the solution value due to the influence of the local environment via electrostatic and hydrophobic effects. Electrostatic effects [19] are observed upon addition of salt to the PE solutions or when an oppositely charged polymer is added to form a complex. Hydrophobic effects occur when the PE experiences hydrophobic moieties or regions that can alter the dielectric environment of the weak ionic group, therefore making it more difficult to achieve an ionized state. Due to the mentioned effects, the pK_a of the PE in the multilayer (pH at which 50% of the polymer functional groups are ionized) depends strongly on the nature of the other polymer used in the assembly process.

This chapter focuses on the LbL assembly of two pairs of biopolymers, poly L-lysine (PLL)/poly L-glutamic acid (PGA) and chitosan (CHIT)/ dextran sulfate (DEX) onto a negatively charged P(NiPAM-co-MAA) NG

(Scheme 1) [20]. In addition, the assembly of biopolymers onto similar hydrogels with micrometer size [18] and biodegradable dextran-hydroxy-ethyl methacrylate (DEX-HEMA) microgels (MG) [21] will be comparatively described. Natural polypeptides and polysaccharides are weak PEs and can adopt multiple conformations in response to changes in the solution pH and temperature. The folding and unfolding of their chains through changes in hydrogen bonding, molecular and electrostatic forces among others, provides a fascinating insight into the properties of natural PEs interactions with NGs via LbL assembly. The chapter analyzes how subtle changes in the internal reorganization of the NG during heating and cooling induce or restrict, through various types of non-covalent interactions, conformational changes in the absorbed PE layers, and how they contribute to the structure and stability of the multilayer films as well as the kinetics of folding in confined geometry. The influence of the nature of the last layer adsorbed on the thermoresponsive behavior, electrophoretic mobility, bilayer thickness and stability over time of coated NGs is described [20]. Further, the mechanisms of interaction between the NG and the PEs are analyzed. The approach illustrated herein is a simple, versatile and interesting method to integrate biopolymers with nanogels for the design of new nanoscale materials that have potential use in biomedical applications.

SCHEME 1 Schematic representation of the LbL assembly of biopolymers onto soft and porous nanogels (NG).

6.2 STRUCTURE AND PROPERTIES OF THE BIOPOLYMERS USED FOR THE LBL ASSEMBLY

PLL and PGA are biopolymers constructed from the basic amino acid monomer building blocks L-lysine and glutamic acid. In contact with water a positive charge forms on the amino group. Scale production of these polymers is carried out via bacterial fermentation; with uses including drug delivery, food preservative etc. to name a few. Both PLL and PGA (Scheme 2) undergo multiple structural conformations such as α-helix, β-sheet or random coil depending on the pH or temperature, and as such are useful model systems for studying protein conformational changes. Poly L-arginine (pARG) and poly L-aspartic acid (pASP) are also biodegradable polyaminoacids. pARG is currently used for the treatment of head and neck cancer, and pASP can be used as a super-swelling material in nappy/feminine-hygiene products and food packaging.

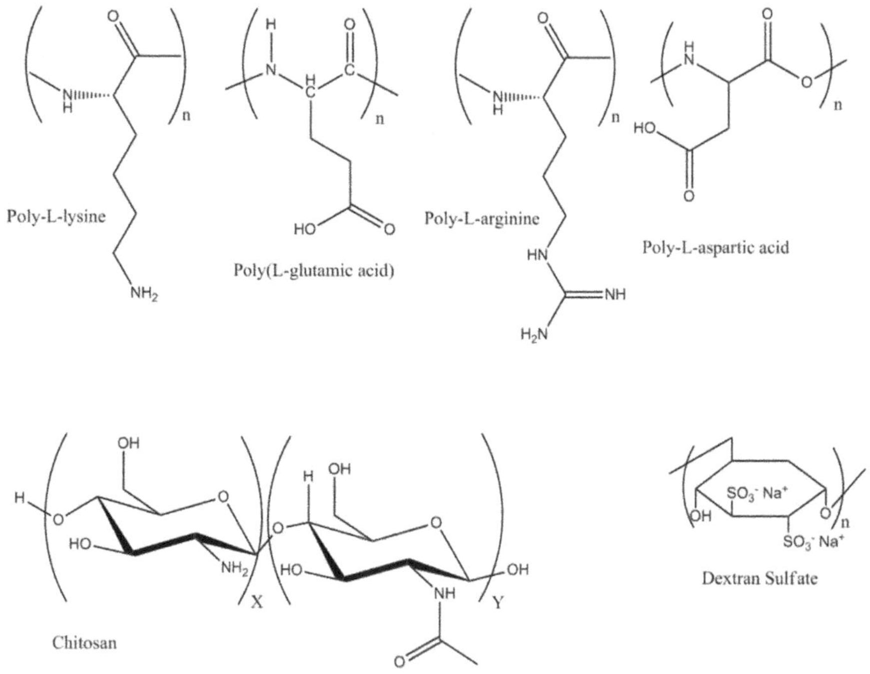

SCHEME 2 Schematic representation of polymeric amino acid and biopolymer/biopolymer derived structures.

Chitin (poly β-1–4-N-acetyl-D-glucosamine) is one of the most abundant biopolymers on the planet. It is mainly found as a structural component of crustaceans and insects or in the cell wall of fungi. Chitosan (CHIT), a deacetylated form of chitin presents far greater solubility in aqueous acids than chitin due to its ability to be protonated to a polycation at low pH. Its ease of further modification, availability and low cost in addition to its biocompatibility all ensure its wide spread use in medical applications, food industry, pollutant adsorbent, and as a natural antimicrobial agent.

DEX, a polyanionic derivative of Dextran is freely soluble and stable in water and salt solutions. DEX is used in many applications such as, an anti-viral agent, stabilizer of proteins and anti-coagulant. CHIT and DEX (Scheme 2) are interesting biocompatible and biodegradable polysaccharides for medical use, since their alternate assembly exhibits anti versus pro-coagulant activity of human whole blood. They are both weak polyelectrolytes, and just like polypeptides, present different structures depending on their structural organization, such as helical or ribbon-like secondary structures. Conformational stability is usually achieved through secondary cooperative interactions between two or more chain segments, usually with long regions of regular covalence sequence. When such regions are interrupted by chain branching or other altered substitution, several chain associations can give rise to higher levels of structure and result in a three-dimensional network.

6.3   PREPARATION OF BIOPOLYMER-COATED GELS

6.3.1   SYNTHESIS OF P(NIPAM-CO-MAA) HYDROGELS

P(NiPAM-co-MAA) NGs can be synthesized by free radical emulsion polymerization [22]. Typically, the NIPAM monomer, the crosslinker N,N'-methylenebis(acrylamide) and the comonomer methacrylic acid (MAA) (9:0.5:1 ratio) are dissolved in water at 70°C and flushed with nitrogen for 40 min while stirring. Polymerization is initiated with a water solution of potassium peroxodisulfate (KPS) and allowed to proceed for about 6 h under a nitrogen stream and constant stirring. The dispersion is passed through glass wool in order to remove particulate matter and further purified by dialysis against deionized and double-distilled water for ~ 1 week. Then the solution is centrifuged for several cycles; between each centrifugation cycle the supernatant is removed and replaced by deionized water to redisperse. Finally the solution is freeze-dried overnight for storage.

Following a similar procedure, P(NiPAM-co-MAA) microgels (MGs) have been recently synthesized in supercritical carbon dioxide (scCO$_2$) [18].

The monomers (5.2 wt.% content relative to CO_2) and perfluoropolyether (10 wt.% content relative to the monomers) were charged into a high-pressure cell at 28.0 MPa and 65°C. All the reactants were soluble in the supercritical medium, and the reaction was allowed to proceed for 24 h under stirring. The obtained polymer was washed continuously with fresh CO_2 for 1 h to remove any residual monomers or crosslinker. Monodisperse crosslinked MGs with an average diameter of 3.0 ± 0.7 μm were obtained.

6.3.2   PREPARATION OF DEX-HEMA MICROGELS

Biodegradable DEX-HEMA MGs can be prepared by an aqueous emulsion technique: DEX-HEMA, fluorescein isothiocyanate (FITC)-labeled DEX solution and dimethyl aminoethyl methacrylate (DMAEMA) are dissolved in water, and subsequently emulsified with an aqueous poly(ethylene glycol) (PEG) solution [21]. Radical polymerization of the pending HEMA groups is initiated by adding N, N, N', N'-tetramethylenediamin (TEMED) and KPS. The reaction is carried out at room temperature for 1 h. Finally, the synthesized MGs (with an average diameter of 7 μm) are washed three times with pure water to remove PEG, KPS, and TEMED, suspended in water and stored at −20°C.

6.3.3   LAYER-BY-LAYER ASSEMBLY OF BIOPOLYELECTROLYTES

The multilayer assembly is typically done by slowly adding 3 mL of an aqueous dispersion of the P(NiPAM-co-MAA) gel (~0.02 wt %) to 12 mL of 1 mg/mL PE solution [20]. PLL and PGA solutions are prepared in buffer (25 mM Tris-(hydroxymethylaminomethane), 10 mM 2-(N-morpholino)-ethanesulfonic acid and 1.15 M NaCl, pH = 7.5). At neutral pH, both PLL ($pK_a \sim 10.5$) and PGA ($pK_a \sim 4.3$) are ionized. CHIT is dissolved in an aqueous solution with 25% acetic acid (pH = 1.3); since the pK_a of the glucosamine units is about 6.0, and the degree of deacetylation of CHIT is ~85%, it is highly positively charged in strong acid conditions. DEX multilayers are assembled from water solution (pH = 6.2); this polysaccharide ($pK_a \sim 4.0$) contains an average of 2.3 sulfate groups per glucozyl residue, hence is fully negatively charged in the conditions of the assembly. Since the gel is negatively charged, the first PE layer deposited is a polycation (PLL or CHIT). The mixture is kept under constant stirring for 4−6 h. After each deposition, the excess of PE and buffer is removed by several ultracentrifugation cycles followed at each step by decantation and redispersion in water by vigorous shaking during at least 4 h. This sequence is repeated until the desired PE layers are deposited. Table 6.1 summarizes the pH of the various components at different stages of the assembly. Before each characterization, the coated microgels are filtered several times in a laminar-flow box.

A similar procedure can be employed to prepare coated DEX-HEMA MGs [21]. A MG aqueous dispersion is mixed with the PE solution, and the PE is allowed to adsorb for 10 min under continuous shaking. The dispersion is then centrifuged, the supernatant is removed and the MG is re-dispersed in water to remove the non-adsorbed PE. As the MG is positively charged, the first layer is a polyanion. The only bio-polycation that coated the DEX-HEMA microgels without aggregation was pARG, which is the only polypeptide with side groups that are charged at almost every pH (the pKa of the side groups is 12.5), and is therefore a strong PE. pASP, PGA and DEX are suitable polyanions for LbL coating of the dex-HEMA MGs. Typically, four biopolymer bilayers are deposited.

TABLE 6.1 Summary of the Solution pH at Different Stages During the Build-Up of the First and Second PE Layers Onto the P(NiPAM-co-MAA) Nanogel (NG)*

Sample	pH	Sample	pH
NG (in water)	4.4	—	—
PLL (in buffer)	7.5	CHIT (in 25% acetic acid)	1.3
PGA (in buffer)	7.5	DEX (in water)	6.2
NG + excess PLL	7.0	NG + excess CHIT	1.9
NG/PLL (1L-coated)	7.3	NG/CHIT (1L-coated)	1.5
NG + excess PGA	7.4	NG + excess DEX	5.7
NG/PGA (2L-coated)	7.5	NG/DEX (2L-coated)	6.6

*The pH of the coated nanogels was determined after washing and redispersion in water.

6.3.4 CHARACTERIZATION TECHNIQUES

6.3.4.1 Electrophoretic Measurements

A parameter commonly used to quantify surface charge on hard and rigid particles is the zeta potential (ζ), a scientific term for electrokinetic potential in colloidal systems. ζ is the potential difference between the dispersion medium and the stationary layer of fluid attached to the dispersed particle. A value of 25 mV (positive or negative) is generally accepted as the arbitrary value that separates low-charged surfaces from highly charged surfaces. ζ can be related to the stability of colloidal dispersions: a high ζ indicates stability, the solution or dispersion will resist aggregation, while colloids with low ζ tend to coagulate or flocculate [23]. Zeta potential is not measurable directly but it can be calculated using theoretical models and an experimentally determined electrophoretic mobility μ_e that is defined as:

$$\mu_e = \frac{\upsilon}{E} \tag{1}$$

where E is the electric field strength and υ the velocity of a dispersed particle. The most widely used equation to calculate ζ was developed by Smoluchowski [24]:

$$\mu_e = \frac{\varepsilon_r \varepsilon_0 \zeta}{\eta} \tag{2}$$

where ε_r is the dielectric constant of the dispersion medium, ε_0 the permittivity of free space ($C^2\ N^{-1}\ m^{-2}$) and η the dynamic viscosity of the dispersion medium (Pa.s). Equation (2) can be applied to dispersed particles of any shape and concentration. This model is valid for most aqueous systems, but fails for nano-colloids in a solution with ionic strength approaching that of pure water. Taking into account that gels are soft, porous, and solvent-penetrable particles that do not conform to the usual hard-sphere model, generally μ_e values are not converted into ζ but directly employed to measure their surface charge density. The LbL assembly of PEs can be monitored by measuring μ_e of diluted aqueous solutions of the gel after each adsorption step as a function of temperature.

6.3.4.2 Dynamic Light Scattering

Dynamic light scattering (DLS) is a non-invasive technique commonly used to measure the particle size of colloidal systems. When monochromatic light from a laser source hits small particles, a time-dependent fluctuation is detected in the scattered intensity, arising from the fact that the small molecules in solutions are undergoing Brownian motion. This scattered light then undergoes either constructive or destructive interferences by the surrounding particles, and this intensity fluctuation contains information about the time scale of movement of the scattering. This information can be derived from the second order autocorrelation curve, which can be expressed as follows:

$$g^2(q;\tau) = \frac{<I(t)I(t+\tau)>}{<I(t)>^2} \tag{3}$$

where q is the wave vector, τ the delay time and I the light intensity. The simplest approach is to treat the first order autocorrelation function as a single exponential decay:

$$g^2(q;\tau) = exp(-\Gamma\tau) \tag{4}$$

where Γ is the decay rate. This is appropriate for a monodisperse population. The translational diffusion coefficient D_t can be obtained at a certain angle using the equations:

$$\Gamma = q^2 D_t \tag{5}$$

$$q = \frac{4\pi n_0}{\lambda} \sin\left(\frac{\theta}{2}\right)$$

where λ is the incident laser wavelength, n_0 is the refractive index of the sample and θ is angle at which the detector is located with respect to the sample cell. Typically, a cumulant analysis is used to determine the particle size [25]. In this method, information can be derived about the variance of the system as follows:

$$g^1(q,\tau) = exp(-\overline{\Gamma}\tau)\left(1 + \frac{\mu_2}{2!}\tau^2 - \frac{\mu_3}{3!}\tau^3 + \cdots\cdots\right) \tag{6}$$

where $\bar{\Gamma}$ is the average decay rate and $\mu_2 \bar{\Gamma}^{-2}$ is the second order polydispersity index (or an indication of the variance). Analogously, the z-averaged translational diffusion coefficient D_z can be calculated at a single angle through the equation:

$$\bar{\Gamma} = q^2 D_Z \tag{7}$$

The hydrodynamic radius R_h can then be obtained from the diffusion coefficient through the Stokes–Einstein equation [26]:

$$D_Z = \frac{K_B T}{6\pi\eta R_h} \tag{8}$$

After each deposition step, DLS experiments are typically performed on highly diluted gel solutions as a function of temperature (i.e., heating from 20 to 40°C and cooling down to 20°C, in steps of 2°C, in order to get a complete cycle).

6.3.4.3 Environmental Scanning Electron Microscopy

The morphology of uncoated as well as LbL-coated ensembles in the hydrated state can be examined using an environmental scanning electron microscope (ESEM). This is a suitable method for morphological characterization of hydrated porous gels to avoid drastic structural changes that might occur during the drying processes. Measurements are typically performed at room temperature under a water vapor pressure of 0.5–0.6 kPa. To prevent damage to the gels or local evaporation, samples are only observed for a short period of time (i.e., 3 min). This microscopic technique images the sample surface by scanning it with a high-energy beam of electrons in a raster scan pattern. The electrons interact with the atoms that constitute the sample producing a variety of signals, which are detected to generate images.

6.3.4.4 Attenuated Total Reflectance FT-IR Spectroscopy

The mechanisms of interaction between the gels and the PEs can be analyzed using ATR-FTIR spectroscopy. The assembly of the PE pairs causes modifications in the position, intensity and shape of the IR bands of the dry assembled gel. Spectra are typically recorded in the wavenumber region between 950 and 1800 cm^{-1}, with a 4 cm^{-1} resolution and no baseline correction. In order to clarify the interactions, a detailed spectral analysis is frequently performed by deconvolution to identify the constituent components of the most relevant bands. A nonlinear least-square fitting method can be used, assuming Gaussian band profiles for the components.

6.3.4.5 Confocal Microscopy and Fluorescence Correlation Spectroscopy

Visual evidence of the LbL assembly on thermoresponsive MGs can be obtained from confocal microscopy images. However, in the case of NGs it is very difficult to visualize the exact location of the LbL coating using fluorescently labeled species. Quantitative evidence of successful LbL assembly on thermoresponsive NGs can be obtained from the auto- and cross-correlation functions of labeled PE using fluorescence correlation spectroscopy (FCS) [27]. This technique was initially designed to measure the diffusion coefficient of fluorescently labeled species at nanomolar concentrations by exploiting the detection of the fluctuation in the fluorescent intensity as the species enter and leave the detection volume – the so-called autocorrelation function, ACF. Commonly, FCS is employed in the context of optical microscopy, in particular confocal microscopy or two-photon excitation microscopy. In these approaches light is focused on a sample and the measured fluorescence intensity fluctuations (due to diffusion, physical or chemical reactions, aggregation, etc.) are analyzed using the ACF. The main disadvantage of using conventional FCS for quantitative analysis is the required information about the size and shape of the excitation/detection volume, which depends strongly on the optical saturation, laser power and refractive index of the sample. Therefore, a calibration is required prior to the sample measurement. To overcome these limitations, Dertinger et al. [28] proposed the dual focus Fluorescence Correlation Spectroscopy (2fFCS), a novel and improved version of FCS that permits determinations of comparable precision to well-established methods such as DLS or small angle neutron scattering (SANS). An external length scale is introduced into the measurement by creating two overlapping excitation foci at a precisely known distance, thus allowing reference free and absolute measurements of the diffusion coefficient close to the infinite dilution limit. 2fFCS data evaluation is modified taking into account correctly finite size effects in colloidal and macromolecular systems. To achieve a precise temperature control, the sample is placed in a cell for inverted microscopes, enabling heating-cooling cycles to be performed with high accuracy. This new method can then be used for qualitative and quantitative experiments on the LbL assembly of PEs. As PEs are able to interdigitate with each other during multilayer formation as well as with the gel, it was not possible to distinguish between PEs adsorbed on or in the gel. Dual color fluorescence cross-correlation spectroscopy measures interactions by cross-correlating two or more fluorescent channels, hence, distinguishing between free labeled PE and those that are bounded to the NG (which have longer diffusion time), with fluorescence imaging providing visual proof.

6.3.4.6 Atomic Force Microscopy

Atomic force microscopy (AFM), which provides 3D images with very high-resolution, is a powerful tool for characterizing materials at the nanoscale by providing morphological information. It consists of a cantilever with a sharp tip (probe) at its end that is used to scan the sample surface. When the tip is brought close to the gel surface, forces between the tip and the gel lead to a deflection of the cantilever according to Hooke's law. The probe deflection is typically measured by a "beam bounce" method, and the measured deflections are used to generate a map of the surface topography. AFM images of hydrogels are typically obtained in the intermittent mode (tapping), where the cantilever oscillates at its resonant frequency, in air at room temperature. Gels are deposited onto mica substrates and dried under a nitrogen stream.

6.4 CHARACTERIZATION OF THE BIOPOLYMER-COATED GELS

6.4.1 MORPHOLOGICAL CHARACTERIZATION

The morphology of uncoated and coated-NGs in the hydrated state can be observed directly by ESEM, as illustrated in Figure 6.1. The image of the neat NG (Figure 6.1a) shows soft, rough and porous nanospheres

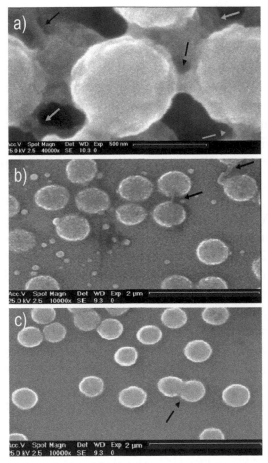

FIGURE 6.1 Typical ESEM micrographs of (a) neat NG, (b) NG/(PLL/PGA) and (c) NG/(CHIT/DEX) in the hydrated state. Reprinted from Ref. [20], © copyright 2010, with permission from Elsevier.

interconnected by sticking bridges. In contrast, the PPL- and CHIT-coated NGs (Figures 6.1b and 6.1c, respectively) are more separated and present more individual particles with few sticking connections; similar morphology has been described for all coated NGs [20]. The comparison of NG/PLL and NG/CHIT images reveals that polypeptide-coated systems partly keep the rough surface of the uncoated NG, while those covered with polysaccharides exhibit a smoother structure. This behavior could be related to differences in layer thickness, as reported for biopolymers assembled onto flat substrates [30]. Polypeptides are more prone to rearrangement and to form complexes with both PLL and PGA in β-sheet antiparallel conformation, with extensive interpenetration and strong ionic interaction between biopolymers from adjacent layers, which led to a highly compact structure that fits easily onto the NG surface, thus retaining its rough morphology. This is consistent with the images reported by De Geest et al. [21] for pARG-coated dex-HEMA MGs showing very rough microspheres attached to each other.

On the other hand, all the systems display spherical shape and low size polydispersity. The mean radius of the neat NG, NG/PLL and NG/CHIT are ~346, 295 and 231 nm, respectively, and those of two layer coated NG/(PLL/PGA) and NG/(CHIT/DEX) are 338 and 269 nm. These values are in good agreement with those obtained from DLS experiments in the swollen state [20] (as will be discussed in a following section), which verifies that the size of the coated NG is strongly influenced by the nature of the PE. Upon adsorption of the first polycation layer there is a strong decrease in the NG size, while the deposition of the second layer (polyanion) causes an important increase in size. It has been observed systematically at temperatures below the VPT that every time the outermost layer is a polycation there is a noticeable decrease in the radius and upon addition of each polyanion the size of the ensemble once again increases. An analogous "odd-even" effect in the NG size depending on the type of PE in the outermost layer has been reported for PLL and PGA coated-P(NiPAM-co-MAA) MG [18] (Figure 6.2) as well as NGs coated with synthetic polymers such as PDACMAC or PPS [15]. The electrostatic attraction between the

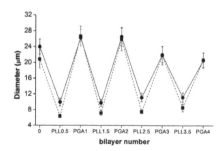

FIGURE 6.2   Diameter of P(NiPAM-co-MAA) MG upon LbL assembly of PLL and PGA at 24°C (circles) and 37°C (squares). Adapted from Ref. [18], © copyright 2012, with permission from the American Chemical Society.

negatively charged NG and the first positively charged PLL or CHIT layer causes the ensemble to contract, leading to a remarkable decrease in size. Upon approach of the negatively charged PGA or DEX layer, the positive charges have to be shared between the NG and the new absorbed PE; this weakens the pulling attraction between the NG and the polycation, hence, the radius increases. Overall, SEM images not only provide the visual proof of the successful LbL assembly but also demonstrate that the morphology of the NGs can be tuned by surface adsorption of the PE layers.

6.4.2   SURFACE CHARGE DENSITY

The sequential deposition of the different PEs onto gels can be monitored using electrophoretic mobility (μ_e) measurements as a function of temperature after each adsorption step. Figure 6.3 compares μ_e of NGs coated with polypeptides (a) and polysaccharides (b) at 20°C vs. the number of layers deposited. Qualitatively similar behavior has been reported for other temperatures tested [20]. The uncoated P(NiPAM-co-MAA) NG is negatively charged in aqueous solution ($\mu_e = -2.7 \times 10^{-8}\, m^2\, V^{-1}s^{-1}$), due to the deprotonation of most of the acidic groups of MAA (pH = 4.4). Upon adsorption of each polycation and polyanion layer, the net surface charge becomes positive and negative, respectively, in agreement with the behavior reported by other authors working with similar biopolymers and different types of substrates [30, 31]. However, in terms of absolute value, the surface charge is not totally reversed upon sequential layering, differing from what is observed for LbL assembly on flat surfaces or rigid particles. This is an indication that interactions between the NG and the LbL films are occurring and may be due to incomplete coverage of the NG surface and/or interpenetration with the PEs. An analogous behavior of charge reversal after each deposition step has been reported for PLL/PGA-coated MGs [18], and demonstrates the feasibility of the LbL confinement of biopolymers onto soft and porous gels. The relatively high values of μ_e obtained are consistent with those reported by Haidar et al. [32] for the assembly of biopolymers onto liposomes, and are a clear indication of the stability of these systems. It is clear from Figure 6.3 that μ_e behavior is qualitatively similar for both kinds of biopolymers, albeit the biggest charge reversal ($+4.3 \times 10^{-8}\, m^2\, V^{-1}s^{-1}$) is achieved when the first

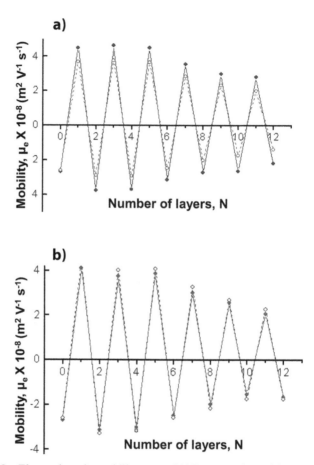

FIGURE 6.3 Electrophoretic mobility μ_e at 20°C vs. number of layers deposited on P(NiPAM-co-MAA) NG: (a) NG/(PLL/PGA) and (b) NG/(CHIT/DEX). Reprinted from Ref. [20], © copyright 2010, with permission from Elsevier.

CHIT layer is deposited compared to PLL ($+3.9 \times 10^{-8}$ m² V⁻¹s⁻¹). This can be explained considering that CHIT presents a larger number of functional groups and higher charge density since it is assembled from strong acid solutions with low ionic strength, and tends to adopt an extended rod-like conformation, resulting in a larger excess of positive charges onto the NG surface. Upon adsorption of the first polyanion layer, the absolute μ_e value is higher for DEX-terminated NG in comparison to PGA-terminated, also related to their different charge density caused by the distinct assembly conditions. As mentioned in the introduction, when these biopolymers are incorporated into the multilayer system, since they are weak PEs, their degree of ionization may change considerably from the solution value due to the influence of the local environment through electrostatic and hydrophobic effects. Thus, when DEX is adsorbed, the presence of abundant oppositely charged cationic groups should increase the degree of ionization of this polyanion in comparison to the solution value, leading to a high surface charge density. In contrast, when PGA is adsorbed, there are fewer cationic groups in the surface, since part of the amine groups of PLL are compensated by salt counterions, hence decreasing the magnitude of the electrostatic effects. Moreover, due to hydrophobic effects, the apparent pK_a of PGA can be increased and that of PLL decreased, which would reduce the degree of ionization of these biopolymers in the multilayer. It has been reported [33] that the pK_a of the carboxylic groups of PGA drops slightly upon incorporation in the multilayer, albeit it shifts to ~4.5 in the presence of salt. Analogously, a pK_a fall from ~10.5 in solution to 9.8 was found for the amine groups of PLL, explained in terms of a proton release from the NH_3^+ due to the interaction with the negatively charged groups of PGA. The aforementioned changes in the degree of ionization of the different groups also influence the charge balance during the LbL assembly.

On the other hand, a systematic decrease in μ_e is found with increasing number of layers deposited onto the NG, irrespective of the type of biopolymer used. This a clear difference with LbL deposition on hard and rigid particles [30], where the magnitude of charge reversal remains almost constant. LbL films are usually considered as ordered non-equilibrium arrangements, and the mobility of the PE chains within the multilayer during and after its construction is well known. The soft and porous nature of the gels allows a certain degree of penetration of the PEs and cooperative interactions that are not feasible for solid and rigid substrates. PEs can diffuse in and out, neutralizing difficult to access charges. Breakage and reformation of ion pairs in all dimensions occur not only between polycations and polyanions but also between polycations and gels; all these facts explain the decrease in μ_e with increasing number of layers. This behavior suggests that during the adsorption of the initial layers, the electrostatic interactions and the intrinsic mechanism of charge compensation are a key factor in the multilayer formation, whereas with increasing number of deposition steps, the secondary interactions become the driving force of the assembly, which is reflected in a decrease in μ_e.

Figure 6.4 compares the temperature dependence of μ_e for the different biopolymer-coated NGs: (a) PLL, (b) PGA, (c) CHIT and (d) DEX [20]. In all cases, the absolute value of μ_e increases with temperature, due to the native thermoresponsive sensitivity of the NG. The gel swelling is entropy driven and depends on the hydrogen bonds between the amide groups of PNiPAM and the water molecules. An increase in temperature breaks the stabilizing H-bonds, hence, the template shrinks, expelling out the charges, thus provoking an increase in the surface charge density. In the collapsed state the NG still retains water inside, and on cooling some trapped immobile segments are able to reorient and reorganize as water flows into the gel. This phenomenon, combined with all the aforementioned effects that cause polymer chains to reorganize, might explain the hysteresis found in μ_e between the heating and the cooling curves.

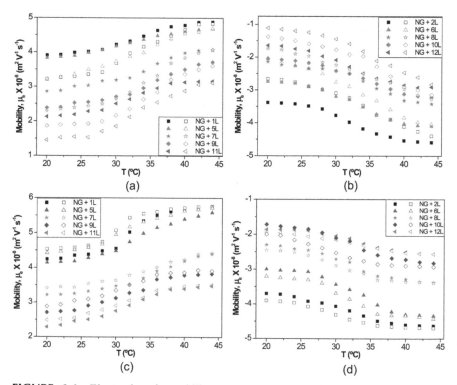

FIGURE 6.4 Electrophoretic mobility μ_e vs. temperature for different biopolymer-terminated NGs: (a) PLL, (b) PGA, (c) CHIT and (d) DEX. Reprinted from Ref. [20], © copyright 2010, with permission from Elsevier.

According to Figure 6.4, this hysteresis is considerably more pronounced for NGs coated with polypeptides, hinting towards a higher degree of internal rearrangement. This behavior is also related to the high ionic strength of the PLL and PGA dipping solutions. As mentioned earlier, salt counter ions occluded within the multilayer swell the ensemble by screening the electrostatic forces that bind the layers together, causing dissociation of individual polycation and polyanion binding sites, thus promoting localized dissociation and increased conformational dynamics. This enhanced mobility of the polymer chains induces strong restructuration of the entanglement of the ensemble, resulting in a remarkable hysteresis. However, for the high M_w polysaccharides assembled from water solutions, less rearranging is allowed, albeit they also experiment small conformational changes between different helical structures induced by the swelling and collapsing of the NG. Comparing identical number of layers and temperature, it is found that polysaccharides exhibit slightly higher μ_e values. For example, at 30°C, μ_e of one and two layer polysaccharide and polypeptide coated microgels are about $+4.5 \times 10^{-8}$, -4.0×10^{-8}, $+4.2 \times 10^{-8}$ and $-3.8 \times 10^{-8}\,m^2\,V^{-1}s^{-1}$, respectively. Polysaccharides, with higher M_w possess longer chains, bearing more functional groups and charges, and tend to adopt a stretched conformation due to the repulsion of charges along the polymer backbone, hence should present more charges at the NG surface. In contrast, polypeptides assembled from solutions with high ionic strength tend to adopt a more coiled conformation due to the screening of the electrostatic forces, exposing fewer charges to the surface. Moreover, the presence of salt decreases the energy barrier between different polymer conformations, and thereby increases the flexibility of the PE chains, enabling them to diffuse more easily and interpenetrate into the template, neutralizing charges otherwise not accessible, thus resulting in lower μ_e values.

6.4.3 HYDRODYNAMIC RADIUS

The thermoresponsive behavior of uncoated and coated NGs can be monitored by DLS. Figure 6.5 compares the evolution of the hydrodynamic radius R_h as a function of temperature for different layers of (a) NG/(PLL/PGA) and (b) NG/(CHIT/DEX) systems [20]. The neat NG in water (pH = 4.4)

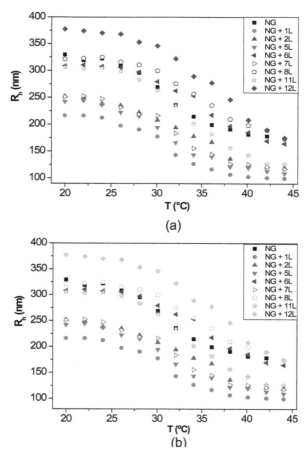

FIGURE 6.5 Hydrodynamic radius R_h vs. temperature for different layers of (a) NG/(PLL/PGA) and (b) NG/(CHIT/DEX). For clarity, only the heating cycles are shown. Reprinted from Ref. [20], © copyright 2010, with permission from Elsevier.

exhibits a R_h of 330 nm at 20°C. During heating, it undergoes a volume phase transition around 32–33°C (VPT) and collapses to ~180 nm; on cooling it recovers the original size. Upon adsorption of the first polycation layer there is a strong decrease in the NG size at 20°C ($R_{h\text{-MG/PLL}}$ ~290 nm; $R_{h\text{-MG/CHIT}}$ ~215 nm.), albeit the coated NGs maintain the thermoresponsive behavior, and on heating they still undergo the transition to a collapsed state. The deposition of the second layer induces an important increase in size ($R_{h\text{-MG/PLL/PGA}}$ ~325 nm; $R_{h\text{-MG/CHIT/DEX}}$ ~255 nm), and the temperature sensitivity of the coated NG is also retained. This "odd-even" effect was corroborated by ESEM images. As pointed out earlier, the decrease in particle size upon adsorption of the first biopolymer layer is considerably stronger for MG/CHIT, where R_h decreases by ~35% to nearly the size of the uncoated NG in the collapsed state, while for MG/PLL the reduction is less significant (~13%). Moreover, the size of one-layer polysaccharide coated NG in the collapsed state is ~100 nm, whereas PLL-terminated system shrinks to ~170 nm. To understand the differences between the behavior of NGs coated with polypeptides and polysaccharides, the PE charge density in the conditions of the assembly and secondary cooperative interactions such as hydrophobic-hydrophobic and hydrogen bonds that contribute to the film formation and stability have to be considered. The polysaccharide displays a high charge density, since it possesses ~85% deacetylated amine groups that are protonated in strong acid solutions (pH = 1.3) with low ionic strength. Furthermore, it presents high M_w and a great number of positively charged functional groups that can shield the negative charges of the NG, causing a rearrangement of the gel in order to adopt a more coiled and compact conformation, resulting in a smaller R_h. In addition, since the NG is porous, PEs and water can diffuse in and out; breakage and reformation of hydrogen bonds take place not only between the PEs and the NG but also with the solvent molecules. Thus, upon deposition of CHIT, strong H-bonds can be formed between its hydroxyl groups and the solvent at the expense of breaking stabilizing interactions between the amide groups of PNiPAM and the water molecules, resulting in a drastic decrease in diameter both in the whole temperature range. Upon approach of the DEX layer, the H-bonds of the confined CHIT are shared between the solvent and the negatively charged biopolymer, weakening their interaction with the solvent. Therefore, some water molecules are rendered mobile again and able to penetrate the NG, causing the ensemble to expand slightly, explaining the fact that NG/(CHIT/DEX) presents the mentioned "odd-even" effect in both swollen and collapsed states.

In contrast, the polypeptides are assembled at neutral pH and in the presence of salt, where the charges of the NG and the PEs are highly screened. Consequently, the electrostatic attraction between the template and the PLL is considerably weaker than the NG-CHIT interaction, and the decrease in size is less drastic. Moreover, secondary cooperative interactions play a very important role in the assembly process. Upon adsorption of the PGA layer, both biopolymers also interact through H-bonds and hydrophobic interactions. To achieve the electrical neutrality required for the multilayer construction, some of the polyanion charges are compensated by salt counterions that get entrapped within the film structure. During the washing steps, the osmotic pressure acts as the driving force for the swelling of the ensemble. The change of the ionic strength leads to a difference in the chemical potential of the ions inside and outside the coated NGs. As a result, an exchange process of ionic pairs occurs between the multilayer and the solution until the chemical potential is equilibrated again, causing an increase in the R_h.

The swelling behavior of coated gels can be characterized through the swelling ratio defined as $(R_h$ swollen$/R_h$ collapsed$)^3$. In the case of P(NiPAM-co-MAA) NGs, this ratio is considerably higher for polysaccharide coated systems (i.e., 8.3 and 13.1 for NG/PLL and NG/(PLL/PGA) vs. 21.6 and 23.3 for MG/CHIT and MG/(CHIT/DEX). This is consistent with the fact that hydrogels containing more hydrophilic groups exhibit higher swelling ratios. The differences between the assembly of polypeptides and polysaccharides are summarized in Figure 6.6, which compares the evolution of the R_h as a function of the number of layers at 20 and 40°C for both types of biopolymers. Clearly, the "odd-even" effect in the R_h is significant for both types of systems in the swollen state, whereas for the collapsed state this effect is only relevant for templates coated with polysaccharides. R_h rises upon increasing layers deposited, which can be interpreted as an increase in the film thickness.

6.4.4 BILAYER THICKNESS

The thickness of bio-polyelectrolyte bilayers can be estimated using AFM according to the method reported by Leporatti et al. [34]. The thickness of dex-HEMA MG coated with 4 bilayers of (DEX/pARG) was estimated from the AFM height profile [21] (Figure 6.7) considering that the measured height is twice the thickness of the whole ensemble. The average thickness obtained was 45 nm, ~11 nm per DEX/pARG bilayer, which is considerably thicker than the values reported for biopolymer multilayers deposited on flat substrates [35].

The bilayer thickness of biopolymer coated NGs has been also estimated from DLS measurements in the swollen state taking the R_h of the NG coated with the first and second layers as reference [20].

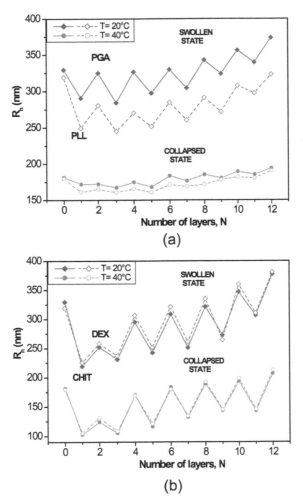

FIGURE 6.6   Hydrodynamic radius R_h vs. number of layers for the swollen (T = 20°C) and collapsed state (T = 40°C) of (a) NG/(PLL/PGA) and (b) NG/(CHIT/DEX). N = 0 corresponds to the neat nanogel. Reprinted from Ref. [20], © copyright 2010, with permission from Elsevier.

Figure 6.8a shows the increment in R_h at 20°C as a function of the number of (PLL/PGA) and (PGA/PLL) bilayers deposited on NG/(PLL/PGA) and NG/PLL, respectively. The thickness of bilayers of CHIT/DEX and DEX/CHIT deposited on NG/(CHIT/DEX) and NG/CHIT are displayed in Figure 6.8b. It is found that PLL- and PGA-terminated films have approximately the same thickness: ~3 and 55 nm for the first and

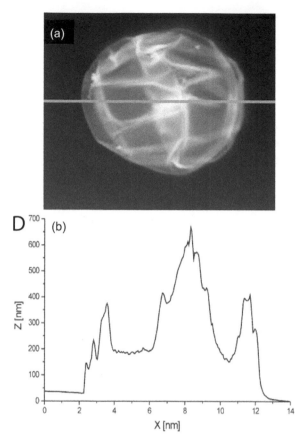

FIGURE 6.7 AFM image of a hollow DEX-HEMA microgel coated with 4 (DEXS/pARG) bilayers. (b) Height profile along the line indicated in (a). Adapted from Ref. [21], with permission from Wiley-VCH.

last bilayers deposited, in agreement with values reported for assembly of these biopolymers onto biological templates [36]. In these systems, values derived from the heating cycles are on average about 16% higher than those obtained from the cooling thermograms, differences probably related to the intermolecular rearrangements that polypeptides undergo when confined on the NG. However, very small differences between data derived from the heating and cooling curves are found for NGs coated with polysaccharides, since their longer chains present a lower degree of freedom for reorganization after adsorption. DEX-terminated bilayers are thicker than CHIT-terminated ones, suggesting that the layer thickness is

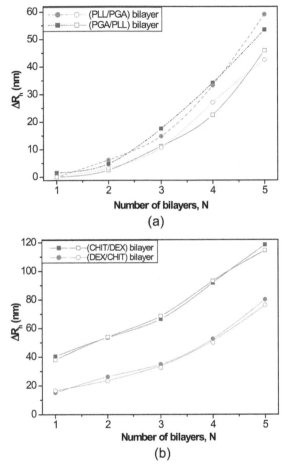

FIGURE 6.8 Bilayer thickness for different polyelectrolyte layers at 20°C: (a) NG/(PLL/PGA) and (b) NG/(CHIT/DEX). The solid and empty symbols correspond to values obtained before and after the heating–cooling cycle is applied, respectively. Reprinted from Ref. [20], © copyright 2010, with permission from Elsevier.

conditioned by the conformation of the biopolymer in the furthest layer. The negatively charged polysaccharide presents a branched backbone and can adopt a more expanded conformation, hence leading to thicker layers, whereas the positively charged one exhibits a linear and flat structure, thereby resulting in thinner layers. These bilayers are thicker than those attained upon assembly of these biopolymers on rigid substrates [35, 37]. The discrepancy can be explained considering the fact that the NG is in an aqueous medium, and the PE layers are fully swollen and bound to a swollen substrate.

Interestingly, polysaccharide bilayers are considerably thicker than polypeptide ones. To understand these differences, it has to be taken into account that the relative layer thickness depends on several factors, mainly charge density, hydrophilicity, M_w, residue size, etc. On the one hand, polysaccharides display higher charge densities, and tend to adopt a stretched conformation to minimize the repulsion between charges, which would lead to thinner layers. On the other hand, they present increased hydrophilicity and consequently adsorb more water, which would result in an increase of the thickness. Furthermore, a rise in the PE M_w and chain length typically causes an increase in the amount of material adsorbed. Overall, it is found that the adsorption of the polysaccharides results in the formation of thicker bilayers.

NGs coated with both types of biopolymers show a non-linear growth of the layer thickness with the number of deposition steps. This type of exponential growth has been previously reported for PLL/PGA [38] and HA (hyaluronic acid)/CHIT [39] multilayer's assembled on solid substrates. Several models have been proposed in the literature to explain the nonlinear build-up mechanism. McAloney et al. [40] suggested that such behavior is caused by the roughness of the film increasing with the number of deposited layers: The macromolecules adsorbed in the initial layers adopt a flat conformation because of the strong interactions with the substrate surface; however, upon increasing layers, the PE chains start to adopt a more loopy conformation, resulting in increased surface area. Another explanation relies on the ability of at least one PE to diffuse "in" and "out" of the whole structure during each bi-layer deposition [35]. When the film is brought in contact with the polycation solution, part of the PE adsorbs on its surface, and a fraction of its chains diffuse within the film. After contact with the polyanion, the remaining free polycation chains diffuse towards the surface and complex with the incoming polyanion chains. These complexes are bound to the film surface and form the new outer layer. Since the amount of free polymer chains that can diffuse out of the film when brought in contact with the PE solution is proportional to the film thickness, the amount of polyanion/polycation complexes formed during this step is also proportional to the film thickness, and the build-up growth becomes exponential. Considering the latter explanation, it would be expected that polypeptides, with lower M_w and shorter chains could more easily diffuse in and out of the multilayer, leading to a more significant increase in the bilayer thickness than polysaccharides. However, fitting of the experimental values reveals that the exponential growth is slightly more pronounced for NGs coated with polysaccharides. Thus, the former model provides a better understanding of the experimental results. Since polysaccharides give rise to thicker films, the interaction with the NG is restricted to the initial layers adsorbed, and the PE would adopt a looser and expanded conformation during the successive dipping cycles, thus deviating strongly from linearity.

6.4.5   TEMPORAL STABILITY

The storage or temporal stability of LbL-coated NGs can be investigated by monitoring the changes in the R_h and in the thermoresponsive behavior vs. time. Studies using fluorescently labeled PLL have demonstrated that there is no desorption with time even after 3 months, demonstrating that biopolymer-coated NGs are very stable [41]. Figure 6.9 shows the DLS curves for various PE terminated systems over different periods of time [20]: (a) NG/PLL (b) NG/(PLL/PGA) (c) NG/CHIT (d) NG/(CHIT/DEX). In the case of multilayers prepared by adsorption of PLL (Figure 6.9a), a significant decrease in R_h with time is found, from 290 to 264, 248, 240, 235 and 229 nm after 3, 6, 10, 18 and 30 days, respectively. However, when the coated NGs are heated, they always collapse to approximately the same size as when they were first coated. Similar trends are found within the R_h of systems where the outermost layer is PGA. This behavior is consistent with the significant decrease in the mean roughness of one-month-old (PLL/PGA) films reported by Pelsöczi et al. [36]. A strong hysteresis is found between the heating and cooling cycles, as well as changes in the VPT values, indicating spatial and temporal reorganization over relatively long time periods. Nevertheless, the differences between the heating and cooling cycles decrease with increasing time (from ~14% to 8% over a period of 1 month), suggesting that the NG-polypeptide interaction eventually stabilizes, and it is expected that an almost reversible steady state would be achieved in an extended period. This indicates that the conformations finally adopted by the polypeptides confined onto the NG,

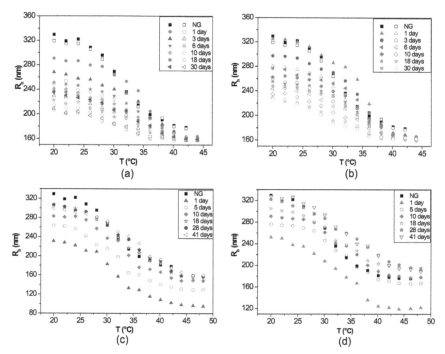

FIGURE 6.9 Temporal stability of LbL-coated nanogels in water over different periods of time: (a) NG/PLL (b) NG/(PLL/PGA) (c) NG/CHIT (d) NG/(CHIT/DEX). Gels coated with polysaccharide layers are reversibly thermoresponsive, and for clarity, only the heating cycles are shown. Reprinted from Ref. [20], © copyright 2010, with permission from Elsevier.

induced by its thermoresponsive behavior, are reversible. Polysaccharide-coated NGs exhibit totally opposite behavior: no hysteresis is found and R_h increases progressively with time to almost the original size of the uncoated template. Thus, for CHIT- terminated NG (Figure 6.9c), it rises by ~15, 23, 30, 33 and 34% after 3, 6, 10, 18 and 30 days, respectively. The NG-polysaccharide interaction seems to stabilize these systems, and after a period of ~20 days acquire similar size as when first coated. When PEs are adsorbed onto the NG and subjected to a heating and cooling treatment, they interpenetrate with the gel and undergo reorganization as water flows in and out of the ensemble. When left standing over time, the system can extend into the aqueous phase due to entropic reasons (i.e., relaxation of loose ends group or collective sliding motion of the NG and loops of the PEs). Considering that polysaccharide-terminated surfaces are more hydrophilic, they absorb more water, which would diffuse through the whole film, leading to a higher degree of disentanglement of the PE chains and the NG network, enabling the entire ensemble to adopt a more expanded conformation.

6.4.6 ATR-FTIR SPECTRA

Costa et al. [18] employed ATR-FTIR spectroscopy to analyze the interactions between biopolymers and P(NiPAM-co-MAA) MGs (Figure 6.10). The spectra are difficult to analyze since the characteristic PE bands overlap with the amide vibrations. Moreover, the exact band positions for the polypeptides depend on their conformation (α-helix, β-sheet, or random coil), due to vibrational interactions between the peptide groups and the

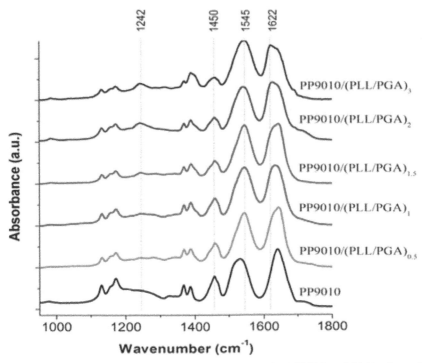

FIGURE 6.10 ATR FT-IR spectra of polypeptide-coated P(NiPAM-co-MAA) microgels. Adapted from Ref. [18], © copyright 2012, with permission from the American Chemical Society.

different forms of hydrogen bonding. The spectra of neat MG is dominated by two strong bands at, 1641 and, 1531 cm^{-1}, corresponding to the amide I and amide II vibrations of the PNiPAM segments, and a weak amide III band that appears at, 1242 cm^{-1}. Upon adsorption of the polypeptides, the amide II band became more intense, and the shoulder corresponding to the C=O stretching of unprotonated carboxylic groups of PMAA at, 1707 cm^{-1} decreases in intensity, indicating that PE layering occurs primarily through ionic interactions. Upon adsorption of the PLL layer, the amide bands shifted significantly to higher energy states, closer to the characteristic vibrations of crosslinked PNiPAM homopolymers (δ_{NH} at, 1543 cm^{-1}, intramolecular H-bonded $v_{C=O}$ at, 1644 cm^{-1} and free $v_{C=O}$ at, 1669 cm^{-1}). The subsequent adsorption of PGA led to a significant shift of the intramolecular H-bonded $v_{C=O}$ (1632 cm^{-1}) and free $v_{C=O}$ (1653 cm^{-1}) back to the lower wavenumbers characteristic of the native NG, whereas the intensity of the free $v_{C=O}$ of PMAA COOH group maintained low, indicating that the PNiPAM segments were strongly involved in H-bonding. The deposition of additional PE layers did not change the spectra significantly. Furthermore, in the high wavenumber region the sequential deposition of PLL and PGA led to a relative increase in intensity and broadening of the band in the N–H stretching region between, 3000 and 3450 cm^{-1} relative to the isopropyl and alkyl vibrations (2850–3000 cm^{-1}), due mainly to the contribution of PLL v_{N-H} and PGA v_{O-H}.

6.4.7 CONFOCAL MICROSCOPY AND FLUORESCENCE CORRELATION SPECTROSCOPY

Further indication of the successful LbL assembly of biopolymers onto gels can be attained by confocal microscopy. Figure 6.11a shows a typical confocal microscopy image of DEX-HEMA MGs coated with four bilayers of (DEX/pARG) [21]. The MGs were fluorescently labeled with FITC-DEX (green color), whereas the pARG was labeled with rhodamine B isothiocyanate (RITC, red color). A clear ring of red fluorescence originating from the labeled pARG surrounding the MGs was observed, indicative of a successful LbL coating. After exposure of the coated MGs to a 0.1 M phosphate buffer (pH 7.4) at 37°C for 5 days the DEX–HEMA microgel core was degraded (Figure 6.11b) and hollow capsules were

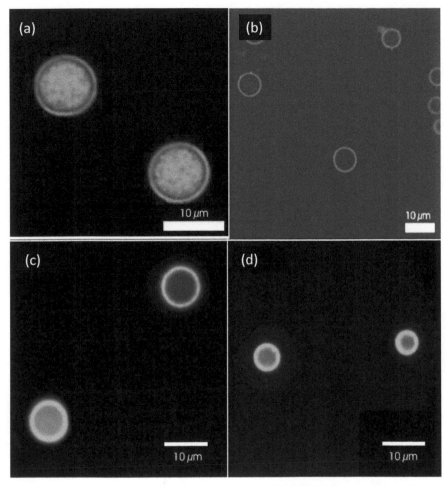

FIGURE 6.11 Top: Confocal microscopy images of DEX-HEMA microgels with four bio-polyelectrolyte bilayers before (a) and after (b) degradation of the MG. Adapted from Ref. [21], with permission from Wiley-VCH. Bottom: Confocal microscopy images of polypeptide P(NiPAM-co-MAA) microgels. Adapted from Ref. [18], © copyright 2012, with permission from the American Chemical Society.

obtained; the microcapsules did not rupture, but released the FITC–DEX as no significant fluorescence was detected within the microcapsules. The authors [21] also reported the influence of the PE molecular weight on the behavior of the core–shell ensemble. Systems with low M_w PGA all ruptured, while those with high M_w PGA did not all explode. This indicates that the increase in molecular weight of PGA changes the interplay between the mechanical properties and the permeability of the bio-poly-electrolyte coating.

Costa et al. [18] also characterized P(NiPAM-co-MAA) coated MGs by confocal microscopy at 24 and 37°C (Figures 6.11c and 6.11d, respectively). The fluorescent labeling shows that these polyions diffuse into and distribute across the MG, with higher fluorescence intensity concentrated on the MG shell, related with the higher concentration of carboxylic acid groups in that area. Another explanation for this phenomenon could be a self-limiting adsorption process that occurs as the polycation adsorbs onto the MG surface even as it diffuses into the gel, thus creating an intrinsic barrier to the diffusion of additional polymer into the gel. The presence of labeled polymer within the gel implies that PEs with higher intrinsic mobility may present more significant "odd-even" effects based on size and responsive behavior, as increased mobility provides a greater access to the acid groups throughout the MG. A progressive increase in fluorescence was observed upon each polycation deposition step, in an apparently linear regime within the bilayer number.

Regarding NGs, it is very difficult to visualize the exact location of the LbL coating using fluorescently labeled species. FCS has been used to demonstrate successful assembly of biopolymers on PNiPAM NGs [41]. Figure 6.12 shows the 2f-FCS curves of the uncoated NG labeled with Rhodamine at (a) 25°C and (b) 40°C. Using the particle size effect model, an R_h of 276 and 174 nm were calculated for the swollen and collapsed

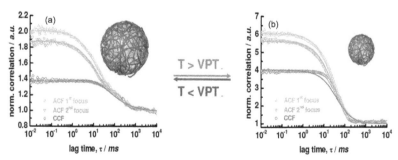

FIGURE 6.12   2f-FCS curves of the uncoated Rhodamine labeled NG detected separately at two different foci (ACF1 and ACF2) and cross-correlated (CCF) at (a) 25°C and (b) 40°C. Adapted from Ref. [41], © copyright 2009, with permission from the American Chemical Society.

state, respectively. These values are in good agreement with the R_h obtained by DLS, which were 265 and 168 nm. When a fluorescently labeled bio-polymer PLL[FITC] was incorporated into the multilayer system, single or auto-correlated experiments performed on each species at different layering stages yield similar curves [41], revealing diffusion times of the same order of magnitude, which confirms the anchoring of the labeled PE to the NG. If the species were moving independently, the cross-correlation curve would have shown only a base-line. Images of the dried PNiPAM/PLL[FITC] reveal indeed that the fluorescence originates from exactly the same species, corroborating that the PLL was anchored to the gel.

6.5 CONCLUDING REMARKS AND FUTURE OUTLOOK

This chapter focuses on the preparation and characterization of biopoly-mer-coated soft and porous gels via the layer-by-layer technique, analyz-ing the influence of the polyelectrolyte nature and the assembly conditions on the behavior of the coated templates. For polysaccharide-terminated gels, the swelling and contraction processes are almost reversible; how-ever, templates coated with polypeptides exhibit strong hysteresis due to temperature triggered conformational changes induced by confinement onto the gel. Further, an "odd–even" effect on the size of the ensembles occurs depending on the type of polyelectrolytes in the outermost layer. Literature data demonstrate that with the proper choice of the bio-poly-electrolytes, the morphology and thermoresponsivity of the coated gels can be finely tuned and their temporal stability improved, making them highly suitable for a wide range of biomedical applications, such as, bio-sensors, encapsulation (storage) and delivery of small molecules, dyes or therapeutics. In particular, the ability of the coated ensembles to release substances in a controlled range of pHs and temperature may serve as basis for the development of new means of in vivo delivery of proteins, enzymes or vaccines. Future challenges include the use of biodegradable gels and shells such as biopolymer-coated DEX-HEMA microgels that are biocompatible, the efficient loading and release of the drug, the incorpo-ration of functionalities to attain site-specifically target delivery and the addition of inorganic nanoparticles to develop hybrid core-shell structures with improved optical, magnetic and electronic properties.

ACKNOWLEDGMENTS

PS gratefully acknowledges the Ministerio de Ciencia e Innovación for the concession of a Juan de la Cierva (JCI-2011–10836). AD wishes to acknowledge the Ministerio de Economía y Competitividad (MINECO) for a "Ramón y Cajal" Senior Research Fellowship co-financed by the EU.

KEYWORDS

- bilayer thickness
- layer-by-layer
- polypeptides
- polysaccharides
- temporal stability
- thermoresponsive gels

REFERENCES

1. Decher, G., Schmitt, J. Fine-tuning of the film thickness of ultrathin multilayer films composed of consecutively alternating layers of anionic and cationic polyelectrolytes. Prog. Colloid Polym. Sci. 1992, 89, 160–164.
2. Decher, G. Fuzzy nanoassemblies: Toward layered polymeric multicomposites. Science, 1997, 277, 1232–1237.
3. Yoo, D., Shiratori, S. S., Rubner, M. F. Controlling bilayer composition and surface wettability of sequentially adsorbed multilayers of weak polyelectrolytes. Macromolecules, 1998, 31, 4309–4318.
4. Shiratori, S. S., Rubner, M. F. pH-dependent thickness behavior of sequentially adsorbed layers of weak polyelectrolytes. Macromolecules, 2000, 33, 4213–4219.
5. Dubas, S. T., Schlenoff, J. B. Factors controlling the growth of polyelectrolyte multilayers. Macromolecules, 1999, 32, 8153–8160.
6. Donath, E., Sukhorukov, G. B., Caruso, F., Davis, S. A., Möhwald, H. Angew. Chem. Int. Ed. 1998, 37, 2202–2205.
7. Caruso, F., Caruso, R. A., Möhwald, H. Nanoengineering of inorganic and hybrid hollow spheres by colloidal templating. Science, 1998, 282, 1111–1114.
8. Caruso, F., Lichtenfeld, H., Donath, E., Möhwald, H. Investigation of electrostatic interactions in polyelectrolyte multilayer films: Binding of anionic fluorescent probes to layers assembled onto colloids. Macromolecules, 1999, 32, 2317–2328.

9. Wang, Y., Yu, A., Caruso, F. Nanoporous polyelectrolyte spheres prepared by sequential coating sacrificial mesoporous silica spheres. Angew Chem Int Ed, 2005, 44, 2888–2892.

10. Sukhorukov, G. B., Volodkin, D. V., Günther, A. M., Petrov, A. I., Shenoy, D. B., Möhwald, H. Porous calcium carbonate microparticles as templates for encapsulation of bioactive compounds. J. Mater. Chem. 2004, 14, 2073–2081.

11. Nayak, S., Lyon, L. A. Soft nanotechnology with soft nanoparticles. Angew. Chem. Int. Ed. 2005, 44, 7686–7708.

12. Calvo, P., Remunan-Lopez, C., Vila-Jato, J. L., Alonso, M. J. Novel hydrophilic chitosan-polyethylene oxide particles as protein carrier. J. Appl. Poly. Sci. 1997, 63, 125–132.

13. Nolan, C. M., Serpe, M. J., Lyon, L. A. Pulsatile release of insulin from layer-by-layer assembled microgel thin films. Macromol. Symp. 2005, 227, 285–294.

14. Shenoy, D. B., Sukhorukov, G. B. Microgel-based engineered nanostructures and their applicability with template-directed layer-by-layer polyelectrolyte assembly in protein encapsulation. Macromol. Biosci. 2005, 5, 451–458.

15. Wong, J. E., Diez-Pascual, A. M., Richtering, W. Layer-by-Layer assembly of polyelectrolyte multilayers on thermoresponsive P(NiPAM-co-MAA) microgel: Effect of ionic strength and molecular weight. Macromolecules, 2009, 42, *1229–1238.*.

16. Fery, A., Schöler, B., Cassagneau, T., Caruso, F. Nanoporous thin films formed by salt-induced structural changes in multilayers of poly(acrylic acid) and poly(allylamine). Langmuir, 2001, 17, 3779–3783.

17. McAloney, R. A., Goh, M. C. In situ investigations of polyelectrolyte film formation by second harmonic generation. J. Phys. Chem. B 1999, 103, 10729–10732..

18. Costa, E., Lloyd, M.M., Chopko, C., Aguiar-Ricardo, A., Hammond. P. T. Tuning smart microgel swelling and responsive behavior through strong and weak polyelectrolyte pair assembly. Langmuir, 2012, 28, 10082−10090.

19. Choi, J., Rubner, M. F. Influence of the degree of ionization on weak polyelectrolyte multilayer assembly. Macromolecules, 2005, 38, 116–124.

20. Diez-Pascual, A. M., Wong, J. E. Effect of layer-by-layer confinement of polypeptides and polysaccharides onto thermoresponsive microgels: A comparative study. J. Colloid Interf. Sci. 2010, 347, 79–89.

21. De Geest, B. G., Déjugnat, C., Prevot, M., Sukhorukov, G. B., Demeester, J., De Smedt, S. C. Self-rupturing and hollow microcapsules prepared from bio-polyelectrolyte coated microgels. Adv. Funct. Mater. 2007, 17, 531–537.

22. Saunders, B. R. On the structure of poly(N-isopropylacrylamide) microgel particles. Langmuir, 2004, 20, 3925–3932.

23. Lyklema, J. Fundamentals of Interface and Colloid Science, Academic Press: San Diego, 1995; Vol. 2, Solid-Liquid Interfaces, p. 3.208.

24. Von Smoluchowski, M. Contribution to the theory of electro-osmosis and related phenomena. Bull. Int. Acad. Sci. Cracovie, 1903, 184–199.

25. Koppel, D. E. Analysis of macromolecular polydispersity in intensity correlation spectroscopy: The method of cumulants. J. Chem. Phys. 1972, 57, 4814–4821.

26. Einstein, A., Fürth, R. Investigations on the Theory of the Brownian Movement; Dover: New York, 1956.

27. Madge, D., Webb, W. W., Elson, E. Thermodynamic fluctuations in a reacting system: Measurement by fluorescence correlation spectroscopy. Phys. Rev. Lett. 1972, 29, 705–708.

28. Dertinger, T., Pacheco, V., von der Hocht, I., Hartmann, R., Gregor, I., Enderlein, J. Two-focus fluorescence correlation spectroscopy: A new tool for accurate and absolute diffusion measurements" Chem. Phys. Chem, 2007, 8, 433–443.

29. Picart, C. Polyelectrolyte Multilayer Films: from physico-chemical properties to the control of cellular processes. Curr. Med. Chem. 2008, 15, 685–69.

30. Zhi, Z. L., Haynie, D. T. Direct evidence of controlled structure reorganization in a nano-organized polypeptide multilayer thin film. Macromolecules, 2004, 37, 8668–8675.

31. Snowden, M. J., Chowdhry, B. Z., Vincent, B., Morris, G.E. Colloidal copolymer microgels of N-isopropylacrylamide and acrylic acid: pH, ionic strength and temperature effects. J. Chem. Soc., Faraday Trans. 1996, 92, 5013–5016.

32. Haidar, Z. S., Hamdy, R. C., Tabrizian, M. Protein Release Kinetics for Core-shell Hybrid Nanoparticles Based on the Layer-by-layer Assembly of Alginate and Chitosan on Liposomes. Biomaterials, 2008, 29, 1207–1215.

33. Gergely, C., Bahi, S., Szalontai, B., Flores, H., Schaaf, P., Voegel, J.-C., Cuisinier, F. J. C. Human serum albumin self-assembly on weak polyelectrolyte multilayer films structurally modified by pH changes. Langmuir, 2004, 20, 5575–5582.

34. Leporatti, S., Voigt, A., Mitlohner, R., Sukhorukov, G., Donath, E., Mohwald, H. Scanning force microscopy investigation of polyelectrolyte nano- and microcapsule wall texture. Langmuir, 2000, 16, 4059–4063.

35. Lavalle, P., Gergely, C., Cuisinier, F. J. C., Decher, G., Schaaf, P., Voegel, J.-C., Picart, C. Comparison of the structure of polyelectrolyte multilayer films exhibiting a linear and an exponential growth regime: an in situ atomic force microscopy study. Macromolecules, 2002, 35, 4458–4465.

36. Pelsöczi, I., Turzo, K., Gergely, C., Fazekas, A., Dékány, I., Cuisinier, F. J. C. Structural Characterization of Self-Assembled Polypeptide Films on Titanium and Glass Surfaces by Atomic Force Microscopy. Biomacromolecules, 2005, 6, 3345–3350.

37. Yuan, W., Dong, H., Li, C. M., Cui, X., Yu, L., Lu, Z., Zhou, Q. pH-Controlled construction of chitosan/alginate multilayer film: Characterization and application for antibody immobilization. Langmuir, 2007, 23, 13046–13052..

38. Lavalle, P., Picart, C., Mutterer, J., Gergely, C., Reiss, H., Voegel, J. C. Modeling the buildup of polyelectrolyte multilayer films having exponential growth. J. Phys. Chem. B 2004, 108, 635–648.

39. Richert, L., Lavalle, P., Payan, E., Zheng, X. S., Prestwich, G. D., Stoltz, J. F., Schaaf, P., Voegel, J.-C., Picart, C. Layer by Layer Buildup of Polysaccharide Films: Physical Chemistry and Cellular Adhesion Aspects. Langmuir, 2004, 20, 448–458.

40. McAloney, R. A., Sinyor, M., Dudnik, V., Goh, C. Atomic Force Microscopy Studies of Salt Effects on Polyelectrolyte Multilayer Film Morphology. Langmuir, 2001, 17, 6655–6663.

41. Wong, J. E., Muller, C. B., Diez-Pascual, A. M., Richtering, W. J. Study of layer-by-layer films on thermoresponsive nanogels using temperature-controlled dual-focus fluorescence correlation spectroscopy. J. Phys. Chem. B 2009, 113, 15907–15913.

CHAPTER 7

USE OF RECYCLED POLYMER IN THE CONSTRUCTION INDUSTRY

DAVI NOGUEIRA DA SILVA,[1] ADALENA KENNEDY VIEIRA,[2] and RAIMUNDO KENNEDY VIEIRA[3]

Faculdade de Tecnologia, Universidade Federal do Amazonas, Av. General. Rodrigo O. J. Ramos, 3000, CEP 69077–000, Manaus – AM, Brazil; [1]E-mail: davi.nogueira@am.senai.br; [2]E-mail: adalenakennedy@gmail.com; [3]E-mail: maneiro01@ig.com.br

CONTENTS

ABSTRACT

The benefits of innovation are already a determinant factor these days. There is no denying that the survival of competitiveness in industry much depends on the innovations. The growth of the concept of sustainability coupled to the development of environmentally friendly products has fostered growing interest of the society and companies on new technologies, new processes and new services. With the growing market, the plastics have taken over the modern world. Products recycled polymer combines the concept of an environmentally friendly product, adding comfort and innovative design. In fact, it becomes necessary since the construction industry is responsible for major environmental impacts before and after project completion. The case study conducted in a School of Professional Education showed consistency in applying the use of recycled polymer in the construction industry. The aim of this chapter turns to a discussion of recycling of polymers and their use in the manufacture of innovative products for the construction industry.

7.1   INTRODUCTION

The growth within the construction industry in Manaus and in all Brazil has shown that this niche is one of the more jobs offers and employs one of the most promising in the field of innovation. According to the Brazilian Chamber of Construction Industry [8], this segment has seen year after year increased consistently, reaching record highs since, 2010.

The recycled polymer products combine the concept of an environmentally friendly product, adding comfort and innovative design that can be incorporated in various ways in the industry, especially in the construction industry.

The development and subsequent use of these products are evidence that this market is potentially strong when it comes to innovation, being partly an exception compared with other segments. In this regard, [7] clarifies that less than 10% of new businesses across the country bring to market a truly unique product, which is not offered by the competition.

The need for innovation becomes necessary since the construction industry is responsible for major environmental impacts. For [13], the construction industry is accountable for consuming 66% of all timber, generates about approximately 40% of urban waste and releases huge amounts of dust.

Industries generally are comprehensive in various industries with specific needs, but the concept of environmentally friendly impact minimization becomes unique and products like recycled polymer is proof of that.

The objective of this work is the study of what is possible for the construction industry to innovate in products through the use of recycled polymer. In this regard, [20] reported that the innovation and knowledge were important factors for the companies' development and competitiveness.

7.2 STRONG CONSTRUCTION INDUSTRY

The construction industry is one of the strongest sectors of Brazilian economy. The Annual Survey of Construction Industry [15] shows that companies in the construction sector performed works in the amount of US$ 112.5 billion; it is a real increase of 23.3% over the previous year. Notwithstanding this growth, the economic crisis has affected the global economy, the Figure 7.1 below shows yearly growth activity within the construction industry.

Despite continuous data growth that comes from the year, 2007, the construction industry has serious environmental problems that cause negative impacts. The construction is an extremely pollutant and with a final product that consumes enormous resources. According to John (2002), these data probably make the construction industry to human activity with the greatest impact of the environment.

The construction industry consumes about 210 million tons per annum of natural aggregates only for the production of concrete, mortar and bricks. The amount of natural resources used by the construction industry

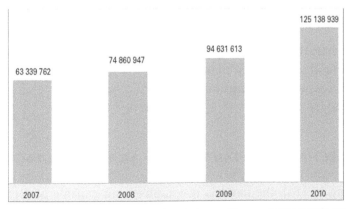

FIGURE 7.1 Annual survey of construction industry (Source: IBGE (2007–2010)).

are approximately one-third of the total consumed annually by the whole society. (John, 2000).

According to [23], the construction industry has the challenge of reconciling their productive activity with alternatives that lead to sustainable development. It is precisely at this point that strategic thinking is stimulating innovation as a differential for the industry. The recycled polymer is presented as one more technological innovation in construction sites with stimulus benefits to the environment, reduced material consumption and increased social satisfaction. According to [9], a positive point to be initially adopted is the use of recyclable materials.

The use of recycled polymers in construction can merge the concept with the correct environmental innovation. Technological innovation is considered by Moreira and Rodrigues (2002) as the main driver of productivity growth, raising the capacity to act in the global competition and conquering new markets and consumers.

7.3   METHODOLOGY

To reach the goal of sustainability, this chapter presents two examples of innovative products made from recycled polymers and intended exclusively for the construction industry. The first product is a recycled polymer block made of PET bottles, intended to replace the conventional bricks. The second product is also made of the base polymer recycled PET; however, it is a blanket lining the walls, designed to replace the glass tiles.

For the development of products, following the methodology manufacturing technique in which the recycled polymer must pass, namely:

(a) technical details of pattern manufacturing;
(b) definition of the raw material;
(c) process of plastic injection;
(d) analyze viability.

7.3.1   TECHNICAL DETAILS OF THE MOLD MANUFACTURING

According to [14], the mold is a complete unit capable of reproducing geometric shapes and dimensions of the desired product. Figure 7.2 shows

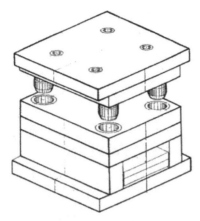

FIGURE 7.2   Example of plastic injection mold (Source: Adapted by Alencar (2002)).

the drawing of a model of an injection mold, especially the upper plate assembly and lower plate assembly.

The mold fabrication process starts with the selection of materials. According to [24], this aspect allows the mold to keep its thermal, mechanical and metallurgical preserved, thus ensuring a smooth operation. The choice of material should be taken into account, since according to [11] the mold confection shall be alloy steel supporting efforts and high temperatures, with coat to acquire a better surface finish of the product, increasing the wear resistance. According to [18] steel is the only material that provides these properties mentioned above, thereby generating reliability in the operation of the mold and long useful lives.

The next step of manufacturing injection molding and machining process, [5] defines machining as the feature that is most important, because it depends on factors such as dimensional tolerances, surface roughness and degree of finish. The product quality will depend integrally from the quality of the machining process. In machining, machine had been using universal turning machines, universal milling machine, flat grinding, cylindrical grinding and CNC machining center. For [19], the molds suffer great efforts during the machining process, affecting the quality of the product. For this reason, it is necessary a final process to finish the molds surface.

To reduce surface imperfections and minimize the degree of irregularities from the machining process should be used bat crystal fibers set for polishing brush, mounted rotating polishing, set tip mounted felt for polishing molds, set cane felt square for polishing molds, support for manual polishing bat, chuck felt disc for polishing disc felt for polishing molds and equipment for ultrasonic polishing.

Figure 7.3 shows the mold for the manufacture of recycled polymer block, made from, 1045 hardened steel, machined surface finished and controlled.

With respect to the product derived from that mold, plates are able to reproduce the form of a block with geometric design characteristics of a brick in the construction.

7.3.2 SELECTION OF RAW MATERIAL

The stage setting for the raw material to be recycled it's not just a mere selection process. Environmental, technological, economic, market and social factors should be part of planning for choosing the best polymer to be recycled. To Barbiere (2004), how the industry is positioned in environmental issues can directly influence the production, planning, competition and sales.

According to [17], the polymers are made from a petrochemical process, giving quality as lightweight, transparency and resistance.

The polymers are present in various sectors, activities, and in our day-to-day. The Brazilian Association of Technical Standards [3] establishes the identification of polymers that can be recycled, such as polypropylene (PP), poly(ethylene terephthalate) (PET), polyethylene (PE), poly (vinyl chloride) (PVC).

Among all polymers available for recycling, PET stands out as the main polymer industry, its high applicability, versatility, and quality jobs.

FIGURE 7.3   Injection mold and recycled polymer block (Source: Senai (2013)).

7.3.2.1   Using PET and Recycled Polymer

PET is a thermoplastic polymer, which is produced on the most strength plastic with high mechanical, thermal, and chemical resistance. PET is made from resin (polymer) synthetic, derived from oil, and its chemical composition is highly resistant to biodegradation. Plastics have significant volume and its separation brings many benefits, because the packaging its can be reused [16].

According to Figure 7.4, the polyethylene (PE) polymer still remains the most commonly found in the urban environment. However, recent data collected in major Brazilian cities show that PET ranks second as municipal waste, with approximately 20% of its use, highlighting and enabling continuous production of new products from recycled polymer.

Furthermore, with respect to Figure 7.4, [25] states that confronting this large consumer market, the PET recycling industry is on the threshold of a new phase, where the post-consumer PET can be applied in various ways, further expanding the consumer market of this material.

With regard to versatility, [12] explains that PET has a huge advantage because it can be reprocessed or recycled several times by the same transformation process. In addition, with respect at this point, [12], states that the recycled polymer emerges as a differentiation because PET is a polymer that can be used in a wide range of applications and have the advantage of being reusable 100%.

On environmental concern, Romao (2005) pointed out that the PET recycling is a viable alternative to minimize the environmental impact caused by the disposal of these materials into sanitary landfills.

On demand issues, [10] states that the production of polymers in the world in 1999 was 168 million tons, with estimated forecast at 210 million tons in 2010.

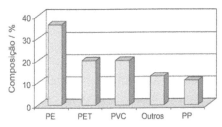

FIGURE 7.4   Polymers mostly found in the urban environment (Source: Aparecida's Adaption (2004)).

Another important point about how the choice of PET polymer to be recycled to create new products for the construction industry has to do with the residual growth of its use in the domestic market. According to the data of economics and statistics of the Brazilian Association of Chemical Industry [2] PET increased by 34.2% in domestic production in the last half of, 2013.

Other historical factors are important, according to [22], PET is the main polymer studied and developed for packaging carbonated beverages. and statistics of the Brazilian Association of Chemical Industry [2] PET increased by 34.2% in domestic production in the last half of, 2013.

7.3.3 PLASTIC INJECTION PROCESS

[21] stresses that the process of plastic injection is a polymer molding technique consisting of breaking into the molten compound into the cavity of a mold. After cooling part, it is extracted and a new molding cycle occurs. This entire process is held in a machine called plastic injection.

According to the Brazilian Association of Technical Standards [4], the machine is used for discontinuous manufacturing injection molded products, consisting essentially of a clamping unit, injection unit and plastic drive system.

During the process, the recycled polymer block, the plastic injector receives the material in the solid state under the form of granules or powder and transports it in pre-established quantities in the mold interior, as in Figure 7.5 highlights:

The plastic injection-manufacturing segment is the most recommended because it allows a high accuracy of shape and dimensional accuracy. This process ensures that the product has a great look, finish and texture.

According to [26], injection molding is a cyclic process of transformation of thermoplastic and includes the following events:

(a) transportation of thermoplastic material;
(b) heating and plasticization of material;
(c) homogenization of the plasticized material;
(d) injection of plasticized material into the cavity of the mold;
(e) cooling and solidification of the material into the cavity;
(f) ejection of the molded part.

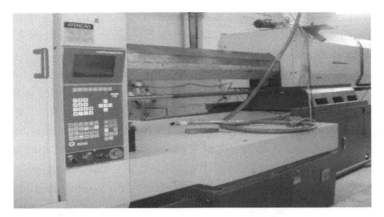

FIGURE 7.5   Plastic injection machine (Source: Senai (2013)).

Additionally, according to the author, the process of injection molding processes is one of the most versatile plastics processing for the processing of polymers, and it is applied in the production of various products.

7.4   RESULTS

The use of recycled polymer becomes an option for waste management in the construction industry. Figure 7.6 shows an example of job where the remains of the PET polymers were used for making plates for covering inner walls.

The plates' production for covering inner walls followed the methodology already mentioned, from mold manufacturing to fixation test with grout and mortar.

FIGURE 7.6   Recycled polymer plates for construction industry (Source: Senai (2013)).

Different colors and shades were achieved extending the applicability for the entire field of construction as homes, schools, businesses, buildings, condominiums, churches, etc.

The esthetic, the designer and the possibility of other forms of decoration are evidence of its viability, opening up a range of utilities within the construction industry.

This process is of extreme relevance for the construction industry because it meets the Article 10 of Resolution 307/2002 of the National Council of Environment (CONAMA), which says that construction waste should be reused or recycled as aggregates being willing to allow their future use or recycling. Figure 7.7 shows another result of success where the PET recycled polymers where it was used for preparation of blocks to replace the conventional bricks.

It was performed the complete assembly of a house with two rooms, measuring 5 x 5 m^2. It was tested the overall alignment of the walls, thermal expansion after several days in the sun, sealing fittings in each block and material strength. Still according to Figure 7.7, tests performed for fixing porcelain for bathrooms and all part of plumbing and electrical.

The definition of the raw material was in partnership with a construction company, where it provided a venue for the separation and collection of PET bottles, as shown in Figure 7.8.

The machine called Pet Pope made the collection, bottles labels separation, lids separation and grinding of raw materials, making a difference in the company. While the mold was machined, Pope Pet machine remained at the construction site of the company storing all the raw material.

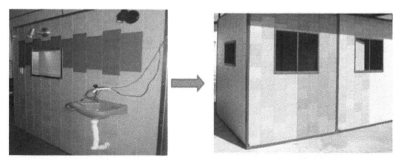

FIGURE 7.7   House made using recycled polymer (Source: Senai (2013)).

FIGURE 7.8   PET bottles collecting machines (Source: Senai (2013)).

The use of recycled polymers in the construction industry satisfies Resolution 307/2002 of National Council of Environment (CONAMA) Article I, which establish guidelines, criteria and procedures for the management of construction waste, regulating the actions needed in order to minimize environmental impacts.

The use of recycled polymers in the construction industry presents another huge benefit is the ability to enjoy the encouragement of an innovative product and increase competitiveness, with a correct environmental position and reduced costs.

7.5   CONCLUSION

The high technological development, coupled with the growing demand for new environmentally friendly products, has produced a remarkable level of search, discovery and use of plastic materials, enabling a wide range of applications, making it a product modern, affordable and increasingly important.

Recycled polymer products are a simple and viable, but needs more economic studies that can foster a whole chain of entrepreneurship, since there are still many questions about the quality of these products and their use.

Reliability is also another factor that needs to be evolved. More technical studies aimed at engineering construction need to be developed.

Testing to ensure reliability needs to be performed where conventional materials may be progressively replaced by polymeric materials.

A transforming society where the recycled polymers are used in a new product again demonstrates the great use of recycled polymer industry and its importance to this.

KEYWORDS

- **construction**
- **environmental**
- **innovation**
- **polymer**
- **recycling**

REFERENCES

1. Aparecida, Márcia. The recycling polymers technology. 1° edição, Campinas SP, 2004.
2. Brazilian Association of Chemical Industry. ABIQUIM. Apparent consumption of thermoplastics increases in the first quarter of, 2013. Disponível em: home Page: http://www.abiquim.org.br/comissao/setorial/resinas-termoplasticas-coplast/sobre-a-comissao. Acesso em: 20 de setembro de, 2013.
3. Brazilian Association of Technical Standards. NBR 13230: Symbology indicative of recyclability and identification of plastic materials. Rio de Janeiro (2008).
4. Brazilian Association of Technical Standards. NBR 13536: injection molding machines for plastics and elastomeric. Technical safety requirements for the design, construction and use. Rio de Janeiro (2008).
5. Batista, Marcelo Ferreira. Roughness study of flat surfaces machined for cherries with spherical top. São Paulo: 2006. Dissertation (Master in Manufacturing) – School of Engineering of São Carlos, University of São Paulo, São Carlos, 2006.
6. BRAZIL. 2002. Resolution CONAMA. 307 – Establishes guidelines, criteria and procedures for the management of construction waste. Official Gazette of the Federative Republic of Brazil from July 17, 2002.
7. Bomfim, Ronaldo. Lack of innovation. Column Follow-up business. Center of Industry of the State of Amazonas (CIEAM). Manaus AM, 2013.
8. Chamber of Brazilian of Construction Industry. Construction Industry. Available in: Home Page: http://www.cbic.org.br/. Access August 20, 2013.

9. Ceotto, Luiz Henrique. The Construction and the Environment. Ed 51–53, 2008, São Paulo SP. Available at: homepage: http://www.sindusconsp.com.br/seções. asp?subcateg=74&categ=16. Accessed on: September 2, 2013.

10. Ferreira, Caio, T., Fonseca, Juliana, B., Saron, Clodoaldo. Recycling of waste poly (ethylene terephthalate) (PET) and polyamide (PA) by means of reactive extrusion for the preparation of the blends. Polymers. 2011, vol.21, n.2, pp. 118–122. ISSN 0104–1428.

11. Ferreira, Nathaniel de Abreu. Processing system for plastic injection. FATEC. In 2012. 1st edition, Sorocaba, São Paul.

12. Goncalves – Days, Sylmara Lopes Francelino and Teodosio, Armindo de Sousa dos Santos. Structure of the reverse chain "paths" and "detours" of PET packaging. Prod. [Online]. 2006, vol.16, n.3, pp. 429–441. ISSN 0103–6513.

13. Hansen, Sandro. Environmental Management: Environment in Construction. 1st edition, Senai / SC, Florianópolis SC, 2008.

14. Herto, Santana de Alencar. Molds Injection. Educational Society of Santa Catarina. 1st edition, Paraná, 2004.

15. Brazilian Institute of Geography and Statistics. IBGE – Annual Survey of Construction Industry (PAIC). 20th edition, IBGE, Rio de Janeiro, 2010.

16. Luiza, Mary; Vilhena, André. Municipal waste: integrated management manual. 2nd edition, IPT / CEMPRE, São Paulo, 2000.

17. Mano, Eloisa; Mendes, Luis Claudio. Introduction to polymers. São Paulo, publisher Edgard bulcher, 2nd ed. 2004.

18. Menges, Georg; Michaeli, Walter; Mohren, Paul. How to make injection molds. 3rd Edition, Hanser. Munich, 2000.

19. Smith, C. E., Rodrigues, F. M. M. Industry and Technological Issues. Ministry of Science and Technology, FINEP, CNI, Brasília – DF, 2002, available at: Home Page: http://www.cni.org.br. Access May 18, 2013.

20. Muccioli, Cristina. Scientific research, innovation and development. Arq Bras. Ophthalmolo. 2007, vol.70, n.3, pp. 383–383. ISSN 0004–2749.

21. Nunes, Joseph Injection Molding. São Paulo, 2008. Available in: Home Page: http://www.carnavale.com.br.html. Access May 16, 2013.

22. NUNES, Edilene de Cassia; Agnelli, José Augusto, Roberto Rossi. Poly (ethylene naphthalate) – PEN: a review of its history and the main trends of its worldwide application. Polymers. 1998 vol.8, n.2, pp. 55–67. ISSN 0104–1428.

23. Pinto, T. P. Environmental management of construction waste: the experience of Sinduscon – SP. 1st edition, Sinduscon – SP, São Paulo, 2005.

24. Pouzada, A. S. Molds for production of polymer parts. Manual designer molds for plastic injection. Volume 1. Marina Grande. Centimfe, 2003.

25. Romao, Wandeson Spinacé, Marcia; PAOLI, Marco. PET: A revision of the processes of synthesis, mechanism of degradation and recycling. Polymers. 2009, vol. 19, n.2, p. 121–132. ISSN 0104–1428.

26. Torres, Jocelito. Computer simulation of the injection process in a virtual environment. Rio Grande do Sul; school Senai professional education Nile Bettanin, 2007. Vol 1.

CHAPTER 8

POLYURONATES AND THEIR APPLICATION IN DRUG DELIVERY AND COSMETICS

ROBIN AUGUSTINE,[1] BHAVANA VENUGOPAL, S. SNIGDHA,[1] NANDAKUMAR KALARIKKAL,[1,2,4] and SABU THOMAS[1,3]

[1]International and Interuniversity Centre for Nanoscience and Nanotechnology, [2]School of Pure and Applied Physics, [3]School of Chemical Sciences, Mahatma Gandhi University, Priyadarshini Hills P.O., Kottayam – 686 560, Kerala, India

[4]Department of Biotechnology, St. Joseph's College, Irinjalakuda, Thrissur – 680 121, Kerala, India; E-mail: Sabu Thomas (sabupolymer@yahoo.com), Nandakumar Kalarikkal (nkkalarikkal@mgu.ac.in)

CONTENTS

ABSTRACT

Polyuronates or polyuronides are polysaccharides containing one or more uronic acid moieties in their molecular structures. Pectins and alginates are the most commonly used polyuronides in everyday applications. Their ability to form highly crosslinked gels in the presence of bivalent or multivalent cations makes them suitable for many applications such as drug delivery and cosmetics. In drug delivery, these substances are used as controlled drug delivery systems. Capability to tune the crosslinking density by varying the crosslinking agents, concentration of crosslinking agents, crosslinking time and pH, make these substances ideal candidates for the controlled delivery of therapeutic agents. The gel forming ability makes them suitable for wide array of cosmetic applications ranging from moisturizing cream to dermal fillers.

8.1 INTRODUCTION

Polyuronates are polysaccharides containing one or more uronic acid residues in their molecular structures. It includes all pectic materials and many plant mucilages, hemicelluloses and some microbial polysaccharides. Most of them are water soluble and all of them are hydrolyzed by hot dilute solutions of strong acids to reducing sugars or aldobionic or free uronic acids. In the polyuronide macromolecule, sugars and uronic acids are joined by glycosidic linkages to form complex acids. The most widely used polyuronides are the pectins and alginates. Alginates are a family of naturally occurring polysaccharides extracted from brown seaweed used by the pharmaceutical industry for specific gelling, thickening, and stabilizing applications. Pectin is a natural, non-starch, linear, heterogeneous polysaccharide that is commercially extracted from citrus fruit peels and apple pomace.

8.2 STRUCTURE OF POLYURONATES

It is generally known that polyuronides (alginates and pectic substances) are natural ion exchangers of outstanding properties. Alginate occurs in the cell wall and intracellular spaces of brown algae [62]. It is an unbranched binary copolymer of 1,4-linked D-mannuronate (ManA) and L-guluronate (GulA) arranged in a blockwise fashion [28]. Pectic substances are present as cell wall constituents in higher plants and consist predominantly of linearly 1,4-linked D-galacturonate (GalA) and its methyl ester. Typical samples of pectin also contain small amounts of neutral sugars [19].

8.2.1 STRUCTURE OF PECTIN

Pectin is a natural, nonstarchy, linear, heterogeneous polysaccharide, comprising a variety of sub-structural elements that varies with the extraction methodology, raw material, location, and other environmental factors [36]. For industrial applications it is extensively extracted from citrus fruit peels and apple pomace, which are then treated with hot dilute mineral acids at a pH of 2 [75]. Pectin mainly consists of D-galacturonic acid (GalA) units [52], joined in chains by means of α (1–4) glycosidic linkages [75, 50] with alternating side chains of α (1–4) D-galactose and D- arabinose [36, 51]. The carboxylate groups of the uronic acids are often present as methyl and carboxamide esters, which are formed as a part of commercial treatment. Thus pectin is widely represented as the methyl ester of polygalacturonic acid [75].

The ratio of esterified galacturonic acid groups to the total galacturonic acid groups is termed as the Degree of Esterification (DE). Pectin with an esterification degree of more than 50% is regarded as High methoxy (HM) pectin and with less than 50% is termed as low methoxy pectin [34]. Neutral sugars, such as rhamnose, are also present as a minor component of the pectin backbone, which creates a kink in the main chain [75, 51]. The presence of several such entities in the structural backbone of pectin, creates two structurally distinct regions within it, viz., Homogalacturonan and Rhamnogalacturonan regions which are also known as "smooth" and "hairy" regions respectively [51, 50, 42]. The chemical structure of pectin is given in Figure 8.1.

8.2.2 STRUCTURE OF ALGINATE

Alginate is a natural unbranched polymer obtained from the marine brown seaweed and also from some soil bacteria as capsular polysaccharide

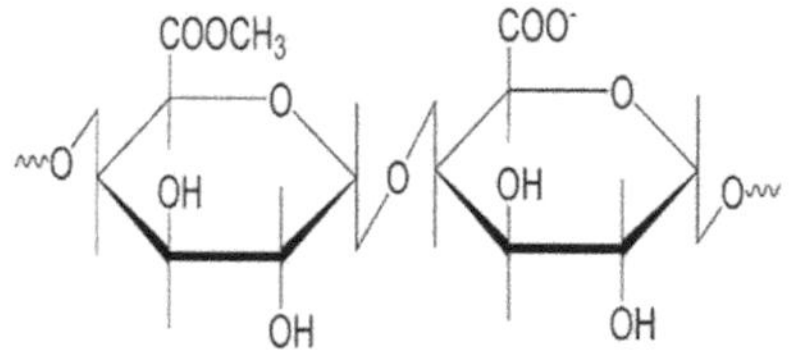

FIGURE 8.1   Structure of pectin.

[33, 10, 25, 12]. The major commercial sources are species of Ascophyllum, Durvillaea, Ecklonia, Laminaria, Lessonia, Macrocystis, Sargassum and Turbinaria. Among these, the most important are Laminaria, Macrocystis and Ascophyllum. Alginate is extracted from the dried and milled algal material after treating with dilute mineral acid to remove or degrade any associated neutral homopolysaccharides such as laminarin and fucoidin. The alginate is then converted from the insoluble protonated form to the soluble sodium salt by the addition of sodium carbonate at a pH below 10. After extraction, the alginate is further purified and converted either to a salt or acid form. Structurally, it is composed of linear copolymers of 1,4-linked D-mannuronic acid (M) and L-guluronic acid (G). Alginate is considered as a true block copolymer because of its homopolymeric regions of M and G block. The M blocks and G blocks are interspersed with alternating areas of M-G blocks which differs in a distribution pattern among various sources [78, 16, 12] for instance, alginates isolated from *L. hyperboea* kelp have a high number of α-L-guluronic acid residues, against the alginates isolated from *A. nodosum* and *L. japonica* [10, 25]. The distribution of monomers along the polymer chain is random and therefore alginates do not have a repeating unit [10]. The carboxylic groups in alginate are capable of forming salt formations such as sodium alginate, with the sodium monovalent ions [12]. The chemical structure of alginate is given in Figure 8.2.

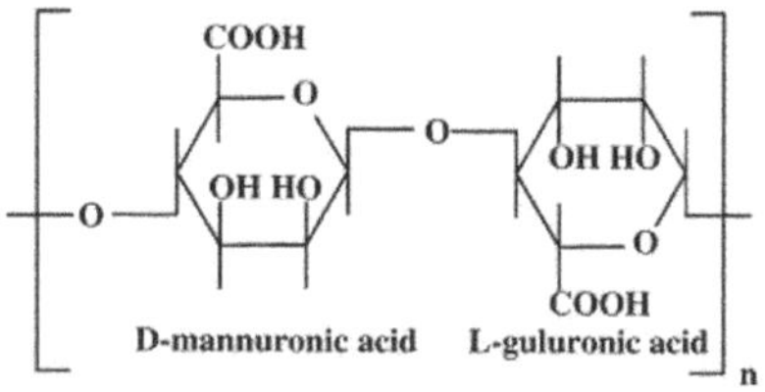

FIGURE 8.2   Structure of alginate.

8.3   PROPERTIES OF POLYURONATES

8.3.1   PROPERTIES OF PECTIN

It has been broadly reported that the solubility of pectin is attributed to the valency of the salt forming cations. Monovalent cationic salts of pectin are highly soluble in water whereas di- or tri-valent cationic salts of pectin are weakly soluble in water. Dilute solutions of pectin show Newtonian behavior but at moderate concentrations they exhibit Non-Newtonian behavior. Further, pectin tends to show a pseudo plastic behavior, which could be related to its concentration in a solution [49]. The viscosity of the pectin on the other hand, is influenced by the molecular weight, degree of esterification, concentration of the preparation and the pH.

8.3.2   PROPERTIES OF ALGINATE

When the monovalent alginate is dissolved in distilled water, form smooth solutions having long flow properties. The solution properties are dependent on both physical and chemical variables. The molecular weight of alginate is found to be in a range $4.6–37 \times 10^4$. They are good at retaining moisture tenaciously. Alginates of alkali metal such as sodium are soluble in water. Divalent metal alginates are insoluble except for magnesium alginate. The flow properties of alginate are concentration dependent such that 2.5% solution of alginate is considered as pseudo plastic with high shear rate whereas a solution by 0.5% has a Newtonian character with a low shear rate. The physical variables, which affect the flow characteristics of monovalent alginate solutions include temperature, shear rate, macromolecular size, concentration in solution and the presence of miscible solvents.

The chemical variables affecting alginate solutions include pH, sequestrants, monovalent salts, polyvalent cations, and quaternary ammonium compounds.

8.4   CROSSLINKING OF POLYURONATES

Polyuronates are anionic polyelectrolyte, which serve as a polymeric coat in drug encapsulation or colloidal delivery system by associating with oppositely charged surfaces [51]. The unique gel forming ability of poly-uronates in the presence of calcium ions makes them ideal for drug delivery [49]. The divalent calcium ions and the negatively charged carboxylate groups of the polyuronates forms intermolecular crosslinks resulting in an "egg-box" model (See Figure 8.3) of gel systems [63]. The chemistry and gel-forming characteristics of polyuronates, have enabled this naturally occurring polymer to be used in pharmaceutical industry, health promotion and treatment. It has also been used potentially in pharmaceutical preparation and drug formulation as a carrier of a wide variety of biologically active agents, not only for sustained release applications but also as a carrier for targeting drugs to the colon for either local treatment or systemic action [76, 43, 74].

8.4.1 PECTIN GELATION

High methoxy (HM) pectins gel in acidic medium in addition of a large quantity of sucrose (>50%), while low methoxy (LM) pectins gel in the presence of divalent cations, such as Ca2+. This ability has been largely exploited to prepare pectinate gel beads by ionotropic gelation of pectin droplets in calcium or zinc solutions [22]. Thus calcium pectinate gel is generally interpreted as an ionic bond between the carboxylate groups of the pectin and the calcium ions. The resulting structure can be represented by the so-called 'eggbox' model [63]. However, high amounts of Ca^{2+} is found to destabilize the solutions, leading to the formation of increasingly heterogeneous thermal gels or precipitation. The use of zinc as the counter ion can lead to stronger pectin beads than the use of a calcium counter ion

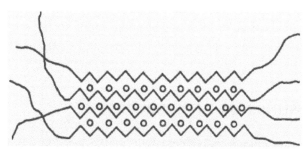

FIGURE 8.3 Egg-box model of polyuronate gelation (the circles represent bivalent or multivalent cations).

due to a higher binding ability with a higher affinity and a higher pecti-nate gel strength [75]. Gel formation and the bioactivity of encapsulated molecules are affected by many parameters such as pectin type, nature and amount of divalent cation, and drying processes [14]. The mechanism of gelation is via the creation of junction zones by the galacturonan region of the pectin with specific sequences of the GalA monomers lying in par-allel or nearby to it is linked electrostatically, or via ionic binding to the carboxyl groups. The interaction of ions and the carboxylate groups in pectin involves intermolecular chelate binding of the cation leading to the formation of macromolecular aggregates (Figures 8.4 and 8.5). Calcium ions take up the interstitial spaces. The gel structure resembles a net like formation of crosslinked pectin molecules.

Pectin is also found to form aggregates of macromolecule at acidic pH whereas at neutral pH it tends to dissociate and shows an expanded net-work. Rate of gelation is also influenced by the DE. When DE is reduced via the treatment with commercial pectin methyl esterase an increase in viscosity and firmer gelling was found in the presence of calcium ions [49, 77]. Furthermore, amidation process increases the gelation ability of LM pectin and the amidated pectin requires less calcium to gel [8, 75, 6, 76].

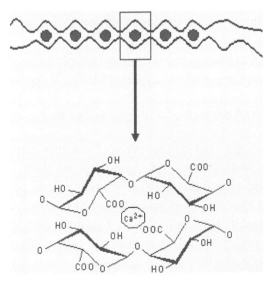

FIGURE 8.4 Coordinate bond formation of bivalent cation with polygalacturonate chains in pectin.

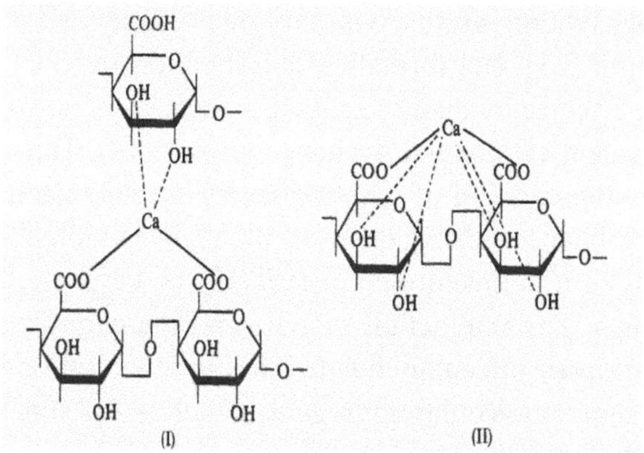

FIGURE 8.5   Chelate bond, (I) Intermolecular, (II) Intermolecular in pectin molecule.

8.4.2 ALGINATE GELATION

Alginate and its derivatives are also able to form gels in the presence of divalent cations such as calcium (Ca^{2+}), through ionic interaction between these cations and the carboxyl groups located on the polymer backbone [12]. The gelation or crosslinking results in the stacking of the guluronic acid blocks of alginate chains [37]. The crosslinks are formed when solution of sodium alginate and the desired substance is extruded as droplets into a divalent solution to encourage crosslinking of the polymers.

This reticulation process consists of the simple replacement of sodium ions with calcium ions, which is expressed in the following reaction:

$$2Na\ (Alginate) + Ca^{2+} \rightarrow Ca\ (Alginate)_2 + 2Na^{+}$$

Binding of calcium to α-L-guluronic acid residues forming dimerizing junctions with other chains and producing insoluble polymeric networks. In biological systems the Ca2+ ion is lost to phosphate, resulting in the breakage of all preformed crosslinks. However, this problem could be prevented by the modification of alginate with long alkyl chains. The functional and physical properties of cation crosslinked alginate beads are controlled by the composition, sequential structure, and molecular size of the polymers.

8.5 POLYURONATES IN DRUG DELIVERY

Drug delivery is the process of administering a pharmaceutical compound to achieve a therapeutic effect in humans or animals. The polyuronates and their low molecular weight fragments are of special interest in human medicine as prophylactic substances and as drugs against intoxication by radioactive strontium and heavy metals. As early as, 1825 Braconnot suggested that pectic substances might be good antidotes for heavy metal poisoning because of the insolubility of the compounds formed [38, 56]. Many scientific institutes, among which those in Canada, the Soviet Union, Great Britain, the USA and Yugoslavia should be mentioned, have successfully investigated in detail the clinical application of alginates and pectic substances.

The goal of drug delivery is to furnish the prescribed drug concentrations to specific sites ensuring a definite drug release profile for a specified period of time [59]. Natural polymers like polyuronates provide greatest advantages in drug delivery since they are biocompatible, non-toxic and biodegradable [33, 12]. The natural polymers, namely, pectin, alginate, hyaluronate, chitosan are used in the form of hydrogels or hydrophilic matrices which is a 3D crosslinked polymeric material and has the ability to absorb and retain water in large amounts from the surroundings [49, 68]. The polyuronates share many similarities with the native tissues or ECM in their mechanical, structural and biological properties, which can be applied in conductive matrices, cellular or biomolecular delivery vehicles or space filling agents. The soft and rubbery nature of the polyuronates helps in minimizing the irritation in the surrounding tissues. The polyuronates also have the advantage of excellent tissue compatibility, easy manipulation and solute permeability that helps in drug delivery [40].

Recently, a great deal of research activity has been undertaken in the development of stimulus-responsive polymeric hydrogels. These hydrogels are designed in such a way that they respond to external or internal stimuli and which can be monitored via changes in the physical nature of the network. Among these systems, pH or temperature responsive hydrogels have been most widely studied in the biomedical field because of the impact of these factors in the in vivo and in vitro application. Tremendous effort has been taken to associate polyuronates with thermo-sensitive macromolecules in an attempt to prepare matrices that present a dual and independent sensitivity to both pH and temperature [13, 87, 15]. Hydrogel materials like polyuronates are often selected since they present synthesis conditions, which are advantageous for maintenance of the native protein conformation can be customized to provide noncovalent functional groups, and it is possible to vary the hydrogel porosity to optimize the cavity formation and diffusion of the macromolecules through the gel. The work of Zhang et al. (2006) was the first attempt at macromolecular imprinting using calcium alginate- based microcapsules *via* an inverse suspension method using hydroxyethyl cellulose to improve the mechanical properties of the microspheres.

8.5.1 PECTIN IN DRUG DELIVERY

Pectin, renowned as the 'miracle polymer, ' has a long and safe history in medical field [49]. It offers several desirable properties of stability under acidic conditions and at higher temperature that is ideal for a drug delivery system. Therefore pectin recently gathered attention as a carrier system and several pectin formulations (hydrogels, films, microspheres and nanoparticles) have been tried so far targeting various proteins and drugs. Unfortunately, pectin is impaired by drawbacks such as, low mechanical strength; low drug loading efficiency and less shear stability. To meet such challenges, pectins are many a time modified chemically and physically to improve its physicochemical properties. As a result binary polymer blends have been developed which were the composite films of natural polysaccharides and synthetic polymers.

Srivastava et al. (2011) carried out a pioneer study in the applicability of orange peel derived pectin as binding agent for the dosage formulation whereas mango peel derived pectin was found to act as a super integrating agent in dosage formulation. Wong et al. (2002) has explored the ability of LM-citrus pectin in making microspheres via emulsification technique. The potential value of HM-pectin in the drug delivery has also come under the study by Sungthongjeen et al. (1999).

Polysaccharide/protein microspheres have been employed in oral drug delivery since long time. These drug carriers from naturally occurring polymers are widely accepted by consumers. A series of microspheres were developed from pectin and corn proteins in the presence of the divalent ions calcium or zinc that showed a high efficiency of drug incorporation and also yield of the microsphere. Since pectin offers protection against the proteolytic attack of the polypeptide drug in the small intestine and from the drastic conditions in the stomach, it is widely suggested as an oral drug delivery (ODD) system.

8.5.1.1   Pectin in Vaginal Drug Delivery

Vaginal route is a major route of drug delivery for both systemic and local diseases. It offers some advantages such as self-insertion, high permeability to drugs, and avoidance of first pass effect. Traditional drug preparations in the form of creams, foams, gels and so forth have short resident time because of the self-cleaning action of vaginal tract. Therefore, mucoadhesive and bioadhesive drug delivery system is extremely appropriate for the retention of the drug in the treatment of local diseases, sexually transmitted diseases and for contraception [81, 18]. The high mucoadhesive strength makes them ideal for the vaginal drug delivery system [3]. It is also an effective site for the delivery of uterine targeted drugs such as terbutaline, progesterone and danazol. A study by Rohan et al. (2009) also indicates the potency of the vaginal drug delivery system in AIDS prevention.

8.5.1.2   Pectin in Anti-Cancer Drug Delivery

Colon cancer is the fourth most fatal cancer. Chemotherapy via injection route is the usual way of reducing tumor growth and thus preventing metastasis. Several studies have been undertaken to find out a possible route of administration of therapeutic drug and have come up with a colon specific drug delivery system via oral administration. Since pectin is only selectively degraded by the colonic micro flora, it can protect the drug degradation from the upper GI tract, thus enhancing maximum bioavailability and lower dosage requirement [84].

Localized anticancer drug delivery and the homotypic cell aggregation is a major issue to confront for the researchers now. Hence a specialized in-situ gellable drug delivery system was put to thought by Takei et al. (2010) based on the periodate oxidized citrus pectin where the drug Doxorubicin (Dox) was coupled by an imine bond. It was then established that the Dox causes the anticancer activity but at the same time the oxidized citrus pectin was found to prevent metastasis.

An investigation has been undertaken by Ashford et al., to assess, in vitro, the potential of several pectin formulations as colonic drug delivery systems. Their findings suggest that either a high methoxy pectin formulation or low methoxy pectin with a carefully controlled amount of calcium should maximize colonic specificity by providing optimal protection of a drug during its passage to the colon and a high susceptibility to enzymatic attack [4].

8.5.1.3   Pectin in Nasal Drug Delivery

The ability of pectin to form weak gels in the presence of calcium ions which are present in the nasal secretions [30] and also their texture makes them patient friendly for nasal drug delivery (NDD) [17, 85].

As an exemplification, a novel system called PecSys (PS), which is a pectin based drug delivery system, is developed to gel on mucosal surfaces when applied. These systems are currently focused in their application in intranasal delivery of drugs where it plays the role of optimizing the absorption of lipophilic drugs into the circulation. The PS based systems were commonly used for the intranasal formulations constituting opioid analgesics for the rapid pain relief. The lead product that uses the PS system is NasalFent, which is a fentanyl nasal spray formulation [82].

8.5.1.4   Pectin in Rheumatoid Arthritis Drug Delivery

Owing to the several advances being made with pectin based colon-specific drug delivery system, Eudragit coated pectin microspheres loaded with aceclofenac for the treatment of rheumatoid arthritis has been developed [65]. The delivery system provided the maintenance of therapeutic concentration of the drug with a complementing anti-inflammatory effect whole day.

8.5.1.5 Pectin in Dental Enamel Protection

The increasing trend in the consumption of acidic soft drinks has led to the dissolution and softening of dental enamel, which is a phenomenon known as erosion. Recently a study has revealed that by the addition of food–approved polymers such as highly esterified pectin (1% w/w) to the citric acid solutions with a pH of typical soft drinks, the effect of citric acid on human dental enamel can be reduced [9]. Similarly, interaction between the pectin coated liposome and dental enamel were also studied to find out the ability of the pectin-coated liposome to mimic the natural protective biofilm on the tooth surfaces. There were no aggregation tendencies for the pectin coated liposome and parotid saliva, which made them a promising device for the dental drug delivery. This ability of pectin-coated liposome to retain on the enamel surfaces also enhances their function as protective structures of the teeth [58].

8.5.1.6 Pectin in Tissue Regeneration

Tissue engineering scaffolds are often used for the delivery of drugs, growth factors and therapeutically useful cells. An ideal scaffold system should have properties that include biocompatibility, suitable microstructure, desired mechanical strength and degradation rate and most importantly the ability to support cell residence and allow retention of metabolic functions. Various natural and synthetic biomaterials have been considered as cell supporting matrices. Several natural polymers have been identified which can mimic the extracellular matrix (ECM), which can promote regeneration in tissues such as bone tissue; soft tissue etc. pectin is one such natural polymer which has the strange ability to mimic as an artificial ECM. Liu et al developed such a composite film, which includes Pectin/ Poly lactide-co-glycolide (PGLA) for the delivery of bioactive compounds in tissue regeneration [44]. Modified pectin with RGD-peptide already found application in bone tissue engineering [54] and similarly pectin/ chitosan microcapsules have been reported for their ability to act as scaffolds for soft tissue regeneration like muscle and cartilage regeneration for the treatment of degenerative pathologies [53].

8.5.1.7 Pectin in Anti-Emetic Drug Delivery

The extent of the role of pectin based drug delivery system also encompasses the treatment of the after effects of chemotherapy such as vomiting and nausea. Ondansteron is one of the anti-emetic drugs which are used for preventing nausea and vomiting caused by chemotherapy. The main problem associated with this drug is that it undergoes first pass effect about 40% in the liver itself. Therefore oral administration is not an advisable route of drug delivery of such drugs. Overall, the ondansteron loaded pectin microspheres were found to be suitable for the effective intranasal delivery of the drug [46].

8.5.1.8 Pectin in Gene Therapy

Pectin was also used as a DNA carrier for gene delivery applications [35]. The rich galactose residues of pectin acted as a potential ligand for the interaction between membrane receptors. DNA being anionic, its complex formation with the anionic pectin is made possible by the modification of pectin with amine groups (cationic). One such modified pectin, Pectin-NH_2– Q (Q=N + $(CH_3)_3$), has proved to be an efficient DNA carrier.

8.5.1.9 Pectin in Miscellaneous Drug Delivery

The emulsifying ability of pectin was exploited for the development of pectin nanoemulsions containing Itraconazole, an anti-fungal agent [11]. The nanoemulsions were prepared by simple homogenization method with chloroform acting as the internal phase. The study also brought to light that a good degree of emulsification is offered by highly esterified pectin. A pectin polymer based transdermal drug delivery system was also developed using meloxicam as a model drug [40]. Meloxicam is a non-steroidal drug with an anti-inflammatory property was found to be compatible with pectin. The transdermal films were then used to achieve controlled release and improved bioavailability of meloxicam. In the same manner, the gelling ability of amidated pectin has also been investigated for their site-specific delivery to the colon as a multiparticulate system.

Several anti-inflammatory drugs were likely incorporated into such bead formulations of pectin providing satisfactory results of drug release [55].

The pectin hydrogels are in addition adopted as a matrix for controlled drug delivery, soft contact lenses, protein separation and matrices for cell encapsulation.

8.5.2 ALGINATE DRUG DELIVERY

In the past three decades, the interest of researchers in experimenting alginate formulations in the field of drug delivery system has been at its heights because of its low cost, pH sensitivity, wide applicability as biomaterial and biocompatibility [33]. Microspheres of alginates have been used for the encapsulation of many bioactive components such as cells, proteins, micronutrients, antibodies etc. Most of the studies undertaken were based on the alginate beads, which are prepared by the gelling of sodium alginate in the presence of calcium ions [21]. Alginates impart a protective role to these bioactive agents by shrinking at the low pH in the stomach and thereby safeguarding the drug from the harsh gastric environment, whereas in the high pH of the intestinal tract, alginate hydrogel disintegrates slowly facilitating the controlled release of the drug. Although, the alginate hydrogels are excellent drug carriers because of its low entrapment efficiency, rapid release and low loading efficiency of water soluble drugs due to high porosity which result in leakage of the drug from the bead. Consequently, to solve these problems, usually it is incorporated with pectin, chitosan, PVP and clay to improve the drug-loading, encapsulation efficiency and release characteristics [33, 72, 47].

For instance, smetic group of clay, often called as Montmorillonite clay, is the common ingredient in pharmaceutical products whose positively charged surfaces interact with the anionic alginate resulting in a unique polymeric material which has exceptional capability to trap drug molecules [29]. Similarly, a combination of alginate and poly (N-acryloylglycine) could lead to a smart drug delivery system with superior pH sensitivity [20]. Micro and nanoparticles were likewise obtained from the alginate by the induction of gelatin with calcium ions. Such gel particles have increased usefulness in the treatment of esophageal reflux treatment and wound healing. Being a natural disintegrator and hemostat, calcium alginate offered an attractive alternative for a sustained release system and wound healing respectively [48]. In another study, the strength of calcium alginate gels to be used as matrices of electrodes for iontophoresis (IOP) was established which a breakthrough in drug delivery systems became. It was found that there was an increase in the order of release at 5 V and 60 min after AC application compared to the usual way of simple passive diffusion [57, 26]. The future much awaits to use alginate based drug systems in the treatment of vascular diseases. It also seems promising as an embolic material for endovascular procedures and also expected to be used in the entrapment of VEGF (Vascular Endothelial Growth Factor) molecule in beads for the angiogenesis therapy [27].

Alginate has also been extensively investigated as a drug delivery device in which the rate of drug release can be varied by varying the drug polymer interaction as well as by chemically immobilizing the drug to the polymer backbone using the reactive carboxylate groups. The encapsulation of proteins and bioactive factors within ionically crosslinked alginate gels are known to greatly enhance their efficiency and targetability and as a result, extensive investigation has been undertaken to develop protein delivery systems based on alginate gels [66]. Recently our group has shown that the applicability of calcium alginate for the immobilization of silver nanoparticles on absorbable surgical gut suture to prevent surgical wound infections and to enhance wound healing. Figure 8.6 shows the ability of the fabricated suture to successfully inhibit the growth of bacteria, *Staphylococcus aureus* and *Escherichia coli.*

Alginate-Carbopol 940 bead formulations were tried by López-Cacho et al. (2012) and shown that the drug release rate from the beads was mainly affected by the Carbopol concentration and the stirring rate during the bead formation process. A scanning Electron Microscopic image of Alginate-Carbopol 940 Bead formulations is shown in Figure 8.7.

8.5.2.1 Alginate in Oral Drug Delivery

Oral drug delivery systems provide a safe route for the drugs to be directly released into the gastrointestinal tract. Nano chitosan blended alginate matrix (CBAM) by grafting with poly methacrylic acid (PMAA) recently being developed for the oral drug delivery. Such a system showed

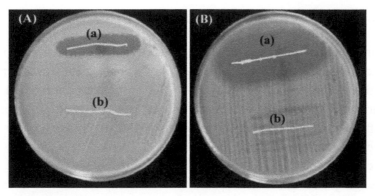

FIGURE 8.6 Plates showing the in vitro antimicrobial activity of the alginate/silver nanoparticle coated suture. Plate (A): On *S. aureus* culture, (a) Ag nano-alginate coated suture, (b) control without coating. Plate (B): On *E. coli* culture, (a) Ag nano-alginate coated suture, (b) control without coating (Augustine et al., 2012).

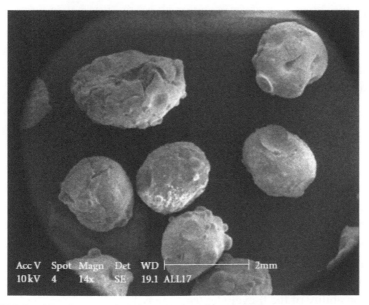

FIGURE 8.7   SEM images of Alginate-Carbopol beads [45].

increased bioadhesivity provided by the chitosan in the matrix and so is studied for the oral delivery of insulin since it protected insulin from the diverse environments in the GI tract [47].

8.5.2.2   Alginate in Anti-Tuberculosis Drug Delivery

The treatment of tuberculosis offers several vexing problem like patient non- compliance, which may ultimately bring about treatment failure and also the rise of multidrug resistance. Alginate gathered attraction from researchers owing to the possession of qualities of a good drug delivery system to co-encapsulate multiple anti-tuberculosis drugs and also complemented controlled release [2, 1].

Reducing the dosage frequency of anti-tuberculosis drugs are quite challenging and this will improve patient compliance. The therapeutic potential of drugs, developed an alginate-based nanoparticulates as a frontline drug delivery system for the anti-tuberculosis drugs such as rifampicin, isoniazid, pyrazinamide and ethambutol was able to achieve the reduced dosage frequency [2]. In a similar way, alginate based micro particulate system is also developed and evaluated [71]. They studied isoniazid, rifampicin and pyrazinamide containing alginate microparticles in guinea pigs. Alginate microparticles containing antitubercular drugs were evaluated for *in vitro* and *in vivo* release profiles. These microparticles exhibited sustained release of isoniazid, rifampicin and pyrazinamide for 3–5 days in plasma and up to 9 days in organs.

8.5.2.3   Alginate in Anti-Emetic Drug Delivery

Antiemetic drugs are used in the treatment of nausea and emesis. Macroscopic mechanical force has been employed for the modification of molecular interactions and molecular reactivity. A designer hydrogel with a blend of properties of alginate and β-cyclodextrin derivative has now been reported in antiemetic drug delivery. Moreover, this is intended for the release of Ondansteron, an anti-emetic drug, controlled by the mild compressions, mimicking the patient's hand. These compressions were the change in the inclusion ability of the β-cyclodextrin moiety [32].

In another study, sodium alginate-xanthan gum based transdermal drug delivery system (TDDS) was used for domperidine delivery to treat nausea and vomiting. The polymer membranes were prepared using xanthan gum (XG) and sodium alginate (SA) by varying the blend compositions. This Drug delivery system has shown a highly effective controlled release and improved bioavailability in in vitro studies [64].

8.5.2.4 Alginate in Anti-Cancer Drug Delivery

The prolonged use of alginate-based hydrogels was never limited now a day, being considerably applied in cancer therapies. Alginate micro particle formulation for the controlled release of the drugs for the eye-cancer treatment has been reported [7]. Cyclophosphane and 5-flurouracil were encapsulated in this system and was revealed as a promising drug delivery system.

The systemic side effects of chemotherapy are a continuing problem in cancer treatment. But it is still preferred after resection to prevent tumor growth in many forms of cancer. In order to lessen the patients suffering from chemotherapy and to reduce the drug dosage, a biopolymer-based delivery system has been widely suggested. Erogğlu et al. developed an alginate based carrier system for the drug mitomycin-C for bladder cancer chemotherapy [23]. The carrier system has proved to be highly cordial with patients.

8.5.2.5 Alginate in Transdermal Drug Delivery

The transdermal drug delivery system provides an efficiently controlled way to delivering the drug into the bloodstream by attaching a medicated patch of membrane on the skin. Crosslinked sodium alginate films were also used as rate controlling membrane (RCM) for the transdermal drug delivery systems. These membranes offered a thin, smooth and pliable surface. It proved less irritating to the skin and safe when studied on the release of diclofenac and diethyl amine as a model drug [39].

8.5.2.6 Alginate in Cardiac Drug Delivery

The Carvedilol (CRV), is a principally prescribed drug for the treatment of mild to severe congestive heart failure (CHF), which has bothered the scientists with its first pass metabolism and therefore poor bioavailability. In order to overcome this, a mucoadhesive alginate microsphere of CRV, for the nasal administration, in the form of a non-aggregated, free flowing powder in spherical shape has been tried [61]. This has greatly reduced the clearance rate of the drug in in vivo studies.

8.5.2.7   Alginate in Anti-Diabetic Drug delivery

An electrospun composite nanofiber based polyvinyl alcohol and sodium alginate transmucosal patch was invented to deliver antidiabetic drug, insulin [70].

It facilitated 99% efficacy of encapsulation and therefore paved way for an ideal carrier for the delivery insulin via sublingual route. Another anti-diabetic drug, Glipizide, is also brought under study for the evaluation of mucoadhesive potential of sodium alginate [60, 86]. These mucoadhesive microspheres are designed in such a way that it confers a sufficient resident time in gastrointestinal tract was maintained for the drug while the treatment of type II diabetes which improved the therapeutic performance of the drug. Similarly, alginate/chitosan based nanoparticles system has been developed for the delivery of insulin via the oral route. These nanoparticles were anionic and had a mean size of 750 nm which is ideal for the uptake within the GI tract [69].

8.5.2.8   Alginate in Antibiotic Delivery

Penicillin-loaded microbeads composed of alginate and octenyl succinic anhydride (OSA) starch prepared by ionotropic pregelation with calcium chloride and evaluated their in vitro drug delivery profile. The results demonstrated that alginate and OSA starch beads can be used as a suitable controlled-release carrier for penicillin G (PenG). SEM micrograph of OSA starch/alginate containing PenG is shown in Figure 8.8.

8.5.2.9   Alginate in Miscellaneous Drug Delivery

The oral administration of liquid dosage forms of suitable consistency and with sustained release characteristics may provide a means of improving the compliance of geriatric patients who experience difficulties in swallowing conventional solid dosage forms.

In situ gelling solutions of sodium alginate and methyl cellulose were formulated by Shimoyama T et al (2012) for the treatment of dysphagic patients. Matrices of 2% methyl cellulose and 0.5% alginate and

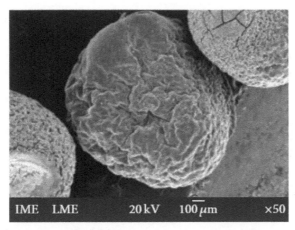

FIGURE 8.8   SEM micrograph of OSA starch/alginate containing PenG [24].

20% d-sorbitol with suitable viscosity when administrated would gel in the stomach under suitable conditions. In a similar way in-vitro release of paracetamol drug was also attempted and found to be an improved way for the sustained release. Liquid formulations of xyloglucan and sodium alginate in appropriate proportions are of suitable consistency for ease of administration to dysphagic patients and form gels in situ in the rat stomach capable of sustaining the release of paracetamol over a 6-hour period [31].

8.6   COSMETIC APPLICATIONS OF POLYURONATES

Cosmetic products are always in great demand in the market since they are believed to transform a normal person to iconic star. There are a huge variety of cosmetics available; each specialized for perfecting one or more of the one-beauty flaws of an individual.

Along with bountiful trends and trademarks, there is also an increased awareness in the utilization of natural ingredients in the cosmetic items. Pectin, alginate, xantha gum, chitosan are some of the natural ingredients included.

The application of natural polymers is contributed by their unique characteristics compared to their chemical counterparts. These include gelation, viscosity, adhesion, pH, swellability, emulsification, moisture absorption, film-forming ability, stabilizing ability and much more.

The present scenario is such that there is no cosmetic product being put out in the market without these polymers. They have become an important component of the cosmetic industry.

The major applications of polyuronates in cosmetics are:

- Emulsifiers
- Thickeners
- Moisturizers
- Surfactants

8.6.1   PECTIN IN COSMETIC APPLICATION

Many of the natural features of pectin are widely being exploited for shaping the cosmetic industry. An ample variety of cosmetic preparations were made with the help of pectin viz., skin and hair care products, foundation, mascara, eyeliner, lipstick etc. Pectins are gelling agents and increase the viscosity in gels and creams. It is extensively used in lotions, creams, hair conditioners and hair styling products. Apart from adding structure through gelation and viscosity build-up, pectin gels on the skin can provide moisture absorption while being skin friendly. Also, pectin gels, when are physically disrupted during processing, are used to formulate lotions and creams without the use of surfactants [66].

In the cosmetic industry, pectin offers great advantage in the form of:

- Gelation
- Viscosity
- Emulsification
- Moisture absorption
- Esterfication
- Adhesion
- Chelation

The unique ability of pectin to form gels and their efficient bio-adhesiveness is a milestone for the development of many skin care products.

8.6.1.1   Pectin as Emulsifier

Gel forms of pectin provide moisture maintenance of the skin. The gellability is also used for the elimination of oil coalescence that in turn improves spreadability and avoids greasiness. It offers a gradual release of oil in water-in-oil emulsion. The viscous nature of the pectin when hydrated is used for attaining an even distribution of the product. This effect has been playing a major role in hair conditioners and hair styling products with an excellent non-slimy spreadability. Furthermore, it also provides a good consistency to water in silicone emulsions, which is used in the preparation of non-wax sticks.

8.6.1.2   Pectin as Moisturizer

The strange character of pectin to swell when hydrated rather than dissolving in it helps them to transform into soft, wet particle system, which imparts a fat like texture to the skin creams. The availability of the carboxyl groups on pectin for esterification/etherification has led to the emergence of the new emulsion based on low molecular weight pectin in skin care products. Often the carboxylic groups are grafted with anti-oxidants for a healing and moisturizing effect.

Oral care markets exploit the film forming ability of low molecular weight pectin to use in breath freshener strips.

Low molecular weight pectin can be easily manipulated to design novel materials to eliminate the allergic response by the enzyme. It can bind to enzymes and can cause immunogenic responses. In a similar way, it binds to keratin, forming a hybrid with more water resistance.

8.6.1.3   Pectin as Chelator

The anionic nature of pectin is yet another important aspect, which prevents the polymerization of organic silicon compounds. Thus pectin always makes crosslinked complexes with proteins under suitable conditions and act as a powerful chelator of allergens such as cobalt, nickel, copper (Table 8.1).

8.6.2   ALGINATE IN COSMETIC APPLICATION

Alginates are tied to an enormous application in the cosmetic industry. Its INCI (International Nomenclature of Cosmetic Ingredient) name is

TABLE 8.1   The Potential Application of Pectin in Cosmetic Industry

S.No.	Products
1.	Facial creams and body lotions
2.	Shaving lotions
3.	Hair care serum
4.	Liquid soap
5.	Feminine hygiene products
6.	Lotionized tissue products
7.	Pro-fragrance products

Algin. The generally recommended proportion of algin in cosmetic items is 0.2–5% and a viscosity of about 1%.

Algin normally functions in thickening, moisture retention, color retention, etc.

8.6.2.1   Alginate as Thickener

Being a natural thickener and bubble stabilizer, it widely used in shampoos to promote good lather.

8.6.2.2   Alginate as Stabilizer

In skin care products it act as emulsion stabilizer thereby provide a soothing and silky feel to the skin. Alginate in occasionally found in tooth products as a bonding agent. Alginate absorbs water easily and therefore is used in facial masks as filmogens.

Dhat Shalaka et al. in 2009 has developed a vitamin E loaded pectin alginate microsphere with an aim to apply in cosmetic field. The microspheres are intended to swell in a cosmetic topical gel to their optimum capacity. While on application they would rupture upon rubbing between and release vitamin E. The FMC biopolymer has updated the cosmetic industry with an Isagel FM alginate in the form of facial masks. The alginate formulation is expected to supply an efficient gelling system, a high rate of gelling, firm and smooth mask and easy peel off.

8.7 CONCLUSIONS

Polyuronates, the most far-flung and versatile compound in the biological world, shares a long history of being an efficacious polymer, turning stone into gold, since time immemorial. The innate ability to absorb and retain water has always astounded the frontiers of modern science. Moreover, this widespread acceptance could be attributed to its innocuous behavior among several other compounds under trial. It has now become a frisson among the researchers to blend skillfully its unique characteristics to formalize the modern biomedical arena. It has surely fortified the pharmacological industry, by helping it to forestall in time with an aim to bring about patient compliance with drug delivery that comprehends disease conditions such as Diabetes, Cancer, HIV, Arthritis's, etc. As a promising drug delivery system, the polyuronates permits an excellent carrier matrix for the drug, protecting it from the adverse environment inside the patient's body. Moreover, the difficulty of controlled drug release is now under command. The food and fashion world are next in the line for being the greatest beneficiaries of Polyuronates. The present cosmetic market is teeming with products based on polyuronates. This hike in polyuronate based cosmetic products could be accredited to the wealth of profitable and much needed traits such as emulsification, thickening, hydration and oleophobicity.

KEYWORDS

- cosmetics
- drug delivery
- pectic materials
- polyuronates
- properties
- structures

REFERENCES

1. Ahmad, Zahoor, G. K. Khuller. "Alginate-based sustained release drug delivery systems for tuberculosis." (2008), 1323–1334.
2. Ahmad, Zahoor, Rajesh Pandey, Sadhna Sharma, G. K. Khuller. "Alginate nanoparticles as antituberculosis drug carriers: formulation development, pharmacokinetics and therapeutic potential." *Indian Journal of Chest Diseases and Allied Sciences* 48, no. 3 (2006), 171.
3. Ahuja, Alka, Roop K. Khar, Javed Ali. "Mucoadhesive drug delivery systems." *Drug Development and Industrial Pharmacy* 23, no. 5 (1997), 489–515.
4. Ashford, Marianne, John Fell, David Attwood, Harbans Sharma, Philip Woodhead. "Studies on pectin formulations for colonic drug delivery." *Journal of Controlled Release* 30, no. 3 (1994), 225–232.
5. Augustine, R., K. Rajarathinam. "Synthesis and characterization of silver nanoparticles and its immobilization on alginate coated sutures for the prevention of surgical wound infections and the in vitro release studies." International Journal of Nano Dimension (IJND) (2012), 205–212.
6. Axelos, M. A. V., J. F. Thibault. *The Chemistry of Low-Methoxyl Pectin Gelation.* Academic Press: New York, 1991.
7. Batyrbekov, Yerkesh O., Dinara Rakhimbaeva, Kuanyshbek Musabekov, Bulat Zhubanov. "Alginate Based Microparticle Drug Delivery Systems for the Treatment of Eye Cancer." In *MRS Proceedings*, vol. 1209. Cambridge University Press, 2009.
8. BeMiller, James N. "An introduction to pectins: structure and properties." *Chemistry and Function of Pectins* 310 (1986), 2–12.
9. Beyer, Markus, Jörg Reichert, Erik Heurich, Klaus D. Jandt, Bernd W. Sigusch. "Pectin, alginate and gum arabic polymers reduce citric acid erosion effects on human enamel." *Dental Materials* 26, no. 9 (2010), 831–839.
10. Brunetti, Michael. "Alginate Polymers for Drug Delivery." PhD diss., Worcester Polytechnic Institute, 2006.
11. Burapapadh, Kanokporn, Mont Kumpugdee-Vollrath, Doungdaw Chantasart, Pornsak Sriamornsak. "Fabrication of pectin-based nanoemulsions loaded with itraconazole for pharmaceutical application." *Carbohydrate Polymers* 82, no. 2 (2010), 384–393.
12. Cao, Ning. Fabrication of alginate hydrogel scaffolds and cell viability in calcium-crosslinked alginate hydrogel. Graduation Thesis, University of Saskatchewan Saskatoon, June, 2011.
13. Cardoso, Susana M., Manuel A. Coimbra, J. A. Lopes da Silva. " Temperature dependence of the formation and melting of pectin–Ca2+ networks: a rheological study." *Food Hydrocolloids* 17, no. 6 (2003), 801–807.
14. Chambin, O., G. Dupuis, D. Champion, A. Voilley, Y. Pourcelot. "Colon-specific drug delivery: influence of solution reticulation properties upon pectin beads performance." *International Journal of Pharmaceutics* 321, no. 1 (2006), 86–93.
15. Chen, Sung-Ching, Yung-Chih Wu, Fwu-Long Mi, Yu-Hsin Lin, Lin-Chien Yu, Hsing-Wen Sung. "A novel pH-sensitive hydrogel composed of N, O-carboxymethyl chitosan and alginate crosslinked by genipin for protein drug delivery." *Journal of Controlled Release* 96, no. 2 (2004), 285–300.

16. Ciofani, Gianni, Vittoria Raffa, Arianna Menciassi, Silvestro Micera, Paolo Dario. "A drug delivery system based on alginate microspheres: mass-transport test and in vitro validation." *Biomedical Microdevices* 9, no. 3 (2007), 395–403.

17. Dale, O., R. Hjortkjaer, E. D. Kharasch. "Nasal administration of opioids for pain management in adults." *Acta Anaesthesiologica Scandinavica* 46, no. 7 (2002), 759–770.

18. Das Neves, J., M. F. Bahia. "Gels as vaginal drug delivery systems." *International Journal of Pharmaceutics* 318, no. 1 (2006), 1–14.

19. De Vries, J. A., C. H. Den Uijl, A. G. J. Voragen, F. M. Rombouts, W. Pilnik. "Structural features of the neutral sugar side chains of apple pectic substances." *Carbohydrate Polymers* 3, no. 3 (1983), 193–205.

20. Deng, K. L., H. B. Zhong, T. Tian, Y. B. Gou, Q. Li, L. R. Dong. "Drug release behavior of a pH/temperature sensitive calcium alginate/poly (N-acryloylglycine) bead with core-shelled structure." Deng et al.–*eXPRESS Polymer Letters* 4 (2010), 773.

21. Dumitriu, Raluca Petronela, Ana-Maria Oprea, Cornelia Vasile. "A drug delivery system based on stimuli-responsive alginate/N-isopropylacryl amide hydrogel." *Cellulose Chemistry Technology* 43, no. 7–8 (2009), 251–262.

22. El-Gibaly, Ibrahim. "Oral delayed-release system based on Zn-pectinate gel (ZPG) microparticles as an alternative carrier to calcium pectinate beads for colonic drug delivery." *International Journal of Pharmaceutics* 232, no. 1 (2002), 199–211.

23. Eroğlu, Muzaffer, Eylem Öztürk, Nalan Özdemúr, Emúr Bakú Denkbapi, Isin Doğan, Abuzer Acar, Murat Güzel. "Mitomycin-C-loaded alginate carriers for bladder cancer chemotherapy: in vivo studies." *Journal of Bioactive Compatible çpolymers* 20, no. 2 (2005), 197–208.

24. Fontes, Gizele Cardoso, Verônica Maria Araújo Calado, Alexandre Malta Rossi, Maria Helena Miguez da Rocha-Leão. "Characterization of Antibiotic-Loaded Alginate-Osa Starch Microbeads Produced by Ionotropic Pregelation." *BioMed Research International*, 2013 (2013).

25. Gombotz, Wayne R., SiowFong Wee. "Protein release from alginate matrices." *Advanced Drug Delivery Reviews* 31, no. 3 (1998), 267–285.

26. Haida, Haruka, Shizuka Ando, Saori Ogami, Ryo Wakita, Hikaru Kohase, Norio Saito, Tomohiko Yoshioka et al. "In vitro evaluation of calcium alginate gels as matrix for iontophoresis electrodes." *Journal of Medical Dental Sciences* 59, no. 1 (2012), 9–16.

27. Hao, Xiaojin, Eduardo A. Silva, Agneta Månsson-Broberg, Karl-Henrik Grinnemo, Anwar J. Siddiqui, Göran Dellgren, Eva Wärdell, Lars Åke Brodin, David J. Mooney, Christer Sylvén. "Angiogenic effects of sequential release of VEGF-A165 PDGF-BB with alginate hydrogels after myocardial infarction." *Cardiovascular Research* 75, no. 1 (2007), 178–185.

28. Haug, Arne, Bjørn Larsen, Olav Smidsrød. "Uronic acid sequence in alginate from different sources." *Carbohydrate Research* 32, no. 2 (1974), 217–225.

29. Iliescu, Ruxandra Irina, Ecaterina Andronescu, Cristina Daniela, Daniela Berger Ghiţulică, Anton Ficai. "Montmorillonite-Alginate Nanocomposite Beads as Drug Carrier for Oral Administration of Carboplatin–Preparation Characterization." U.P.B. Sci. Bull. 2011, 73(3), 2–15.

30. Illum, Lisbeth, Peter James Watts. "Compositions for nasal administration." U.S. Patent 6,342,251, issued January 29, 2002.

31. Itoh, Kunihiko, Reina Tsuruya, Tetsuya Shimoyama, Hideki Watanabe, Shozo Miyazaki, Antony D'Emanuele, David Attwood. "In situ gelling xyloglucan/alginate liquid formulation for oral sustained drug delivery to dysphagic patients." *Drug Development Industrial Pharmacy* 36, no. 4 (2010), 449–455.
32. Izawa, Hironori, Kohsaku Kawakami, Masato Sumita, Yoshitaka Tateyama, Jonathan P. Hill, Katsuhiko Ariga. "β-Cyclodextrin-crosslinked alginate gel for patient-controlled drug delivery systems: regulation of host–guest interactions with mechanical stimuli." *J. Mater. Chem. B* 1, no. 16 (2013), 2155–2161.
33. Jaya, S., Durance, T. D., Wang, R. "Physical characterization of drug loaded microcapsules and controlled in vitro release study." *The Open Biomaterials Journal* 2 (2010), 9–17.
34. Judd, Patricia A., Truswell, A. S. "Comparison of the effects of high-and low-methoxyl pectins on blood and fecal lipids in man." *Br J Nutr* 48, no. 3 (1982), 451–458.
35. Katav, Tali, LinShu Liu, Tamar Traitel, Riki Goldbart, Marina Wolfson, Joseph Kost. "Modified pectin-based carrier for gene delivery: cellular barriers in gene delivery course." *Journal of Controlled Release* 130, no. 2 (2008), 183–191.
36. Khandelwal, Mohit, Ankit Ahlawat, Ram Singh. "Polysaccharides Natural Gums for Colon Drug Delivery." Pharma innovations, 2012, 8–12.
37. Kim, Chong-Kook, Eun-Jin Lee. "The controlled release of blue dextran from alginate beads." *International Journal of Pharmaceutics* 79, no. 1 (1992), 11–19.
38. Kohn, Rudolf. "Ion Binding On Polyuronats—Alginate Pectin." (1975); www.iupca.org.
39. Kulkarni, Raghavendra V., Yogesh J. Wagh. "Crosslinked Alginate Films as Rate Controlling Membranes for Transdermal Drug Delivery Application." *Journal of Macromolecular Science, Part A: Pure Applied Chemistry* 47, no. 7 (2010), 732–737.
40. Kumar, Manish, Abhishek Kr Chauhan, Sachin Kumar, Arun Kumar, Sachin Malik. "Design Evaluation of Pectin Based Metrics for Transdermal Patches of Meloxicam." Asian Journal of Pharmaceutical Science Healthcare 2, no 3 (2010) 244–247.
41. Kumar, Manoj, Rakesh Kumar Mishra, Ajit K. Banthia. "Development of pectin based hydrogel membranes for biomedical applications." *International Journal of Plastics Technology* 14, no. 2 (2010), 213–223.
42. Liu, Lin Shu, Joseph Kost, Fang Yan, Robert C. Spiro. "Hydrogels from biopolymer hybrid for biomedical, food, and functional food applications." *Polymers* 4, no. 2 (2012), 997–1011.
43. Liu, Linshu, Marshall L. Fishman, Joseph Kost, Kevin B. Hicks. "Pectin-based systems for colon-specific drug delivery via oral route." *Biomaterials* 24, no. 19 (2003), 3333–3343.
44. Liu, LinShu, Young Jun Won, Peter H. Cooke, David R. Coffin, Marshal L. Fishman, Kevin B. Hicks, Peter X. Ma. "Pectin/poly (lactide-co-glycolide) composite matrices for biomedical applications." *Biomaterials* 25, no. 16 (2004), 3201–3210.
45. López-Cacho, J. M., Pedro L. González-R, B. Talero, A. M. Rabasco, M. L. González-Rodríguez. "Robust optimization of alginate-carbopol 940 bead formulations." The Scientific World Journal, 2012 (2012).
46. Mahajan, Hitendra S., Bhushankumar V. Tatiya, Pankaj P. Nerkar. "Ondansetron loaded pectin based microspheres for nasal administration: In vitro and in vivo studies." *Powder Technology* 221 (2012), 168–176.
47. Mahkam, Mehrdad. "Modification of nano alginate-chitosan matrix for oral delivery of insulin." *Nature Science* 7, no. 8 (2009), 1–7.

48. Malesu, Vijay Kumar, Debasish Sahoo, P. L. Nayak. "Alginate Nanocomposites Blended With Cloisite 30b As A Novel Drug Delivery System For Anticancer Drug Curcumin," International Journal Of Applied Biology And pharmaceutical Technology. 2011, 2(3), 1–9.
49. Mishra, R. K., A. K. Banthia, A. B. A. Majeed. "Pectin based formulations for biomedical applications: A review." *Asian Journal of Pharmaceutical Clinical Research* 5, no. 4 (2012), 1–7.
50. Morris, Gordon A., Samil M. Kök, Stephen E. Harding, Gary G. Adams. "Polysaccharide drug delivery systems based on pectin and chitosan." *Biotechnology Genetic Engineering Reviews* 27, no. 1 (2010), 257–284.
51. Muhiddinov, Z., D. Khalikov, T. Speaker, R. Fassihi. "Development and characterization of different low methoxy pectin microcapsules by an emulsion-interface reaction technique." *Journal of Microencapsulation* 21, no. 7 (2004), 729–741.
52. Mukhiddinov, Z. K., D. Kh Khalikov, F. T. Abdusamiev, Ch Avloev. "Isolation and structural characterization of a pectin homo and ramnogalacturonan." *Talanta* 53, no. 1 (2000), 171–176.
53. Munarin, F., P. Petrini, S. Farè, M. C. Tanzi. "Structural properties of polysaccharide-based microcapsules for soft tissue regeneration." *Journal of Materials Science: Materials in Medicine* 21, no. 1 (2010), 365–375.
54. Munarin, F., S. G. Guerreiro, M. A. Grellier, M. C. Tanzi, M. A. Barbosa, P. Petrini, P. L. Granja. "Pectin-based injectable biomaterials for bone tissue engineering." *Biomacromolecules* 12, no. 3 (2011), 568–577.
55. Munjeri, O., J. H. Collett, J. T. Fell. "Amidated pectin hydrogel beads for colonic drug delivery-An in vitro study." *Drug Delivery* 4, no. 3 (1997), 207–211.
56. Muzzarelli, Riccardo AA, Joseph Boudrant, Diederick Meyer, Nicola Manno, Marta DeMarchis, Maurizio G. Paoletti. "Current views on fungal chitin/chitosan, human chitinases, food preservation, glucans, pectins and inulin: A tribute to Henri Braconnot, precursor of the carbohydrate polymers science, on the chitin bicentennial." *Carbohydrate Polymers* 87, no. 2 (2012), 995–1012.
57. Narra, Kishore, Unnikrishnan Dhanalekshmi, Govindaraj Rangaraj, Devendiran Raja, Celladurai Senthil Kumar, Pully Neelakanta Reddy, Asit Baran Mandal. "Effect of Formulation Variables on Rifampicin Loaded Alginate Beads." *Iranian Journal of Pharmaceutical Research* 11, no. 3 (2011), 715–721.
58. Nguyen, S., M. Hiorth, M. Rykke, G. Smistad. "Polymer coated liposomes for dental drug delivery–interactions with parotid saliva and dental enamel." *European Journal of Pharmaceutical Sciences 50* (2013), 78–85.
59. Oh, Jung Kwon, Ray Drumright, Daniel J. Siegwart, Krzysztof Matyjaszewski. "The development of microgels/nanogels for drug delivery applications." *Progress in Polymer Science* 33, no. 4 (2008), 448–477.
60. Patel, Jayvadan K., Rakesh P. Patel, Avani F. Amin, Madhabhai M. Patel. "Formulation and evaluation of mucoadhesive glipizide microspheres." *AAPS Pharm Sci Tech* 6, no. 1 (2005), E49-E55.
61. Patil, Sanjay B., Krutika K. Sawant. "Development, optimization and in vitro evaluation of alginate mucoadhesive microspheres of carvedilol for nasal delivery." *Journal of Microencapsulation* 26, no. 5 (2009), 432–443.
62. Percival, E., R. H. McDowell. "Algal walls—composition and biosynthesis." In *Plant Carbohydrates II*, pp. 277–316. Springer Berlin Heidelberg, 1981.

63. Perić-Hassler, Lovorka, Philippe H. Hünenberger. "Interaction of alginate single-chain polyguluronate segments with mono-and divalent metal cations: a comparative molecular dynamics study." *Molecular Simulation* 36, no. 10 (2010), 778–795.
64. Rajesh, N. "Feasibility of xanthan gum–sodium alginate as a transdermal drug delivery system for domperidone." *Journal of Materials Science: Materials in Medicine* 20, no. 10 (2009), 2085–2089.
65. Ramasamy, Thiruganesh, Hima Bindu Ruttala, Suresh Shanmugam, Subbiah Kandasamy Umadevi. "Eudragit-coated aceclofenac-loaded pectin microspheres in chronopharmacological treatment of rheumatoid arthritis." *Drug Delivery* 20, no. 2 (2013), 65–77.
66. Robin Augustine, Rajakumari Rajendran, Uroš Cvelbar, Miran Mozetič, Anne George. "Biopolymers for Health, Food, Cosmetic Applications." Handbook of Biopolymer-Based Materials: From Blends Composites to Gels Complex Networks: (2013), 801–849.
67. Rohan, Lisa Cencia, Alexandra B. Sassi. "Vaginal drug delivery systems for HIV prevention." *The AAPS Journal* 11, no. 1 (2009), 78–87.
68. Sadeghi, Mohammad. "Pectin-Based Biodegradable Hydrogels with Potential Biomedical Applications as Drug Delivery Systems." *Journal of Biomaterials Nanobiotechnology* 2, no. 1 (2011), 36–40.
69. Sarmento, B., A. Ribeiro, F. Veiga, P. Sampaio, R. Neufeld, D. Ferreira. "Alginate/chitosan nanoparticles are effective for oral insulin delivery." *Pharmaceutical Research* 24, no. 12 (2007), 2198–2206.
70. Sharma, A., A. Gupta, G. Rath, A. Goyal, R. B. Mathur, S. R. Dhakate. "Electrospun composite nanofiber-based transmucosal patch for anti-diabetic drug delivery." *Journal of Materials Chemistry B* (2013).
71. Sharma, Sadhna, G. K. Khuller, S. K. Garg. "Alginate-based oral drug delivery system for tuberculosis: pharmacokinetics and therapeutic effects." *Journal of Antimicrobial Chemotherapy* 51, no. 4 (2003), 931–938.
72. Shi, Jun, Natalia M. Alves, Joao F. Mano. "Chitosan coated alginate beads containing poly (N-isopropylacrylamide) for dual-stimuli-responsive drug release." *Journal of Biomedical Materials Research Part B: Applied Biomaterials* 84, no. 2 (2008), 595–603.
73. Shimoyama, Tetsuya, Kunihiko Itoh, Michiya Kobayashi, Shozo Miyazaki, Antony D'Emanuele, David Attwood. "Oral liquid in situ gelling methylcellulose/alginate formulations for sustained drug delivery to dysphagic patients." *Drug Development Industrial Pharmacy* 38, no. 8 (2012), 952–960.
74. Shukla, S., D. Jain, K. Verma, S. Verma. "Pectin-based colon-specific drug delivery." *Chronicles of Young Scientists* 2, no. 2 (2011), 83.
75. Sriamornsak, Pornsak, Jurairat Nunthanid. "Calcium pectinate gel beads for controlled release drug delivery: I. Preparation and in vitro release studies 1 paper presented in part at the Eleventh International Symposium on Microencapsulation held at Bangkok, Thailand, 27–29 August, 1997.1." *International Journal of Pharmaceutics* 160, no. 2 (1998), 207–212.
76. Sriamornsak, Pornsak. "Chemistry of pectin and its pharmaceutical uses: A review." *Silpakorn University International Journal* 3, no. 1–2 (2003), 206–228.
77. Srivastava, Pranati, Rishabha Malviya. "Sources of pectin, extraction and its applications in pharmaceutical industry–An overview." *IJNPR* 2, no. 1 (2011), 10–18.

78. Suksamran, Tittaya, Praneet Opanasopit, Theerasak Rojanarata, Tanasait Ngawhirun-pat, Uracha Ruktanonchai, Pitt Supaphol. "Biodegradable alginate microparticles developed by electrohydrodynamic spraying techniques for oral delivery of protein." *Journal of Microencapsulation* 26, no. 7 (2009), 563–570.

79. Sungthongjeen, S., T. Pitaksuteepong, A. Somsiri, P. Sriamornsak. "Studies on pectins as potential hydrogel matrices for controlled-release drug delivery." *Drug Development Industrial Pharmacy* 25, no. 12 (1999), 1271–1276.

80. Takei, Takayuki, Mitsunobu Sato, Hiroyuki Ijima, Koei Kawakami. "In situ gellable oxidized citrus pectin for localized delivery of anticancer drugs and prevention of homotypic cancer cell aggregation." *Biomacromolecules* 11, no. 12 (2010), 3525–3530.

81. Valenta, Claudia. "The use of mucoadhesive polymers in vaginal delivery." *Advanced Drug Delivery Reviews* 57, no. 11 (2005), 1692–1712.

82. Watts, Peter, Alan Smith. "PecSys: in situ gelling system for optimized nasal drug delivery." (2009), 543–552.

83. Wong, T. W., L. W. Chan, H. Y. Lee, P. W. S. Heng. "Release characteristics of pectin microspheres prepared by an emulsification technique." *Journal of Microencapsulation* 19, no. 4 (2002), 511–522.

84. Wong, Tin Wui, Gaia Colombo, Fabio Sonvico. "Pectin matrix as oral drug delivery vehicle for colon cancer treatment." *AAPS PharmSciTech* 12, no. 1 (2011), 201–214.

85. Yadav, Neha, Gordon Morris, S. E. Harding, Shirley Ang, G. G. Adams. "Various non-injectable delivery systems for the treatment of diabetes mellitus." *Endocrine, Metabolic Immune Disorders-Drug Targets* 9, no. 1 (2009), 1–13.

86. Yaswanth Allamneni, B. V., V. K. Reddy, P. Dayananda Chary, Venkata Balakrishna Rao N, Kumar, S. Chaitanya, Arun Kumar Kalekar. "Performance Evaluation of Mucoadhesive Potential of Sodium Alginate on Microspheres Containing an Anti-Diabetic Drug: Glipizide." International Journal of Pharmaceutical Sciences Drug Research, 2012; 4(2), 115–122.

87. Yu, Cui-Yun, Bo-Cheng Yin, Wei Zhang, Si-Xue Cheng, Xian-Zheng Zhang, Ren-Xi Zhuo. "Composite microparticle drug delivery systems based on chitosan, alginate and pectin with improved pH-sensitive drug release property." *Colloids Surfaces B: Biointerfaces* 68, no. 2 (2009), 245–249.

88. Zhang, Feng Ju, Guo Xiang Cheng, Xiao Guang Ying. "Emulsion and macromolecules templated alginate based polymer microspheres." *Reactive Functional Polymers* 66, no. 7 (2006), 712–719.

CHAPTER 9

BIOLOGICAL DELIGNIFICATION OF BIOMASS

JITHIN JOY,[1,2] CINTIL JOSE,[1,2] P. LOVELY MATHEW,[1]
SABU THOMAS,[2] and MOAYAD N. KHALAF[3]

[1]*Department of Chemistry, Newman College, Thodupuzha, Kerala, India*

[2]*International and Inter University Centre for Nanoscience and Nanotechnology, Mahatma Gandhi University, Kottayam, Kerala, India*

[3]*Chemistry Department, College of Science, University of Basrah, P.O. Box 773, Basrah, Iraq*

CONTENTS

9.1 INTRODUCTION

9.1.1 THE COMPOSITION OF LIGNOCELLULOSIC MATERIAL

The annual production of lignocellulosic matters has been estimated about, 1010 metric tons worldwide [2]. The main constituents of the lignocellulosic biomass are cellulose (a homopolymer of glucose), hemicellulose (a heteropolymer of pentoses and hexoses) and lignin (a polymer of phenyl propanoid units) [82]. These three polymers are strongly intermeshed and chemically bonded by non-covalent forces [112]. The polysaccharides (cellulose and hemicellulose) on average accounts for 55–75% on the dry weight basis in plant cell wall and can be deconstructed into simple sugars [142], organic acids, acetone and glycerol [21, 150] (Figure 9.1).

It is very important to note that only a small amount of the cellulose, hemicellulose and lignin produced as by-products in agriculture or forestry is used, the rest being considered as waste. Many microorganisms are capable of degrading and using cellulose and hemicellulose as carbon and energy sources. However, a much smaller group of filamentous fungi has evolved with the ability to break down lignin, the most recalcitrant component of plant cell walls. These are known as white-rot fungi (WRF), which possess the unique ability of efficiently degrading lignin to CO_2. Other lignocellulose degrading fungi are brown-rot fungi that rapidly depolymerize cellulosic materials while only modifying lignin. Collectively, these wood and litter-degrading fungi play an important role in the carbon cycle.

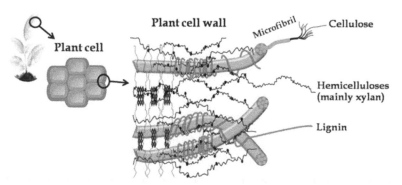

FIGURE 9.1 Structure of lignocellulosic plant biomass [149].

9.1.1.1   Cellulose

Cellulose, the most abundant biopolymer on earth, is a structural component biosynthesized as microfibrils by a number of living organisms [49, 11]. The chains of poly β-(1, 4)-D-glucozyl residues aggregate to form fibrils, which are long thread-like bundles of molecules stabilized by intermolecular hydrogen bonds [4, 31, 140]. Each microfibril can be considered as a string of cellulose crystals linked along the microfibril axis by amorphous domains [6], for example, twists and kinks. Individual cellulose microfibrils have diameters in the range 2–20 nm [6, 95]. Because of the high potential of cellulose nanoparticles or nanocellulose in different applications such as reinforcement of polymers nanocomposites and increasing interest to develop nanoscale materials, researchers have developed different routes to produce cellulosic nanoparticles. The cellulose in a plant consists of parts with a crystalline (organized) structure, and parts with not well-organized amorphous structure. These cellulose fibrils are mostly independent and weakly bound through hydrogen bonding [86]. Cellulose appears in nature to be associated with other plant compounds, and this association may affect its biodegradation.

9.1.1.2   Hemicellulose

Hemicellulose is a complex carbohydrate structure that consists of different polymers like pentoses (like xylose and arabinose), hexoses hexose's hexoses (like mannose, glucose and galactose), and sugar acids. The dominant component of hemicellulose from hardwood and agricultural plants, like grasses and straw, is xylan, while this is glucomannan for softwood [129] Hemicellulose serves as a connection between the lignin and the cellulose fibers and gives the whole cellulose–hemicellulose–lignin network more rigidity [86].

Hemicellulose is a polysaccharide with a lower molecular weight than cellulose. It is formed from D-xylose, D-mannose, Dgalactose, D-glucose, L-arabinose, 4-O-methyl-glucuronic, D-galacturonic and D-glucuronic acids. Sugars are linked together by β-1,4- and sometimes by β-1, 3-glycosidic bonds. The main difference between cellulose and hemicellulose is that hemicellulose has branches with short lateral chains consisting of different sugars, and cellulose consists of easily hydrolyzable oligomers.

9.1.1.3   Lignin

Lignin is, after cellulose and hemicellulose, one of the most abundant polymers in nature and is present in the cellular wall. The main purpose of lignin is to give the plant structural support, impermeability, and resistance against microbial attack and oxidative stress. The amorphous heteropolymer is also non-water soluble and optically inactive; all this makes the degradation of lignin very tough. The solubility of the lignin in acid, neutral or alkaline environments depends, however, on the precursor (p-coumaryl, coniferyl, sinapyl alcohol or combinations of them) of the lignin [44].

Lignin is a high-molecular, hydrophobic polymer and thus not soluble in aqueous solution. The situation is complicated by the fact that lignin does not serve as a sole growth substrate for WRF [3, 77]. It was therefore suggested that the purpose of lignin removal is to expose cellulose and hemicellulose fibers for consumption and further fungal growth [76]. On the other hand, the hemicellulose also physically restrict the access of enzymes to lignin. The close association of lignin and hemicellulose suggests that the primary attack on the wood cell wall require enzymatic degradation of the hemicellulose prior to lignin degradation. The covalent bonds of hemicellulose to lignin may be hydrolyzed, resulting in lignin-carbohydrate complexes of lower molecular weight capable of diffusing from the matrix. Alternatively, the hemicellulose could strip their substrates off the cell wall, rendering the residual lignin exposed to attack by LiP or MnP. Sound wood cells cannot be invaded by wood decay enzymes [139, 14]. Hydroxylation and depolymerization of the hemicellulose component would render lignin accessible for LiP and MnP.

Hardwood and softwood are distinguished by structural elements building the phenyl propane backbone of the lignin component. Lignin is a three dimensional, optically inactive phenyl propanoid polymer randomly synthesized from coniferyl, p-coumaryl and sinapyl alcohol precursors [132]. Softwood lignin is referred to as guaiacyl lignin, containing more than 95% coniferyl alcohol (4-hydroxy- 3-methoxy-cinnamyl alcohol) units. A structural model of spruce lignin was proposed by Adler (1977). The remaining elements are mainly p-coumaryl (6hydroxycinnamyl) alcohols with trace amounts of sinapyl (3,5-dimethoxy-4-hydroxy-cinnamyl alcohol) alcohols. Typical hardwood lignins, also classified as guaiacyl-syringyl lignins, contain coniferyl- and sinapyl alcohol-derived subunits (Figure 9.2).

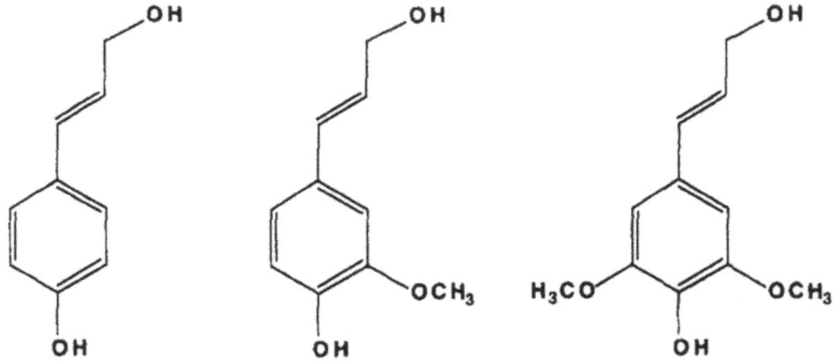

p-coumaryl alcohol   coniferyl alcohol   sinapyl alcohol

FIGURE 9.2   Precursor alcohols of lignin.

Various plants differ in aromatic constituents and in the linkages of these aromatics to other cell wall components. Grasses contain fewer lignin than woody plants and have more p-hydroxyphenyl units along with smaller amounts of guaiacyl and syringyl units [146]. Grass cell walls also have high amounts of ester-linked p-coumaric and ferulic acids, while woody and legume plants have been little of these constituents [19, 65]. These esters are linked to xylans through arabinose [51]. Hardwood lignins typically contain 1.2–1.5 methoxyl groups per phenyl propane unit. Although classified as guaiacyl-syringyl lignin, grass lignin also contains additional small amounts of p coumaryl alcohol-derived subunits. Spruce (softwood) lignin contains 92–96 methoxyl and 15–20 free phenolic functional groups per 100 phenylpropane units, whereas the corresponding numbers for birch (hardwood) lignin are 139–158 and 9–13, respectively [132, 1] (Table 9.1).

9.2   DELIGNIFICATION

Lignin component of biomass is recalcitrant. Therefore, pretreatment is an essential step for removal of lignin so that cellulose may be saccharified by cellulose enzyme. During enzymatic hydrolysis, the availability of cellulose is hindered by the presence of lignin, which is removed or modified by pretreatment methods. There are several methods of pretreatments,

TABLE 9.1 Composition of Some Lignocellulosic Materials

	Lignocellulosic residues	Lignin (%)	Hemicellulose (%)	Cellulose (%)	Reference
1.	Hardwood stems	18–25	24–40	40–55	Howard et al. (2003), Malherbe and Cloete (2002)
2.	Softwood stems	25–35	25–35	45–50	Howard et al. (2003), Malherbe and Cloete (2002)
3.	Nut shells	30–40	25–30	25–30	Howard et al. (2003)
4.	Corn cobs	15	35	45	Howard et al. (2003), Prassad et al. (2007), McKendry (2002)
5.	Paper	0–15	0	85–99	Howard et al. (2003)
6.	Rice straw	18	24	32.1	Howard et al. (2003), Prassad et al. (2007), McKendry (2002)
7.	Sorted refuse	20	20	60	Howard et al. (2003)
8.	Leaves	0	80–85	15–20	Howard et al. (2003)
9.	Cotton seeds hairs	0	5–20	80–95	Howard et al. (2003)
10.	Newspaper	18–30	25–40	40–55	Howard et al. (2003)
11.	Waste paper from chemical pulps	5–10	10–20	60–70	Howard et al. (2003)
12.	Primary waste water Solids	24–29	NA	8–15	Howard et al. (2003)
13.	Swine waste	NA	28	6	Howard et al. (2003)
14.	Solid cattle manure	2.7–5.7	1.4–3.3	1.6–4.7	Howard et al. (2003)
15.	Coastal Bermuda grass	6.4	35.7	25	Howard et al. (2003)
16.	Switch grass	12.0	31.4	45	Howard et al. (2003)

TABLE 9.1 Continued

	Lignocellulosic residues	Lignin (%)	Hemicellulose (%)	Cellulose (%)	Reference
17.	S32 rye grass (early leaf)	2.7	15.8	21.3	Howard et al. (2003
18.	S32 rye grass (seed setting)	7.3	25.7	26.7	Howard et al. (2003)
19.	Orchard grass (medium maturity)	4.7	40	32	Howard et al. (2003)
20.	Grasses (average values for grasses)	10–30	25–50	25–40	Howard et al. (2003), Malherbe and Cloete (2002)
21.	Sugar cane bagasse	19–24	27–32	32–44	Rowell (1992)
22.	Wheat straw	16–21	26–32	29–35	Rowell (1992), Prassad et al. (2007),
23.	Barley straw	14–15	24–29	31–34	Rowell (1992)
24.	Oat straw	16–19	27–38	31–37	Rowell (1992)
25.	Rye straw	16–19	27–30	33–35	Rowell (1992), Stewart et al. (1997), Reguant and Rinaudo (2000), Hon (2000)
26.	Bamboo	21–31	15–26	26–43	Rowell (1992), Stewart et al. (1997), Reguant and Rinaudo (2000), Hon (2000)
27.	Grass Esparto	17–19	27–32	33–38	Rowell (1992), Stewart et al. (1997), Reguant and Rinaudo (2000), Hon (2000)
28.	Grass Sabai	22	23.9	NA	Rowell (1992), Stewart et al. (1997), Reguant and Rinaudo (2000), Hon (2000)

TABLE 9.1 Continued

	Lignocellulosic residues	Lignin (%)	Hemicellulose (%)	Cellulose (%)	Reference
29.	Grass Elephant	23.9	24	22	Rowell (1992), Stewart et al. (1997), Reguant and Rinaudo (2000), Hon (2000)
30.	Bast fiber Seed flax	23	25	47	Rowell (1992), Stewart et al. (1997), Reguant and Rinaudo (2000), Hon (2000)
31.	Bast fiber Kenaf	15–19	22–23	31–39	Rowell (1992), Stewart et al. (1997), Reguant and Rinaudo (2000), Hon (2000)
32.	Bast fiber Jute	21–26	18–21	45–53	Rowell (1992), Stewart et al. (1997), Reguant and Rinaudo (2000), Hon (2000)
33.	Leaf Fiber Abaca (Manila)	8.8	17.3	60.8	Rowell (1992), Stewart et al. (1997), Reguant and Rinaudo (2000), Hon (2000)
34.	Leaf Fiber Sisal (agave)	7–9	21–24	43–56	Rowell (1992), Stewart et al. (1997), Reguant and Rinaudo (2000), Hon (2000)
35.	Leaf Fiber Henequen	13.1	4–8	77.6	Rowell (1992), Stewart et al. (1997), Reguant and Rinaudo (2000), Hon (2000)
36.	Banana waste	14	14.8	13.2	John et al. (2006)

among which chemical pretreatment is the most effective and practiced for industrial applications. Alkali pretreatments can be performed at room temperature [2] and causes minimal sugar loss and more effective on agricultural residues than woody materials [84]. On the other hand, acid pretreatment solubilizes the hemicellulosic fraction of the biomass and thus makes the cellulose more accessible to enzymes. This method appears as a favorable method for industrial applications, and it can be performed at high and low temperatures. Saccharification of dilute sulfuric acid pretreated lignocellulosic biomass results in high-sugar yield [2].

Lignin content can be easily determined by Fourier transform near infrared (FT-NIR) [32, 41, 105, 135, 17]. Schwanninger et al., 2001 developed a partial least-squares regression model for the determination of the lignin content of spruce wood based upon the absorbance bands of aromatic C–H vibrations, which are characteristic for units of the lignin macromolecule, in the spectral region between, 6102 and, 5762 cm^{-1}. Furthermore, the lignin content of spruce wood meals is linearly correlated to the intensity of the valley of the second derivative of the NIR spectrum near, 5980 cm^{-1} [133].

9.2.1 CHEMICAL DELIGNIFICATION

Lignin is crossed linking with hemicellulose, and cellulose makes it highly recalcitrant. Therefore, the partial or complete removal of lignin is necessary for the production of cellulose free of hemicellulose and lignin. Several methods such as steam explosion, steam treatment with diluted acid or alkali, organosolv extraction and ammonia fiber expansion have been largely used to improve the yield of cellulose production [2]. However, most of these pretreatments are environments unfriendly and commercially uncompetitive and these processes are intensive, require costly equipment's and often generate toxic compounds, which make the [2].

9.2.2 BIOLOGICAL DELIGNIFICATION

Due to the three-dimensional polymer interconnected through diverse carbon–carbon and other bonds in lignin made it very difficult to degrade, and they are not hydrolysable under biological conditions [107]. Compared to the other pretreatment methods biological delignification of biomass by solid-state cultivation is considered to be an environmentally friendly process. Among microorganisms, the WRF are the most capable lignin degraders and thus have been potential for the biological breakdown of plant materials [29]. These delignification processes are advocated by researchers since it provides the additional advantages of no use of severe chemicals, reduced energy input, no requirement for pressurized and corrosion-resistant reactors, no waste stream generated, and reduced or no inhibitor to fermentation [70, 47, 102].

The ability of White Rot Fungus (WRF) to delignification of the biomass have been receiving extensive attention for applications in industries processing lignocellulosic materials to produce cellulose, bio-fuels, or cellulose-enriched forage for ruminants [20]. Numerous WRF has been used for delignification of lignocellulose [70, 144, 8, 154, 159, 47, 131, 130, 148]. Some SRF appeared to be selective in their degradation of lignocellulose, specifically degrading lignin and hemicellulose and leaving a major part of the cellulose [68]. Lignin degrading enzymes produced by the WRF breaks the lignin binding with the hemicellulose and cellulose in the wood matrix [125].

9.2.2.1 Classification of Wood Rotters

Biomass degrading fungi are classified into three specific rotting groups depending upon the macroscopic differences. They are white-rot, brown-rot and soft-rot fungi. Generally, wood rotting fungi are the ability to degrade all the composition of the wood. However, the rates of degradations are different for cellulose, hemicellulose and lignin. Brown-rot fungi (BRF) are mainly soft wood degrades found in coniferous plants. BRF is more competent in gaining energy from wood for growth and reproduction than WRF [42].

Lignin is a randomly constructed polymer of phenyl propanoid substructures, and degradation of this structure requires a non-specific enzymatic system [34, 53]. WRF is so far unique in their ability to completely degrade all components of lignocellulosic materials. However, the rates of degradations are different for cellulose, hemicellulose and lignin. The ability to degrade lignin in the biomass is dissimilar for different WRF. Based on these criteria WRF has been divided into classes namely selective delignification and simultaneous delignification [34, 12]. Classification according to these groups proved to be difficult since many WRF show both types of decay while acting on the same substrate, or they attack lignin and polysaccharides sequentially [34]. Biological pretreatment of lignocellulosic materials with WRF are reported [104, 144].

The wood-decaying *basidiomycetes* of white-rot type are the most efficient microorganisms in depolymerization and lignin mineralization. [157, 68, 69, 62]. The enzymes and lignin degradation by WRF has been broadly studied in liquid cultivations [120], and in solid-state cultivations.

Based on the nature of the polymers that enzymes break down, plant cell wall degrading enzymes have been classified into cellulolytic, ligninolytic and xylanolytic. They contain extracellular nonspecific and non stereoselective enzyme system composed by laccases (EC 1.10.3.2, benzenediol:oxygen oxidoreductase), lignin peroxidase (EC 1.11.1.14), manganese peroxidase (EC 1.11.1.13), versatile peroxidase and H_2O_2 producing oxidases and secondary metabolites [122, 96, 116]. Biochemical research delignification turned into a new extent by the discovery of ligninase (lignin peroxidase (Lip)) [43, 147], Tien and Kirk, 1983) and Mn-peroxidase (MnP; [85]) from *Phanerochaete chrysosporium*. Lip's are strong oxidants that interact directly with non-phenolic lignin structures to cleave them, but cannot penetrate the small pores in sound lignocellulose. Manganese-dependent peroxidases produce small diffusible strong oxidants that can penetrate the substrate.

Irpex lacteus CD2 produced LiP for delignification of corn stover [161]. Wheat straw incubated with *T. versicolor* observed that F. fomentarius had a MnP activity of 0.17 U/mL and B. adusta laccase has an activity of 0.22 U/mL [67]. *Ceriporiopsis subvermispora* produced MnP and laccase for delignification of corn stover [154]. No laccase was detected, and MnP plays a major role in the process of sugarcane bagasse pretreatment with *C. subvermispora* [26]. Only MnP was detected in the process of *Eucalyptus grandis* delignification with *Phellinus flavomarginatus* [37]. Songulashvili evaluated 18strains of basidiomycetes in submerged fermentation of mandarin peeling and showed that ligninolytic activities for *T. versicolor* varied from 17.1 to 20.4 U/mL for laccase and 0.06 to 0.71 U/mL for MnP [137]. Phanerochaete chrysosporium, Pycnoporus cinnabarinus, Crinipellissp. RCK-1, Pleurotus ostreatus and Trametes versicolor have been tested for their lignin degradation abilities, when grown under solid-state fermentation (SSF) [47, 83, 136]. Tuomela studied the capacity of nine white-rot fungi, namely Abortiporus biennis, Bjerkandera adusta, Dichomitus squalens, Phanerochaete chrysosporium, Phanerochaete sordida, Phlebia radiata, Pleurotus ostreatus, Trametes hirsuta, and Trametes versicolor, to mineralize synthetic 14C-labeled lignin in soil, in order to reveal the effect that soil has on the action of different ligninolytic systems.

Pleurotus sp. degraded 6.8% lignin in water hyacinth after 22 days [109], while P. cinnabarinus degraded 8.5% lignin in sugarcane bagasse [104]. The biological delignification efficiency in Prosopisjuliflora and Lantana camara by white rot fungus, P. cinnabarinus by13.13 and 8.87%, respectively [47]. The degradation in terms of substrate weight loss, lignin and cellulose degradation increased with an increase in incubation time, and this could directly be correlated with the enzyme production profile by these three fungi.

The extracellular oxidative enzymes ligninase, manganese peroxidase and lactase may be defined as phenol oxidases. Both ligninase (EC No.1.11.1.14 Diarylpropane: oxygen, hydrogen-peroxideoxidoreductase) and manganese peroxidize (EC No. 1.11.1.13 Mn(I1): hydrogen-peroxide oxidoreductase) belong to the class of peroxidizes and oxidize their substrates by two consecutive one-electron oxidation steps with intermediate cation radical formation. Due to its high redox potential, the preferred substrates for LIP are nonphenolic methoxyl-substituted lignin subunits, whereas MnP acts exclusively as a phenol oxidase on phenolic substrates using Mn^{2+}/Mn^{3+} as an intermediate redox couple. Oxidation of phenolic substrates by ligninase leads to their polymerization [58, 110]. Ring-cleavage of aromatic rings is a key step of lignin mineralization. Non-phenolic syringyl and biphenyl model compounds are oxidized by lignin peroxidase and subsequent ring cleavage. In contrast, oxidation of the corresponding phenoliccompounds by ligninase did not yield ring-opened products [54]. Alkoxylgroups activate aromatic rings towards oxidation by ligninase, partly explaining why syringyl lignin is easier to degrade than guaiacyl lignin (Erikssonet al, 1990). Lactase (EC No. 1.10.3.2. (benzenediol:0, oxidoreductase) is a true phenoloxidase with broad substrate specificity. It oxidizes phenols and phenolic lignin substructures byone-electron abstraction with formation of radicals that can repolymerize or lead to depolymerization [59]. Demethylation reactions of terminal phenolic units

catalyzed by lactase may therefore be of importance for native lignin degradation, and the utilization of lactase or laccase-producing WRF for bio
pulping has become a focus of interest recently [120, 103].

In view of these findings it seems likely that structural and chemical
differences in the very inhomogeneous lignin substrate should lead to a
specialization in the respective degrading microorganisms, particularly
the oxidative enzymes expressed by each organism. Bacteria also attack
softwood and hardwood cell walls. They have been described as primary
wood colonizers [92].

9.3   MECHANISM OF LIGNIN DEGRADATION

The enzymatic hydrolysis of cellulose, particularly hydrogen-bonded and
ordered crystalline cellulose is a complex process [55].

9.3.1   CELLULASE

The widely accepted mechanism for cellulose hydrolysis suggests that
three different types of enzyme activities work synergistically in a complete cellulase system during this process [57, 93]. Based on their structural
properties, cellulases can be divided into three groups: (1) endoglucanases
or β-1,4-endoglucanases (EG, EC 3.2.1.4) which randomly hydrolyze
accessible intramolecular β-1,4-glucosidic bonds in cellulose chains, generating oligosaccharides of various lengths and consequently, new chains
ends [93, 158]; (2) exoglucanases (cellobiohydrolases, CNHs; EC 3.2.1.91)
acting on the chain terminal to release soluble cellobiose (cellobiohydro-
lase) or glucose (glucanohydrolase) as major products [152, 158] and (3)
β-glucosidases (BGL, EC 3.2.1.21) which hydrolyze cellobiose to glucose,
in order to eliminate cellobiose inhibition [93, 158]. The use of cellulases
for applications in the industry has been extensively reported [7, 138, 153,
71]. However, only limited works have been published on the use of enzymatic hydrolysis for nanocellulose production [111, 56, 6].

There is growing experimental evidence that reductive processes play
a pivotal role in lignin biodegradation and are also part of the ligninolytic system. A range of monomeric and dimeric aromatic aldehydes and

acids are reduced to the corresponding alcohols in ligninolytic cultures of *P. chrysosporium,* and the responsible enzymes have been purified and characterized [3, 33, 87, 108]. In addition, the enzymes cellobiose:quinone oxido reductase [155] and NADH:quinone oxidoreductase [24], which are involved in quinone reduction, have been isolated and purified from the same fungus.

The lignin degradation by *P. chrysosporium* must involve both oxidative and reductive conversions. Oxidative fragmentation and cleavage reactions are required for the chemical breakdown of the polymer. Lignin depolymerization, therefore, can be considered as an oxidative process, whereas the metabolism of lignin fragments involves a combination of oxidations and reductions. The initial steps of lignin degradation by *P. chrysosporiunz* seem to involve a combination of oxidative conversions catalyzed by MnP and LIP, possibly in conjunction with the action of Mn^{+3} ions, veratryl alcohol or other compounds, as redox mediators. This primary attack would lead to C_α-oxidation in the lignin polymer or to $C_\alpha–C_\beta$ cleavage and alkyl-phenyl cleavage under release of small lignin fragments. The C_α-carbonyl groups already present or formed in the process above are linked to both phenolic and non-phenolic aromatic rings. For further degradation, the latter have to undergo demethylation to yield phenols or can be separated from the polymer by oxidation of an adjacent $_\beta$-ring by LIP, followed by Cp-ether bond cleavage [79]. Phenolic lignin substructures with a $C_{\alpha –}$ carbonyl function, both natives or stemming from oxidation of phenolic substructures by MnP or oxidation of non-phenolic structures by LIP and subsequent demethylation, could in contrast be attacked by MnP. This process would lead to the liberation of lignin fragments from the polymer via C_α-C_β cleavage and alkyl-phenyl cleavage. Importantly, these results suggest that no further enzymes capable of cleaving C_α-oxo substructures need to be contemplated. Lignin biodegradation by WRF is an oxidative process and phenol oxidases are the key enzymes [82, 89, 118]. Of these, LiP, MnP and laccases from WRF (especially Botrytis cinerea, *P. chrysosporium*, Stropharia coronilla, P. ostreatus and Trametes versicolor) have been studied [61, 100] (Figures 3–5).

LiP and MnP oxidize the substrate by two consecutive one-electron oxidation steps with intermediate cation radical formation. LiP degrades non-phenolic lignin units (up to 90% of the polymer), whereas MnP generates Mn^{3+}, which acts as a diffusible oxidizer on phenolic or non-phenolic

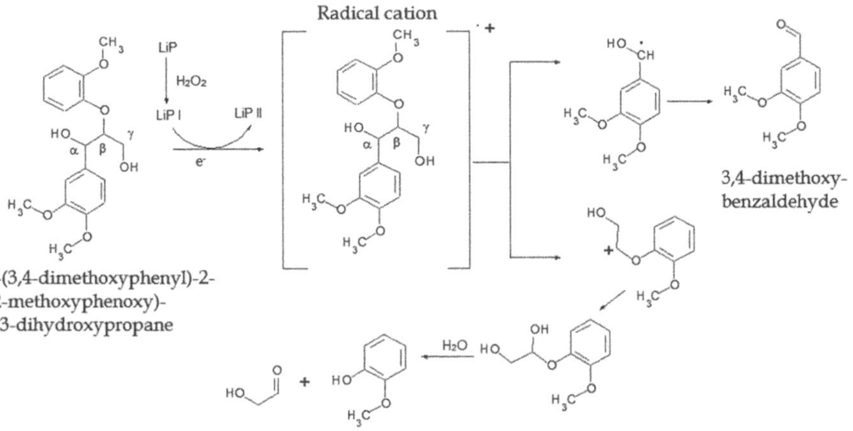

FIGURE 9.3   LiP-catalyzed oxidation of non-phenolic β-O-4 lignin model compound (Wong, 2009).

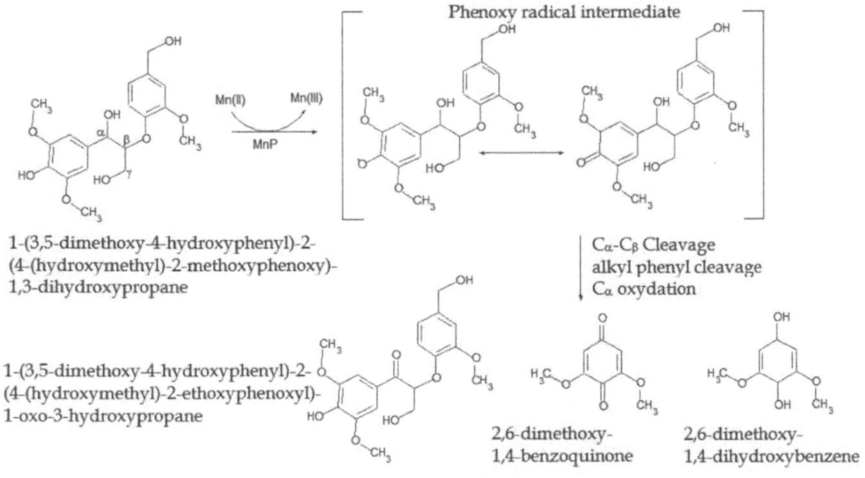

FIGURE 9.4   MnP-catalyzed oxidation of phenolic aryglycerol β-aryl ether lignin model compound (Wong, 2009).

lignin units via lipid peroxidation reactions [63, 27]. Laccase are blue copper oxidases that catalyze the one-electron oxidation of phenolics and other electron-rich substrates [50]. These include arylalcohol oxidase (AAO) described in Pleurotus eryngii [74] and other fungi, and *P. chrysosporium* glyoxal oxidase [74]. Fungal aryl-alcohol dehydrogenases (AAD) and quinone reductases (QR) are also involved in lignin degradation [46, 48].

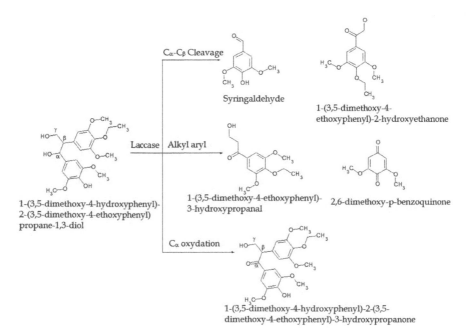

FIGURE 9.5 Laccase-catalyzed oxidation of phenolic β-1 lignin model compound (Wong, 2009).

Laccases or ligninolytic peroxidases (LiP and MnP) produced by WRF oxidize the lignin polymer, thereby generating aromatic radicals (a). These evolve in different non-enzymatic reactions, including C-4-ether breakdown (b), aromatic ring cleavage (c), Cα–Cβ breakdown (d), and demethoxylation (e). The aromatic aldehydes released from Cα–Cβ breakdown of lignin, or synthesized de novo by the fungus (f, g), are the substrates for H_2O generation by AAO in cyclic redox reactions also involving AAD. Phenoxy radicals from C4-ether breakdown (b) can repolymerize on the lignin polymer (h) if they are not first reduced by oxidases to phenolic compounds (i). The phenolic compounds formed can be again reoxidized by laccases or peroxidases (j). Phenoxy radicals can also be subjected to Cα–Cβ breakdown (k), yielding p-quinones. Quinones from g and/or k contribute to oxygen activation in redox cycling reactions involving oxygen activation in redox cycling reactions with QR, laccases, and peroxidases (l, m). This results in reduction of the ferric iron present in wood (n), either by superoxide cation radical or directly by the semiquinone radicals, and its reoxidation with concomitant reduction of H_2O_2 to a hydroxyl free radical ($OH\times$) (o). The latter is a very mobile and very strong oxidizer that can initiate the attack on lignin (p) in the initial stages of wood decay, when the small size of pores in the still-intact cell wall prevents the penetration of ligninolytic enzymes. Then, lignin degradation proceeds by oxidative attack of the enzymes described above. In the final steps, simple products from lignin degradation enter the fungal hyphae and are incorporated into intracellular catabolic routes [100]. Fungal feruloyl and p-coumaroyl esterases are capable of releasing feruloyl and p-coumaroyl units and play an important role in biodegradation of recalcitrant cell walls in grasses [82]. These enzymes act synergistically with xylanases to disrupt the hemicellulose-lignin association, without mineralization of lignin per se [15, 38]. Therefore, hemicellulose degradation is required before efficient lignin removal can commence. In *P. chrysosporium*, a co-metabolizable carbon source is essential for lignin degradation [77], and it is produced in response to nitrogen starvation [75]. This indicates that the ligninolytic system is formed as part of secondary metabolism in this organism. Carbohydrate starvation likewise leads to a rapid but transient onset of ligninolytic activity. Elevated oxygen levels increase the rate of lignin biodegradation through the production of hydrogen peroxide as the extracellular oxidant and the subsequent induction of ligninolytic activity [79, 78, 35].

Ligninolytic activity in *P. chrysosporium* and other WRF is associated with multiple isoenzymes. At least 21 heme peroxidases are produced in liquid cultures of *P. chrysosporium* [88, 147]. The physical and kinetic characteristics of the isozymes are very similar, but differences in stability, quantity and catalytic properties have been described [36]. They may be the consequence of three different structural genes or post-translational modifications. *Coriolus versicolor* [106], *Panus tigrinus* [98], *Rigidoporus lignosus* [40], and *Ceriporiopsis subcermispora* [94] secrete lactase isoenzymes.

9.4 FACTORS AFFECTING THE EFFICIENCY OF ENZYMES

Environmental conditions such as temperature, humidity, microclimate and nitrogen content of the substrate may also govern the selectivity of lignin biodegradation in vivo.

9.4.1 EFFECT OF CARBON AND NITROGEN SOURCES

An increase in production of laccase and MnP by T. trogii was observed by the addition of easily available carbon and nitrogen sources to the culture medium, such as malt-extract and peptone [90]. This is due to the presence of aromatic amino acids tryptophan and tyrosine in the Malt-extract broth. Same observation in Phlebia radiate and Phlebia fascicularia with laccase and MnP activity in *T. versicolor* of about 2.40 and 2.00 U/mL, respectively was observed [5]. A large increase in LiP production when adding tryptophan to the cultures of *T. versicolor*, *P. chrysosporium* and *Chrysosporium lignorum* was observed [23].

9.4.2 EFFECT OF HEAVY METALS

Heavy metals have a great important role in the regulation of ligninolytic and cellulolytic enzymes at the transcription level as well as during their catalytic action [9]. The presence of Mn^{2+} is known to induce the production of MnP in many WRF, but Mn2+ lowers LiP titers [126]. Cu^{2+} has been testified to being a strong laccase inducer in several species, among them *T. pubescens* [39] and *T. versicolor* [22]. Production of additional laccase isoenzymes not present under natural conditions, was observed after Cu^{2+} addition in *P. chrysosporium* [30] and Marasmius quercophilus [81], but not in T. trogii [91]. In T. trogii, addition of copper increased the activities of MnP and glyoxal oxidase. The presence of heavy metals affects the growth of WRF, the decrease of fungal growth rate is sometimes accompanied with a prolonged lag phase. However, high Cu^{2+} concentration could inhibit LiP production by T. trogii. Cu^{2+} inhibited the growth of Ganoderma lucidum at concentrations less than 1 mM, while 150 ppm of Cu^{2+} decreased the growth rate of *P. chrysosporium*. In decaying wood, manganese promotes selective removal of lignin by WRF [13, 72, 73]. For some white-rot fungi, a high content of manganese in the culture medium results in an increase of activity of ligninolytic enzymes [115, 114, 128] Probably, the Mn(III) ions, oxidized by manganese peroxidases to produce free radicals, preferentially degrade the aromatic structures present in the lignocellulose. Kerem and Hadar observed a higher degradation level when the lower amount of $MnSO_4$ was added to wheat straw, resulting in degradation of 50% of the lignin and only a slight increase in cellulose degradation [72, 73].

9.4.3   EFFECT OF OXYGEN/MOISTURE

Since lignin degradation is strictly an oxidative process, the concentration of oxygen in the gas phase has a major role in the delignification process. Delignification process requires the presence of oxygen at a partial pressure equal to that in the natural atmosphere. In fact, lignin is degraded much faster in the presence of pure oxygen than in the presence of air [121] and ligninolytic activity is not observed in low partial pressures (5.07 Pa) of oxygen [80]. The lignin degradation is affected by oxygen at two stages, firstly stimulating the transcription of the ligninolytic enzyme system and secondly at the time of oxidation of lignin [10]. The efficiency of gas exchange between the gaseous phase and the solid substrate also influences the fermentation. The moisture content in the substrate must be controlled at an optimum level [66]. Excessive water hinders the exchange of gases and creates anaerobic conditions inside the substrate, while insufficient water does not allow optimal fungal activity. The optimum ratio of solid-to-liquid in solid-state fermentation depends upon the quality, particle size and water-holding capacity of the substrate [156].

9.4.4   EFFECT OF PH

Optimum pH for the biological delignification observed to be occur at lower pH. Different species differs in their optimum pH for maximum delignification. *P. chrysosporium* degrades lignin maximum at pH 4.0 with decreasing activity towards lower pH values [80]. Boyle noticed that lignin degradation by *P chrysosporium* and *P. sajor-caju* increased with decreasing pH down to pH 3.0 [16]. The enzymes purified from *P. chrysosporium* have their activity optimum at pH 3.0 (Lip; [147]) and pH 4.5 (MnP; [43]). The pH optimum of lactase from *Phellinus noxius* is at pH 4.6 [40]. The pH of the natural substrate varies between 4 and 6 for both hardwoods and softwoods and corresponds to the growth optimum for most wood rotting fungi. Furthermore, fungi are able to maintain a low pH environment at the hyphal level by secreting a polysaccharide slime layer [18]. In order to promote the delignification of a lignocellulosic substrate, it is also essential to maximize the rate as well as the specificity of lignin molecule degradation, avoiding polysaccharide consumption [73].

Solid-state fermentation of lignocellulosic by-products has an advantage over submerged fermentation in that it requires only one-tenths of the fermenter size as compared with the latter. The capacity of lignin degradation by fungi is influenced by the penetration of hyphae into the substrate and the extent of close physical contacts between the substrate and the degrading fungus. The fungal species investigated may also express gene products differently in agitated or stationary cultures than during growth on solid substrates in natural environments. Contrary to liquid culture, *P. chrysosporium* produces MnP as major peroxidase when grown on wood [28]. *C. subvermispora* and *P. brevispora* tested negative for LIP production in agitated culture, whereas the use of southern hybridization technique with a cDNA probe from *P. chrysosporium* revealed the presence of LiP genes in both fungi [123]. In the case of *P. brevispora*, LiP production in liquid culture was stated earlier [113].

9.5 CONCLUSION

The pulp and paper industry is emerging as one of the potential large markets for enzyme application. Microbial enzymes: cellulases, xylanases and ligninases, create new technologies for pulp and paper processing. Xylanases reduce the amount of chemicals required for bleaching; cellulases smooth fibers. Lignin-degrading enzymes remove lignin from biomass. The most important application of enzymes in the pulp and paper industry is in the prebleaching of Kraft pulp. Xylanase prebleaching technology is now in use at several mills worldwide [7, 145]. Reducing the cost for enzyme production is still needed in order to develop enzymatic treatment processes for different industrial and environmental applications, which might be more competitive than conventional and other novel treatment technologies. The fungal production of lignin modifying enzymes through the bioconversion of lignocellulosic residues has been widely investigated in recent years. This approach is attractive because of foreseeable effects on cost reduction, waste reuse and enhanced enzyme production. Solid state fermentation (SSF) processes have shown to be particularly suitable for the production of enzymes by filamentous fungi, since they reproduce the natural habitats of such fungi [25]. Therefore, this strain could be an attractive and alternative source of these enzymes, which have gained renewed interest in recent years, mainly due to their applications in paper industries for pulp treatment, improving the effectiveness of conventional bleaching. Full biological processes are advantageous due to high selectivity and mild reaction conditions, but they are not available yet. Recent studies promote the production of genetically engineered multitask microorganisms able not only to lignocelluloses delignification but also saccharification and fermentation. Additionally, these microorganisms should be marginally inhibited by substrate and product concentration. While this technological dream becomes a scalable economic reality, pretreatments may be carried out by chemical means. In this sense, there is a wide range of possibilities.

KEYWORDS

- biomass
- biosynthesized
- cellulose
- composition
- lignocellulosic material
- structural component

REFERENCES

1. Adler, E. (1977). Lignin chemistry – past, present and future. Wood Sci. Technol. 11, 169–218.
2. Alvira, P., Tomas-Pejo, E., Ballesteros, M., Negro, M. J. (2010). Pretreatment technologies for an efficient bioethanol production process based on enzymatic hydrolysis: A review. Bioresource Technology, 101, 4851–4861.
3. Ander, P., Hatakka, A., Eriksson, K.-E. (1980). Vanillic acid metabolism by the white-rot fungus Sporotrichum pulverulentum. Arch. Microbial., 125, 189–202.

4. Andresen, M., Johansson, L. S., Tanem, B. S., Stenius P (2006). Properties and characterization of hydrophobized microfibrillated cellulose. Cellulose 13:665–677.
5. Arora, D. S., Gill, P. K. (2001). Effects of various media and supplements on laccase production by some white rot fungi, Biores. Technol. 77, 89–91.
6. Azizi, Samir, M. A. S., Alloin, F., Dufresne, A. (2005). A review of recent research into cellulosic whiskers, their properties and their application in nanocomposite field. Biomacromolecules, 6, 612–626.
7. Bajpai, P. (1999). Application of enzymes in the pulp and industry, Biotechnol. Prog. 5 147–157.
8. Bak, J. S., Ko, J. K., Choi, I. G., Park, Y. C., Seo, J. H., Kim, K. H. (2009). Fungal pretreatment of lignocellulose by Phanerochaete chrysosporium to produce ethanol from rice straw. Biotechnology and Bioengineering, 104, 471–482.
9. Baldrian, P. (2003). Interactions of heavy metals with white-rot fungi, Enzyme Microb. Technol. 32, 78–91.
10. Barlev, S. S., Kirk, T. K. (1981). Effects of molecular oxygen on lignin degradation by Phanerochaete chrysosporium. Biochem. Biophys. Res. Commun. 99, 373–378.
11. Bhat, M. K., Bhat S (1997). Cellulose degrading enzymes and their potential industrial applications. Biotechnol Adv 15, 583–620.
12. Blanchette, R. A. (1995). Degradation of the lignocellulose complex in wood, Can. J. Bot., 73, 999–1010.
13. Blanchette, R. A. (1984). Manganese accumulation in wood decayed by white-rot fungi. Phytopatholgy 74, 725–730.
14. Blanchette, R. A., Abad, A. R., Farrell, R. L., Leathers, R. L. (1989). Detection of lignin peroxidase and xylanase by immunocytochemical labeling in wood decayed by basidiomycetes. Appl. Environ. Microbial. 55, 1457–1465.
15. Borneman, W. S., Hartley, R. D., Morrison, W. H., Akin, D. E., Ljungdahl, L. G. (1990). Feruloyl and p-coumaroyl esterase from anaerobic fungi in relation to plant cell wall degradation. Appl Microbiol Biotechnol. 3, 345–51.
16. Boyle, C. D., Kropp, B. R., Reid, I. D. (1992). Solubilization and mineralization of lignin by white rot fungi. Appl. Environ. Microbial. 58, 3217–3224.
17. Brinkmann, K., Blaschke, L., Polle, A. (2002). Comparison of different methods for lignin determination as a basis for calibration of near infrared reflectance spectroscopy and implications of lignoproteins. J Chem Ecol (United States), 28, 2483–501.
18. Buchala, A. J., Leisola, M. S. A. (1987). Structure of the P-glucan secreted by Phanerochaete chrysosporium in continuous culture. Carbohydr. Res. 165, 146–149.
19. Buxton, D. R., Russell, J. R. (1988). Lignin constituents and cell-wall digestibility of grass and legume stems, Crop Sci., 28, 553–558.
20. Cardona, C. A., Sanchez, O. J. (2007). Fuel ethanol production: process design trends and integration opportunities. Bioresour. Technol. 98, 2415–2457.
21. Celi'nska, E., Grajek, W. (2009). Biotechnological production of 2,3-butanediol – Current state and prospects. Biotechnology Advances, 27, 715–725.
22. Collins, P. J., Dobson, A. D. W. (1997). Regulation of laccase gene transcription in Trametes versicolor, Appl. Environ. Microbiol. 63 3444–3450.
23. Collins, P. J., Field, J. A., Teunissen, P., Dobson, A. D. (1997). Stabilization of lignin peroxidases in white rot fungi by tryptophan, Appl. Environ. Microbiol. 63, 2543–2548.

24. Constam, D., Muheim, A., Zimmermann, W., Fiechter, A. (1991). Purification and characterization of an intracellular NADH:quinone oxidoreductase from Phanerochaete chrysosporium. J. Gen. Microbial. 137, 2209–2214.
25. Couto, S. R., Sanrom'an, M. A. (2005). Application of solid-state fermentation to ligninolytic enzyme production—review. Biochem. Eng. J. 22, 211–219.
26. Costa, S., Gonçalves, A., Esposito, E. (2005). Ceriporiopsis subvermispora used in delignification of sugarcane bagasse prior to soda/anthraquinone pulping. Applied Biochemistry and Biotechnology, 122, 695–706.
27. Cullen, D., Kersten, P. J. (2004). Enzymology and molecular biology of lignin degradation. In: Brambl, R., Marzluf, G. A., editors. The Mycota, I. I. I., Biochemistry and molecular biology Berlin-Heidelberg: Springer-Verlag;., 249–273.
28. Datta, A., Bettermann, A., Kirk, T. K. (1991). Identification of a specific manganese peroxidase among ligninolytic enzymes secreted by Phanerochaete chrysosporium. Appl. Environ. Microbial. 57, 1453–1460.
29. Dias, A. A., Freitas, G. S., Marques, G. S. M., Sampaio, A., Fraga, I. S., Rodrigues, M. A. M. (2010). Enzymatic saccharification of biologically pre-treated wheat straw with white-rot fungi. Bioresource Technology, 101, 6045–6050.
30. Dittmer, J. K., Patel, N. J., Dhawale, S. S. (1997). Production of multiple laccase isoforms by Phanerochaete chrysosporium grown under nutrient sufficiency, FEMS Microbiol. Lett. 149, 65–70.
31. Dufresne, A., Cavaille' J. Y., Vignon, M. R. (1997). Mechanical behavior of sugar beet cellulose microfibrils. J Appl Polym Sci 64:1185–1194.
32. Easty, D. B., Berben, S. A., DeThomas, F. A., Brimmer, P. J., Near-infrared spectroscopy for the analysis of wood pulp: quantifying hardwood softwood mixtures and estimating lignin content. Tappi J Wood Pulp Anal, 1990;73:257–261.
33. Enoki, A., Goldsby, G. P., Gold, M. H. (1981). P-Ether cleavage of the lignin model compound 4-ethoxy3- methoxyphenylglycerol P-guaiacyl ether and derivatives by Phanerochaete chrysosporium. Arch. Microbial., 129, 141–145.
34. Eriksson, K.-E. L., Blanchette, R. A., Ander, P. (1990). Microbial and Enzymatic Degradation of Wood and Wood Components, Springer, Berlin, 407.
35. Faison, B. L., Kirt, T. K. (1983). Relationship between lignin degradation and production of reduced oxygen species by Phanerochaete chrysosporium. Appl Environ Microbiol, 46, 1140–1145.
36. Farrell, R. L., Murtagh, K. E., Tien, M., Mozuch, M. D., Kirk, T. K. (1989). Physical and enzymatic properties of lignin peroxidase isoenzymes from Phanerochaete chrysosporium. Enzyme Microb. Technol. 11, 322–328.
37. Fernandes, L., Loguercio-Leite, C., Esposito, E., Menezes Reis, M. (2005). In vitro wood decay of Eucalyptus grandis by the basidiomycete fungus Phellinus flavomarginatus. International Biodeterioration & Biodegradation 55, 187–193.
38. Fillingham, I. J., Kroon, P. A., Williamson, G., Gilbert, H. J., Hazlewood, G. P. (1999). A modular cinnamoyl ester hydrolase from the anaerobic fungus Piromyces equi acts synergistically with xylanase and is part of a multiprotein cellulose-binding cellulase-hemicellulase complex. Biochem, J., 343, 215–224.
39. Galhaup, C., Wagner, H., Hinterstoisser, B., Haltrich, D. (2002). Increased production of laccase by the wood-degrading basidiomycete Trametes pubescens, Enzyme Microb. Technol. 30, 529–536.

40. Geiger, J. P., Huguenin, B., Nicole, M., Nandris, D. (1986). Laccases of Rigidoporus lignosus and Phellinus noxius, I. I., Effect of, R., lignosus lactase Ll on thioglycolic lignin of hevea. Appl. Biochem. Biotechnol. 13, 97–110.

41. Gierlinger, N., Jacques, D., Schwanninger, M., Wimmer, R., Paques, L. E. (2004). Heartwood extractives and lignin content of different larch species (Larix sp.) and relationships to brown-rot decay-resistance. Trees, 18, 230–236.

42. Gilbertson, R. L. (1980). Wood-rotting fungi of north America. Mycologia 72, 1–49.

43. Glenn, J. K., Morgan, M. A., Mayfield, M. B., Kuwahara, M., Gold MH (1983). An extracellular H2O2-requiring enzyme preparation involved in lignin biodegradation by the white-rot basidiomycete Phanerochaete chrysosporium. Biochem. Biophys. Res. Commun. 114: 1077–1083.

44. Grabber, J. H. (2005). How do lignin composition, structure, and crosslinking affect.

45. degradability, A review of cell wall model studies. Crop Sci., 45, 820–831.

46. Guillén, F., Martínez, A. T., Martínez, M.,J. (1992). Substrate specificity and properties of the aryl-alcohol oxidase from the ligninolytic fungus Pleurotus eryngii. Eur J Biochem, 209, 603–611.

47. Gupta, R., Mehta, G., Khasa, Y. P., Kuhad, R. C. (2011). Fungal delignification of ligno-cellulosic biomass improves the saccharification of cellulosics. Biodegradation, 22, 797–804.

48. Gutiérrez, A., Caramelo, L., Prieto, A., Martínez, M. J., Martínez, A. T. (1994). Anisaldehyde production and aryl-alcohol oxidase and dehydrogenase activities in ligninolytic fungi of the genus Pleurotus. Appl Environ Microbiol, 60, 1783–1788.

49. Habibi, Y., Goffin, A. L., Schiltz, N., Duquesne, E., Dubois, P., Dufresne A (2008). Bionanocomposites based on poly (e-caprolactone)-grafted cellulose nanocrystals by ring opening polymerization. J Mater Chem 18, 5002–5010.

50. Hammel, K. E. (1997). Fungal degradation of lignin. In: Cadisch, G., Giller, K. E., editors. Plant litter quality and decomposition. CAB-International, 33–46.

51. Hartley, R. D., Ford, C. W. (1989). Phenolic constituents of plant cell walls and wall biodegradability. In: N. G. Lewis, M.G. Paice (Editors), Plant Cell Wall Polymers: Biogenesis and Biodegradation. American Chemical Society, Washington, DC, 137–145.

52. Hatakka, A. (1999). Transformation of wheat straw in the course of solid-state fermentation by four ligninolytic basidiomycetes. Enzyme and Microbial Technology 25, 605–612.

53. Hatakka, A. (2001). Biodegradation of lignin. In: Hofrichter, M., Steinbu¨chel, A. (Eds.), Biopolymers. Lignin, Humic Substances and Coal, vol. 1. Wiley–VCH, Germany, pp. 129–180.

54. Hattori, T., Higuchi, T. (1991). Degradation of phenolic and nonphenolic syringyl and biphenyl lignin model compounds by lignin peroxidase. Mokuzai Gakkaishi 37, 542–547.

55. Hayashi, N., Sugiyama, J., Okano, T., Ishihara M (1998). The enzymatic susceptibility of cellulose microfibrils of the algal-bacterial type and the cotton-ramie type. Carbohydr Res 305:261–269.

56. Henriksson, M., Henriksson, G., Berglund, L. A., Lindstrom. T. (2007). An environmentally friendly method for enzyme assisted preparation of microfibrillated cellulose (MFC) nanofibers. Eur Polym, J., 43, 3434–3441.

57. Henrissat, B. (1994). Cellulases and their interaction with cellulose. Cellulose, 1, 169–196.

58. Higuchi, T. (1986). Catabolic pathways and role of ligninases for the degradation of lignin substructure models by white rot fungi. Wood Res. 73, 58–81.

59. Higuchi, T. (1989). Mechanisms of lignin degradation by lignin peroxidase and lactase of white-rot fungi. In: Lewis, N. G., Paice, M. G. (Eds.), Biogenesis and Biodegradation of Plant Cell Polymers. ACS Symposium Series 399, 482–502.

60. Hon, D. N. S. (2000). Pragmatic approaches to utilization of natural polymers: Challenges and opportunities. In: Frollini, E., Leao, A. L., Mattoso, L. H. C., editors. Natural polymers and agrofibers composites. New York: Marcel Dekker Inc;. p. 1–14.

61. Howard, R. L., Abotsi, E., Jansen, van, Rensburg, E. L., Howard, S. (2003). Lignocellulose biotechnology: issues of bioconversion and enzyme production. Afr J Biotechnol, 2, 602–19.

62. Jal´c, D., Siroka, P., Fejes, J., Ceresn'akov'a, Z. (1999). Effect of three strains of Pleurotus tuber-regium (Fr.) Sing. on chemical composition and rumen fermentation of wheat straw., J. Gen. Appl. Microbiol. 6, 277–282.

63. Jensen, Jr, K.,A., Bao, W., Kawai, S., Srebotnik, E., Hammel, H. E. (1996). Manganese-dependent cleavage of nonphenolic lignin structure by Ceriporipsis subvermispora in the absence of lignin peroxidase. Appl Environ Microbiol, 62, 3679–86.

64. John, F., Monsalve, G., Medina, P. I. V., Ruiz, C. A. A., (2006). Ethanol production of banana shell and cassava starch. Dyna Universidad Nacional de Colombia., 73, 21–27.

65. Jung, H. G., Valdez, F. R., Abad, A. R., Blanchette, R. A., Hatfield, R. D. (1992). Effect of white rot basidiom ycetes on chemical composition and in vitro digestibility of oat straw and alfalfa stems, J. Anim. Sci., 70, 1928–1935.

66. Kamra, D. N., Zadraz´il, F. (1988). Microbiological improvement of lignocellulosics in animal feed production: a review. In: Zadraz´il, F., Reiniger, P. (Eds.), Treatment of Lignocellulosics with White-Rot Fungi. Elsevier, Essex, UK, pp. 56–63.

67. Kapich, A. N., Prior, B. A., Lundell, T., Hatakka, A. (2005). A rapid method to quantify pro-oxidant activity in cultures of wood-decaying white-rot fungi. J. Microbiol. Meth., 61, 261–271.

68. Karunanandaa, K., Varga, G. A. (1996a). Colonization of crop residues by white-rot fungi: cell wall monosaccharides, phenolic acids, ruminal fermentation characteristics and digestibility of cell wall fiber components in vitro. Anim. Feed Sci. Technol., 63, 273–288.

69. Karunanandaa, K., Varga, G. A. (1996b). Colonization of crop residues by white-rot fungi (Cyathus stercoreus): effect on ruminal fermentation pattern, nitrogen metabolism, and fiber utilization during continuous culture. Anim. Feed Sci. Technol., 61, 1–16.

70. Keller, F. A., Hamilton, J. E., Nguyen, Q. A. (2003). Microbial pretreatment of biomass: potential for reducing severity of thermochemical biomass pretreatment. Applied Biochemistry and Biotechnology 105, 27–41.

71. Kenealy, W. R., Jeffries, T. W. (2003). Enzyme processes for pulp and : a review of recent developments. In: Barry, G., Nicholas, D. D., Schultz TP (eds) Wood deterioration and preservation: advances in our changing world. Oxford University Press, Washington, 210–239.

72. Kerem, Z., Hadar, Y. (1993). Effect of manganese on lignin degradation by Pleurotus ostreatus during solid-state fermentation. Appl. Environ. Microbiol. 59, 4115–4120.

73. Kerem, Z., Hadar, Y. (1995). Effect of manganese on preferential degradation of lignin by Pleurotus ostreatus during solid-state fermentation. Appl. Environ. Microbiol. 61, 3057–3062.

74. Kersten, P., Cullen, D. (2007). Extracellular oxidative systems of the lignin-degrading Basidiomycete Phanerochaete chrysosporium. Forest Genet Biol, 44, 77–87.

75. Keizer, P., Kirt, T. K., Zeikus, J. G. (1978). Ligninolytic enzyme system of Phanerochaete chrysosporium: synthesized in the absence of lignin in response to nitrogen starvation. J Bacteriol, 135, 790–7.

76. Kirk, T. K., Fenn, P. (1982). Formation and action of the ligninolytic system in basidiomycetes. In: Swift, M. J., Frankland, J., Hedger, J. N. (Eds.1, Decomposer basidiomycetes Br. Mycol. Sot. Symp. 4. Cambridge University Press, Cambridge, pp. 67–90.

77. Kirk, T. K., Connors, W. J., Zeikus, J. G. (1976). Requirement for a growth substrate during lignin decomposition by two wood-rotting fungi. Appl. Environ. Microbial. 32, 192–194.

78. Kirk, T. K., Cullen, D. (1998). Enzymology and molecular genetics of wood degradation by white rot fungi. In: Young, R. A., Akhtar, M., editors. Environmentally friendly technologies for the pulp and industry. New York: John Wiley & Sons, 273–308.

79. Kirk, T. K., Farell, R. L. (1987). Enzymatic "combustion": the microbial degradation of lignin. Annu Rev Microbiol, 41, 465–505.

80. Kirk, T. K., Schultz, E., Conors, W. J., Lorenz, L. F., Zeikus, J. G. (1978). Influence of culture parameters on lignin metabolism by Phanerochaete chrysosporium. Arch. Microbiol. 117, 277–285.

81. Klonowska, A., Le Petit, J., Tron, T. (2001). Enhancement of minor laccases production in the basidiomycete Marasmius quercophilus C30, FEMS Microbiol. Lett. 200, 25–30.

82. Kuhad, R. C., Singh, A., Eriksson, K. E. L. (1997). Microorganisms and enzymes involved in the degradation of plant fiber cell walls. Advanced Biochemical Engineering Biotechnology, 57, 46–125.

83. Kuhar, S., Nair, L. M., Kuhad, R. C. (2008). Pretreatment of lignocellulosic material with fungi capable of higher lignin degradation and lower carbohydrate degradation improves substrate acid hydrolysis and the eventual conversion to ethanol. Canadian Journal Microbiology, 54, 305–313.

84. Kumar, R., Mago, G., Balan, V., Wyman CE (2009). Physical and chemical characterizations of corn stover and poplar solids resulting from leading pretreatment technologies. Bioresour Technol 100, 3948–3962.

85. Kuwahara, M., Glenn, J. K., Morgan, M. A., Gold, M. H. (1984). Separation and characterization of two extracellular HzO, -dependent oxidases from ligninolytic cultures of Phanerochaete chrysospvtium. FEBS Lett. 169, 247–250.

86. Laureano-Perez, L., Teymouri, F., Alizadeh, H., Dale, B. E. (2005). Understanding factors that limit enzymatic hydrolysis of biomass. Appl. Biochem. Biotechnol., 124, 1081–1099.

87. Leisola, M. S. A., Fiechter, A. (1985). New trends in lignin biodegradation. In: Mizrahi, A., van Wezel, A. L. (Eds.), Advances in Biotechnological Processes. Alan, R., Liss Inc., New York, 59–89.

88. Leisola, M. S. A., Kozulic, B., Meussdoerffer, F., Fiechter, A. (1987). Homology among multiple extracellular peroxidases from Phanerochaete chrysosporium. J. Biol. Chem.262, 419–424.

89. Leonowicsz, A., Matuszewska, A., Luterek, J., Ziegenhagen, D., Wojtas-Wasilewska, M., Cho, N.,S. (1999). Biodegradation of lignin by white rot fungi. Fungal Genet Biol, 27, 175–85.

90. Levin, L., Herrmann, C., Papinutti, V. L. (2008). Optimization of lignocellulolytic enzyme production by the white-rot fungus Trametes trogii in solid-state fermentation using response surface methodology. Biochemical Engineering Journal, 39(1), 207–214.

91. Levin, L., Ramos, A. M. Forchiassin, F. (2002). Copper induction of lignin modifying enzymes in the white rot fungus Trametes trogii, Mycologia, 94, 377–383.

92. Liese, W., Greaves, H. (1975). Micromorphology of bacterial attack. In: Liese, W. (Ed.), Biological Transformation of Wood by Microorganisms. Springer, New York, 74–88.

93. Liu, H., Fu, S. Y., Zhu, J. Y., Li, H., Zhan, H. Y. (2009). Visualization of enzymatic hydrolysis of cellulose using AFM phase imaging. Enzyme Microb Technol, 45, 274–281.

94. Lobos, S., Larrain, J., Salas, L., Cullen, D., Vicuna, R. (1994). Isozymes of manganese-dependent peroxidase and lactase produced by the lignin-degrading basidiomycete Ceriporiopsis subuermispora. I. Gen. Microbial. 140, 2691–2698.

95. Lu, J., Askel, P., Drzal LT (2008). Surface modification of microfibrillated cellulose for epoxy composite applications. Polymer 49:1285–1296.

96. Mai, C., K'ues, U., Militz, H. (2004). Biotechnology in the wood industry, Appl. Microbiol. Biotechnol. 63 477–494.

97. Malherbe, S., Cloete, T. E. (2002). Lignocellulose biodegradation: fundamentals and applications. Re/Views Environ Sci Bio/Technol., 1, 105–114.

98. Maltseva, O. V., Niku-Paavola, M.-L., Leontievsky, A. A., Myasoedova, N. M., Golovleva, L. A. (1991). Ligninolytic enzymes of the white-rot fungus Panus tigrinus. Biotechnol. Appl. Biochem. 13, 291–302.

99. Martin, C., López, Y., Plasencia, Y., Hernández, E. (2006). Characterization of agricultural and agroindustrial residues as raw materials for ethanol production. Chem Biochem Eng, 20, 443–7.

100. Martínez, A. T., Speranza, M., Ruiz-Dueñas, F. J., Ferreira, P., Camarero, S., Guillén, F, (2005). Biodegradation of lignocellulosics: microbial, chemical, and enzymatic aspects of the fungal attack of lignin. Int Microbiol, 8, 195–204.

101. McKendry, P. (2002). Energy production from biomass: overview of biomass. Bioresour Technol, 83, 37–43.

102. Menon, V., Rao, M. (2012). Trends in bioconversion of lignocellulose: Biofuels, platform chemicals and biorefinery concept. Progress Energy Combustion Science, 38, 522–550.

103. Messner, K., Srebotnik, E. (1994). Biopulping: An overview of developments in an environmentally safe-making technology. FEMS Microbial. Rev. 13, 351–364.

104. Meza, J. C., Sigoillot, J. C., Lomascolo, A., Navarro, D., Auria, R. (2006). New process for fungal delignification of sugarcane bagasse and simultaneous production of laccase in a vapor phase bioreactor. Journal of Agricultural and Food Chemistry, 54, 3852–3858.

105. Michell, A. J., Schimleck, L. R. (1996). NIR spectroscopy of woods from Eucalyptus globulus. Appita, 49, 23–26.

106. Morohoshi, N. (1991). Laccases from the ligninolytic fungus Coriolus versicolor. In: Leatham. G., Himmel, M. E. (Eds.), Enzymes in Biomass Conversion. ACS Symposium Series 460, pp. 204–207.

107. Mosier, N., Wyman, C., Dale, B., Elander, R., Lee, Y. Y., Holtzapple, M., Ladisch, M. (2005). Features of promising technologies for pretreatment of lignocellulosic biomass. Bioresource Technology. 96, 673–686.

108. Muheim, A., Waldner, R., Leisola, M. S. A., Fiechter, A. (1990). An extracellular aryl-alcohol oxidase from the white-rot fungus Bjerkandera adusta. Enzyme Microb. Technol., L, 204–209.

109. Mukherjee, R., Nandi, B. (2004). Improvement of in vitro digestibility through biological treatment of water hyacinth biomass by two Pleurotus species. Inter-national Biodeterioration Biodegradation, 53, 7–12.

110. Odier, E., Mozuch, M. D., Kalyanaraman, B., Kirk, T. K. (1988). Ligninase-mediated phenoxy radical formation and polymerization unaffected by cellobiose: quinone oxidoreductase. Biochimie 70, 847–852.

111. Paakko, M., Ankerfors, M., Kosonen, H., Nykanen, A., Ahola, S., Osterberg, M., Ruokolainen, J., Laine, J., Larsson, P. T., Ikkala, O., Lindstrom, T. (2007). Enzymatic hydrolysis combined with mechanical shearing and high-pressure homogeneization for nanoscale cellulose fibrils and strong gels. Biomacromolecules, 8, 1934–1941.

112. Pérez, J., Muñoz-Dorado, J., De-la-Rubia, T., Martínez, J. (2002). Biodegradation and biological treatments of cellulose, hemicellulose and lignin: an overview. Int Microbiol, 5, 53–63.

113. Perez, J., Jeffries, T. W. (1990). Mineralization of 14C-ringlabeled synthetic lignin correlates with the production of lignin peroxidase, not of manganese peroxidase or lactase. Appl. Environ. Microbial. 56, 1806–1812.

114. Perez, J., Jeffries, T. W. (1992). Roles of manganese and organic acid chelators in regulating lignin degradation and biosynthesis of peroxidases by Phanerochaete chrysosporium. Appl. Environ. Microbiol. 58, 2402–2409.

115. Perie, F. H., Gold, M. H. (1991). Manganese regulation of manganese peroxidase expression and lignin degradation by the white-rot fungus Dichomitus squalens. Appl. Environ. Microbiol. 57, 2240–2245.

116. Pointing, S. B. (2001). Feasibility of bioremediation by white-rot fungi, Appl. Microbiol. Biotechnol., 57, 20–33.

117. Prassad, S., Singh, A., Joshi, H. C. (2007). Ethanol as an alternative fuel from agricultural, industrial and urban residues. Resour Conserv Recycl., 50, 1–39.

118. Rabinovich, M. L., Bolobova, A. V., Vasil'chenko. (2004). Fungal decomposition of natural aromatic structures and xenobiotics: a review. Appl Biochem Microbiol, 40, 1–17.

119. Reguant, J., Rinaudo, M., Fibers Lignocellulosiques. En Iniciation á la Chimie et á la Physico- Chimie Macromoleculares. Les polymères naturels: Structure, modifications, applications. Groupe Français d'études et d' applications des polymères, France; 2000. p. 13.

120. Reid, I. D., Paice, M. G. (1994). Biological bleaching of kraft pulps by white-rot fungi and their enzymes. FEMS Microbial. Rev., 13, 369–376.

121. Reid, I. D., Seifert, K. A. (1980). Lignin degradation by Phanerochaete chrysosporium in hyperbaric oxygen. Can. J. Microbiol. 26, 11658–11671.

122. Ren, X., Buschle-Diller, G. (2007). Oxidoreductases for modification of linen fibers. Colloids Surf. A: Physicochem. Eng. Aspects, 299, 15–21.
123. Riittimann, C., Salas, L., Vicuna, R. (1992). Studies on the ligninolytic system of the white-rot fungus *Ceriporiopsis subcermispora*. In: Kuwahara, M., Shimada, M. (Eds.), Biotechnology in the Pulp and Industry. Uni Publishers Ltd., Tokio, pp. 243–248.
124. Rodr'ıguez Couto, S., Sanrom'an, M. A. (2005)Application of solid-state fermentation to ligninolytic enzyme production, Biochem. Eng. J. 22, 211–219.
125. Rodrigues, M. A. M., Pinto, P., Bezerra, R. M. F., Dias, A. A., Guedes, C. V. M., Cardoso, V. M. G., Cone, J. W., Ferreira, L. M. M., Colaco, J., Sequeira, C. A. (2008). Effect of enzyme extracts isolated from white-rot fungi on chemical composition and in vitro digestibility of wheat straw. Anim. Feed Sci. Technol., 141, 326–338.
126. Rothschild, N., Levkowitz, A., Hadar, Y., Dosoretz, C. G. (1999). Manganese deficiency can replace high oxygen levels needed for lignin peroxidase formation by Phanerochaete chrysosporium, Appl. Environ. Microbiol. 65, 483–488.
127. Rowell, M. R., Opportunities for lignocellulosic materials and composites. Emerging technologies for material and chemicals from biomass: Proceedings of symposium. Washington, DC: American Chemical Society; 1992. p. 26–31.
128. Ruttiman-Johnson, C., Cullen, D., Lamar, R. T. (1994). Manganese peroxidases of the white-rot fungus Phanerochaete sordida. Appl. Environ. Microbiol. 60, 599–605.
129. Saha, B. C. (2003). Hemicellulose bioconversion. J. Ind. Microbiol. Biotechnol. 30, 279–291.
130. Saritha, M., Arora, A., Nain, L. (2012). Pretreatment of paddy straw with Trameteshirsuta for improved enzymatic saccharification. Bioresource Technology, 104, 459–465.
131. Saritha, M., Arora, A., Singh, S., Nain, L. (2012). Streptomyces griseorubens mediated delignification of paddy straw for improved enzymatic saccharification yields. Bioresour Technol, 135, 12–17.
132. Sarkanen, K. V., Ludwig, C. H. (1971). Definition and Nomenclature. In: Sarkanen, K. V., Ludwig, C. H. (Eds.), Lignins: Occurrence, Formation, Structure and Reactions. John Wiley & Sons, New York, 1–18.
133. Schwanninger, M., Hinterstoisser, B., Gradinger, C., Messner, K., Fackler, K. (2004). Examination of spruce wood biodegraded by Ceriporiopsis subvermispora using near and mid infrared spectroscopy. J Near Infrared Spectrosc, 12, 397–409.
134. Schwanninger, M., Hinterstoisser, B. (2001). Determination of the lignin content in wood by FT-NIR. In: 11th International Symposium on Wood and Pulping Chemistry, Nice, Centre Technique Papeterie.3, 641–644).
135. Shenk, J. S., Workman, J. J., Westerhaus, M. O. (2001). Application of NIR spectroscopy to agricultural products. In: Burns, D. A., Ciurczak, E. W., editors. Handbook of near-infrared analysis. New York: Dekker Inc;. p. 419–74.
136. Shrivastava, B., Thakur, S., Khasa, Y. P., Gupte, A., Puniya, A. K., Kuhad, R. C. (2011). White rot fungal conversion of wheat straw to energy rich cattle feed. Biodegradation, 22, 823–831.
137. Songulashvili, G., Elisashvili, V., Wasser, S. P., Nevo, E., Hadar, Y. (2007). Basidiomycetes laccase and manganese peroxidase activity in submerged fermentation of food industry wastes. Enzyme Microb. Technol., 41, 57–61.
138. Spiridon, I., Popa, VI. (2000). Application of microorganisms and enzymes in the pulp and industry. Cellul Chem Technol, 34, 275–285.

139. Srebotnik, E., Messner, K., Foisner, R. (1988). Penetrability of white-rot degraded pine wood by the lignin peroxidase of Phanerochaete chrysosporium. Appl. Environ. Microbiol. 54, 2608–2614.

140. Stenstad, P., Andresen, M., Tanem, B. S., Stenius P (2008). Chemical surface modifications of microfibrillated cellulose. Cellulose 15:35–45.

141. Stewart, D., Azzini, A., Hall, A., Morrison, I., (1997). Sisal fibers and their constituent non-cellulosic polymers. Ind Crops Prod, 6, 17–26.

142. Swana, J., Yang, Y., Behnam, M., Thompson, R. (2011). An analysis of net energy production and feedstock availability for biobutanol and bioethanol. Bioresource Technology, 102, 2112–2117.

143. Sweet, M. S., Winandy, J. E. (1999). Influence of degree of polymerization of cellulose and hemicellulose on strength loss in fire-retardant-treated southern pine. Holzforschung, 53, 3, 311–317.

144. Taniguchi, M., Suzuki, H., Watanabe, D., Sakai, K., Hoshino, K., Tanaka, T. (2005). Evaluation of pretreatment with Pleurotus ostreatus for enzymatic hydrolysis of rice straw. Journal of Bioscience and Bioengineering, 100, 637–643.

145. Tengerdy, R. P., Szakacs, G. (2003). Bioconversion of lignocellulose in solid substrate fermentation, Biochem. Eng. J. 13, 169–179.

146. Terashima, N., Fukushima, K., He, L.-F. Takabe, K. (1993). Comprehensive model of the lignified plant cell wall. In: H. G. Jung, D. R. Buxton, R. D. Hatfield, J., Ralph (Editors), Forage Cell Wall Structure and Digestibility. American Society of Agronomy, Madison, WI, 247–270.

147. Tien, M. (1987). Properties of ligninase from, P. chrysosporium and their possible applications. CRC Crit. Rev. Microbial. 15, 141–168.

148. Tiwari, R., Rana, S., Singh, S., Arora, A., Kaushik, R., Agrawal, V. V., Saxena, A. K., Nain, L. (2013). Biological delignificaton of paddy straw and Parthenium sp. using a novel micromycete Myrothecium roridum LG7 for enhanced saccharification. Bioresour Technol, 135, 7–11.

149. Tomme, P., Warren, R. A. J., Gilkes, N. R. (1995). Cellulose Hydrolysis by Bacteria and Fungi, Advances in Microbial Physiology, 37, 1–81.

150. Tran, H. M. T., Cheirsilp, B., Hodgson, B., Umsakul, K. (2010). Potential use of Bacillussubtilis in a co-culture with Clostridium butylicum for acetone-butanol-ethanol production from cassava starch. Biochemical Engineering Journal, 48, 260–267.

151. Tuomela, M., Oivanen, P., Hatakka, A. (2002). Degradation of synthetic C-lignin by various white-rot fungi in soil. 34, 1613–1620.

152. Turbak, A. F., Snyder, F. W., Sandberg, K. R. (1983). Microfibrillated cellulose, a new cellulose product: properties, uses and commercial potential. J Appl Polym Sci: Appl Polym Symp 37, 815–827.

153. Viikari, L., Alapuranen, M., Puranen, T., Vehmaanpera, J., Siika-Aho, M. (2007). Thermostable enzymes in lignocellulose hydrolysis. Adv Biochem Eng/Biotechnol, 108, 121–145.

154. Wan, C., Li, Y. (2010). Microbial delignification of corn stover by Ceriporiopsis subvermispora for improving cellulose digestibility. Enzyme and Microbial Technology, 47, 31–36.

155. Westermark, U., Eriksson, K.-E. (1975). Purification and properties of cellobiose: quinone oxidoreductase from Sporotrichum pulverulentum. Acta Chem. Stand. Ser. B 29, 419–424.

156. Zadraz˘il, F., Brunnert, H., Grabbe, K. (1983). Edible mushrooms. In: Rehm, H.-J., Reed, G. (Eds.). Biotechnology: A Comprehensive Treatise, Vols. 1–8. Weinheim, Germany, pp. 145–187.
157. Zadrazil, F., Galleti, G. C., Piccaglia, R., Chiavari, G., Francioso, O. (1991). Influence of oxygen and carbon dioxide on cell wall degradation by white-rot fungi. Anim. Feed Sci. Technol., 32, 137–142.
158. Zhang, Q., Jang, L., Lu, J., Hou, L., Jin, H., Pu, J. (2006 a). Research progress of alcoholic fermentation of corn stover. Science and Technology of Food Industry, 10, 198–201.
159. Zhang, X., Yu, H., Huang, H., Liu, Y. (2007). Evaluation of biological pretreatment with white rot fungi for the enzymatic hydrolysis of bamboo culms. International Biodeterioration and Biodegradation, 60, 159–164.
160. Zhang, Y. H. P., Himmel, M. E., Mielenz, J. R. (2006 b). Outlook for cellulase improvement: screening and selection strategies. Biotechnol Adv, 24, 452–481.
161. Zhu, L., O'Dwyer, J. P., Chang, V. S., Granda, C. B., Holtzapple, M. T. (2008). Structural features affecting biomass enzymatic digestibility. Bioresource Technology, 99, 3817–3828.

CHAPTER 10

RECENT RESEARCH IN THE APPLICATIONS OF CHITIN, CHITOSAN AND OLIGOSACCHARIDES

P. N. SUDHA, T. GOMATHI, and S. AISVERYA

Department of Chemistry, D.K.M. College for Women, Thiruvalluvar University, Vellore, Tamilnadu, India, Tel: (+91) 98429 10157; E-mail: drparsu8@gmail.com

CONTENTS

ABSTRACT

As functional materials, chitin and chitosan offer a unique set of characteristics: biocompatibility, biodegradability to harmless products, nontoxicity, physiological inertness, antibacterial properties, heavy metal ions chelation, gel forming properties and hydrophilicity, and remarkable affinity to proteins. In this article, an effort has been made to review the available literature information on chitin and chitosan. The purpose of this chapter is to present a review of about 30 years of research in our team in the context of a precise scientific strategy. Thus, these years were devoted to improve the production of chitin and chitosan, produce series of co-polymers and co-oligomers, improve their characterizations, reveal a general law of behavior, generate nano-particles, physical gels and derived forms, show a continuum of structure from solutions to other physical states, propose the concept of materials decoys of biological media, etc., The chapter ends with a review of the applications of chitin and chitosan in medicine, pharmacy, agriculture, the food industry, cosmetics, among others.

10.1   INTRODUCTION

Biopolymers are synthesized by an enormous number of living organisms. The biopolymers oligosaccharides and polysaccharides offer greater potential chemical diversity orders of magnitude with multiple functions, which make them relevant to almost any area of research. Both oligosaccharides and polysaccharides are the types of carbohydrate molecules, but they differ in size and complexity. Oligosaccharides are much smaller and simpler carbohydrate molecules consist of carbon, oxygen and hydrogen. An oligosaccharide can contain two to six linked monosaccharide units such as glucose, fructose or galactose. The smallest are called monosaccharides, which are sugars containing just a few carbons, while the largest are called polysaccharides. Polysaccharides are the most common type of carbohydrate found in nature. Polysaccharides can have either linear or branched structures.

In particular, the marine polysaccharides especially chitin and chitosan have wide range of applications in various fields. Chitin is believed to be a major structural component of animal skeletons since at least the Cambrian Period, more than 550 million years ago, although it probably originated in eukaryotic protozoans sometime in the Proterozoic Eon [111]. Chitin was first discovered in mushrooms by the French Professor, Henrni Braconnot, in 1811. In 1820s chitin was also isolated from insects. There is a worldwide market for chitin and for certain derived products such as chitosan (the deacetylated form of chitin) and glucosamine.

10.1.1 CHITIN AND CHITOSAN

Chitin is the most abundant natural polymer next to the cellulose and is similar to cellulose in many respects. Chitin contains 2-acetamido-2-deoxy-β-D-glucose through a β (1→4) linkage (Figure 10.1). The most abundant source of chitin is the shell of crab and shrimp. Chitin is obtained by alkaline deproteinization, acid demineralization and decoloration by organic solvents contact of crustaceans' wastes. Chitin is rapidly biodegraded [212]. The annual worldwide chitin production has been estimated to be, 10^{11} tons, and industrial use has been estimated to be 10,000 metric tones [78]. Chitin have excellent bio-compatibility, non-toxicity and wound healing properties, so it has been widely applied in medical and healthcare fields for applications such as release capsules for drugs, man-made kidney membranes, anticoagulants and immunity accelerants [18, 108, 117, 121, 124, 170]. However, chitin is not soluble in common solvents due to the existence of intra- and inter-molecular hydrogen bonds in chitin and its highly crystalline structure. This strongly restricts many applications of chitin.

The deacetylated form of chitin is chitosan. Chitosan was discovered in 1859 by Professor C. Rouget. Chitosan contains 2-acetamido-2-deoxy-β-D-glucopyranose and 2-amino-2-deoxy-β-D-glucopyranose residues [3] (Figure 10.2). Compared with chitin, chitosan have lower crystallinity.

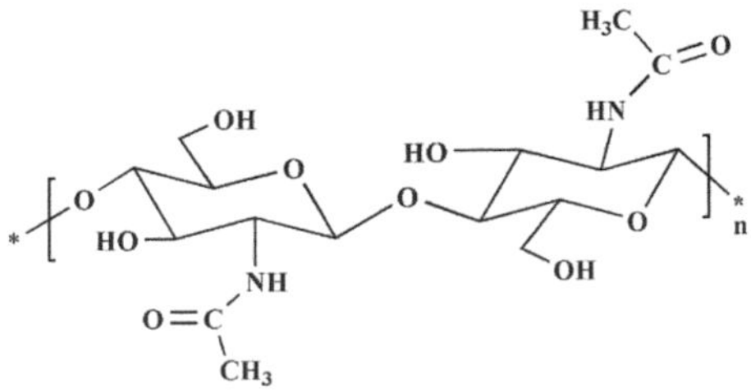

FIGURE 10.1   Structure of Chitin.

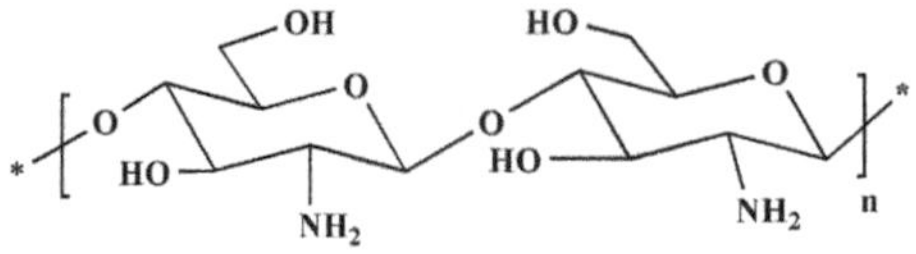

FIGURE 10.2 Structure of Chitosan.

The crystallinity is decreased during the deacetylation process, due to the removal of a portion of the acetyl groups from the chitin structure and thus, to the greater presence of primary amine groups [130]. These facts change the intra and intermolecular interactions, conformed mainly by hydrogen bonds [136], and generate amorphous zones in the biopolymer, with swelling capacity higher than that of crystalline zones, owing in part to the great affinity of primary amine groups to water. The lower crystallinity of chitosan increases the accessibility to the adsorption sites of the biopolymer. The surface of chitosan contains several positively charged amino groups therefore it can easily react with negatively charged biomolecules like DNA, proteins, phospholipids which makes chitosan as bioactive in nature. And the biocompatibility tendency of chitosan is due to the monomeric building blocks of glucosaminoglycans or glycoprotein. In addition to biocompatibility, chitosan exhibits biodegradability, non-toxicity, non-bacterial, non-allergenic, anti-fungal, anti-acid, anti-viral, anti-ulcer and adsorptive properties of chitosan exhibits the promising biomaterial with wide range of applications in various fields [24] like textile and printing, ion exchange chromatography [148], manufacture of pharmaceuticals [99], cosmetics [137], food and packaging industry, wound healing [175], dental [115] biotechnology [105] and agricultural sciences.

Being considered to be materials of great futuristic potential with immense possibilities for structural modifications to impart desired properties and functions, research and development work on chitin and CS have reached a status of intense activities in many parts of the world [142, 71, 133]. Starting in 1982, the study of the physicochemical properties of chitosan in solution through the investigations on various interaction mechanisms and the elaboration of new materials with applications in numerous fields. Much research has been focused on chitosan as a source of bioactive material during past few decades. Chitin/chitosan and their carboxymethyl derivatives have received much attention in the recent decades due to their abundance [146] yet, chitin and chitosan exist in many forms. It is well known that some of the structural characteristics of chitin/chitosan and their derivatives such as degree of acetylation (DA), degree of substitution (DS) and molecular weight (MW) greatly influence various properties such as solubility, physiological activities [53, 166]. Fundamental knowledge of the interactions between chitin and proteins, polysaccharides, calcium carbonate, enzymes, drugs, cells and synthetic materials is not only important for elucidating bio-logical processes associated with chitin, but also for designing novel chitin-based biomaterials. Model chitin surfaces and the development of surface characterization techniques provide a convenient way to study and quantify these interactions [13].

10.1.2 OLIGOSACCHARIDES

Oligosaccharides are the low molecular weight polysaccharides possess a large variety of biological activities on numerous organisms [180, 33]. Oligosaccharides found on the surface of cells as part of glycoproteins and glycolipids play key roles in the control of various normal and pathological processes in living organisms, such as protein folding, cell–cell communication, bacterial adhesion, viral infection, masking of immunological epitopes, fertilization, embryogenesis, neural development and cell proliferation and organization into specific tissues [181, 179, 27].

Consequently, the most active oligosaccharides should be anionic or neutral with degree of polymerization between 4 and 30 partially pyruvate, phosphate, sulfate or acetylate. Nevertheless, very low amount of specific bioactive structures have been reported. Many researches lead to an impact that oligosaccharides and their derivatives are in demand. Thus the production of more and more bioactive oligosaccharides are looked for in the development of oligosaccharide-based therapies and are required to efficiently improve enzymatic and chemical synthesis processes [74, 55, 134].

As for the biological activities of chitosan, LMWC or COSs show versatile biological activities, and these activities are dependent on their molecular weight and DD. Unlike high-molecular weight chitosan, COSs are easily absorbed through the intestine, quickly enter the blood flow, and have systemic biological effects in organisms. Therefore, in this chapter, the recent applications of COSs (chitosan oligosaccharides) are discussed.

Bioactive Chitosan oligosaccharides have significant applications especially in the field of food and biomedical industries. Chitosanase is the key enzyme required for the preparation of biologically active COS from chitosan. The use of chitosanase for the biocontrol of phyto pathogens and for developing transgenic plants is one of the major areas of research. The success in using chitosanase for diverse applications depends on the production of highly active enzyme at a reasonable cost. Research is also focused on developing thermostable chitosanases from microorganisms and modifying them genetically to acquire favorable chattels in the enzyme. Chitosanase producing microorganisms can also be employed in valorization of abundant crustacean bio waste/byproducts.

Microbial chitosanases with different biological roles have been found in nature. Chitosan degrading microorganisms are widely distributed in nature and microorganisms secrete chitosanase extracellularly to degrade chitosan for their nutritional purpose [159]. Numerous processes have been employed for the production of COSs, which usually involve hydrolysis in acidic, alkaline, or oxidizing conditions. In general, many acids including hydrochloric acid, nitrous acid, phosphoric acid, and hydrogen fluoride have been used to obtain COSs [31, 56, 123, 128]. In early studies, COSs were produced by partial hydrolysis of chitosan with concentrated HCl.

10.2   APPLICATIONS OF CHITIN, CHITOSAN AND OLIGOSACCHARIDES

Since chitin was discovered, it has been object of numerous studies in order to understand its properties and find its application in very diverse fields. Chitin and chitosan offer a unique set of characteristics: biocompatibility, biodegradability to harmless products, nontoxicity, physiological inertness, antibacterial properties, heavy metal ions chelation, gel forming properties and hydrophilicity, and remarkable affinity to proteins. Due to these advantages chitin and chitosan has been widely used in various fields such as treating water, biomedical, cosmetic and agricultural or food industrial.

10.2.1   BIOMEDICAL APPLICATIONS

In biomedical applications, chitin and chitosan acts as a promising biomaterial due to their physicochemical properties like biodegradability, non-toxicity, non-bacterial, non-allergenic, anti-fungal, anti-acid, anti-viral, anti-ulcer and adsorptive properties. These properties, find several biomedical applications in tissue engineering [63, 100], wound healing [101], as excipients for drug delivery [61] and also in gene delivery [43, 59].

The antimicrobial activity of chitin, chitosan and their derivatives against different groups of microorganisms, such as bacteria, yeast, and fungi, has received considerable attention in recent years [70, 90, 95], emphasized the increased antimicrobial activity of carboxymethyl chitosan, which is due to the essential transition metal ions unavailable for bacteria or binds to the negatively charged bacterial surface to disturb the cell membranes. Chitosan has strong antimicrobial and antifungal activities which effectively controls fruit decay [1]. Chitosan coating on fruit and vegetable, adjust the permeability of carbon dioxide and oxygen thus reducing the respiration rate [37]. Also it was used in many post harvest fruits and vegetables, such as grape, berry, jujube and fresh-cut lotus root [129, 203, 191].

Jayakumar et al. (2009) developed α- and β-chitin membranes using α- and β-chitin hydrogel for tissue engineering and wound dressing applications. Li et al. (2009) reported the possibility of making films of mPEG-g-chitosan by preparing a composite film with suitable hollow and high capacity of water adsorption, which could have potential application in wound healing and tissue engineering.

Chitin and chitosan were both found to have an accelerating effect on wound healing [113, 98, 161, 175, 189]. When used in wound management, chitosan and its derivatives turn into gel, when they come into contact with body fluids and reduce friction between the dressing material and the wound. They also accelerate wound healing with their haemostatic properties and stimulate macrophage. The wide panel of chitosan properties and processed materials gives to this biosourced polymer a quite promising future as biomaterial as demonstrated by emerging products on the market notably in the wound dressing field.

Ribeiro et al. (2009) evaluated the applicability of a chitosan hydrogel as a wound dressing. They isolated fibroblast cells from rat skin to assess the cytotoxicity of the hydrogel. The results showed that chitosan hydrogel was able to promote cell adhesion and proliferation. Chitosan sponges also find application in bone tissue engineering, as a filling material [23].

Chen et al. (2008) reported composite nanofibrous membrane of chitosan/collagen, which is known for its beneficial effects on wound healing. To make an effective wound healing accelerator, water-soluble chitosan/heparin complex was prepared using water-soluble chitosan with wound healing ability and heparin with ability to attract or bind growth factor related to wound healing process [79]. Xu et al. (2007) fabricated a novel wound dressing material of chitosan/hyaluronic acid composite films. The results demonstrate that the chitosan mixed hyaluronic acid may produce inexpensive wound dressing with desired properties. Recently, Madhumathi et al. (2010) developed novel α-chitin/nanosilver composite scaffolds for wound healing applications.

Tissue engineering has emerged as a major area of regenerative medicine, with the capability of surpassing some of the disadvantage of using pure synthetic materials in conventional replacement procedures. Chitosan and its derivatives have been reported as attractive candidates for scaffolding materials because they are expected to degrade as new tissues are being formed while minimizing inflammatory reactions and toxic degradation products [89, 107, 174, 171, 46, 28, 172].

Liver tissue engineering requires a perfect ECM for primary hepatocytes culture to maintain a high level of liver-specific functions and desirable mechanical stability [38]. Lee et al. (2010) developed a microfluidic-based pure chitosan microfiber for liver tissue engineering applications without the use of any chemical additives. Periodontal regeneration is of utmost importance in the field of dentistry that essentially reconstitutes and replaces the lost tooth supporting structures. Alveolar bone loss is a common finding associated with periodontal degeneration. Various treatment modalities have been used to regenerate or fill bony defects using different biomaterials such as bioglass and hydroxyapatite.

Biodegradable materials are most extensively used in cardiovascular tissue engineering. Polymers such as collagen, gelatin, fibrin, HA, alginate, and decellularized matrices [211, 20] can be produced from biological sources and no toxic degradation or inflammatory reactions are expected [76]. Cardiac patches of silk fibroin (SF) combined with microparticles of chitosan or HA were fabricated by Yang et al. (2009).

Chitosan is also mucoadhesive [29]. Mucus is a blend of molecules including salts, lysozyme, and mucins, which are highly hydrated glycoproteins primarily responsible for the viscoelastic properties of mucus. Sialic acid residues on mucin have a pKa of 2.6, making them negatively charged at physiological pH [29, 187]. The microspheres with a mucoadhesive property can offer additional advantages that may help to prolong residence time and improve uptake of vaccines incorporated with them. Chowdary and Rao (2004) well documented the advantages of mucoadhesive drug delivery systems such as bioavailability improvement of drugs, absorption enhancement of macromolecules and prolonged residence time at the site of application.

Chitosan microspheres have potential application in drug delivery systems because they can be used to enable the controlled release of many drugs and to improve the bioavailability of degradable substances or to enhance membrane permeability. The role of chitosan microspheres as a drug delivery system has been widely studied in a variety of drugs, such as proteins, peptides, and vaccines [41, 57, 58, 69, 178, 182].

Chitosan is suitable for nerve regeneration based on its biocompatibility and biodegradability. Haipeng et al. (2000) reported that neurons cultured on the chitosan membrane can grow well and that chitosan tube can promote repair of the PNS. Yuan et al. (2004) found that chitosan fibers supported the adhesion, migration, and proliferation of SCs, which provide a similar guide for regenerating axons to Bungner bands in the nervous system [162]. The blood compatibility of chitin and chitosan remains a prime aspect almost 20 years after Hirano and Naishila (1985) reported his studies on chitosan and its acyl derivatives. Kuen et al. (1995) confirmed the blood compatibility properties of N-acylchitosans. An N-acylation of 20–50% was achieved and their susceptibility to lysozyme degradation was found to be comparable to acetyl-chitosan. Anti-thrombogenic activity was obtained for collagen coated on N-acyl derivative fibers similar to the results obtained in 1985. The chitosan fibers offer the potential of being fabricated into blood vessels and their blood compatibility results demonstrated in the recent work augurs well for applications where hemocompatibility is sought. Wang et al. (2005) showed a superior blood compatibility through chitosan/collagen/heparin matrix in implantable bioartificial liver (IBL) applications.

The functional oligosaccharides of various origins (viruses, bacteria, plants and fungi) have been shown to exert potent immunomodulatory activities [164, 194, 208, 210]. The functional oligosaccharides also improve immunity which innate defense responses activated through the interaction of sugar moieties with innate receptors on the plasma membrane of host cells, in particular in macrophages and dendritic cells [4, 119]. Oligosaccharide libraries clearly have value for drug discovery research in a wide range of diseases [150]. Limited numbers of such libraries currently exist and they are largely based on mammalian polysaccharides and glycoconjugates. The unique structures and sulphation patterns of marine derived sugars are being increasingly recognized and investigated [81].

10.2.2 COSMACEUTICAL APPLICATIONS

Chitin and chitosan derivatives have wide applications in the field of cosmetics due to their high water solubility and positive anti-microbial characteristics. Jayakumar et al. (2010) emphasis the relationship between molecular structure and moisture-retention ability of chitin and chitosan derivatives has been reported in skin care cosmetics. Using chitosan of different MWs, it was concluded that the water-holding capacity of chitosan was superior to that of conventional methylcellulose. The patents for the use of chitin and chitosan derivatives in skin care cosmetics cover strong moisturizing activity, photo-protection from UV and sunlight, anti-aging, skin elasticity-increasing activity antibacterial and anti-inflammatory effect. Skin hydration and moisture mask effects of chitosan were examined based on viscosity measurements of various molecular weights and swelling degree (DS) of carboxymethyl chitin (CMC) and O-carboxymethyl chitosan (O-CMCS). Hyaluronic acid was used as the control. These results revealed that the 6-carboxymethyl groups in the molecular structure of chitin and chitosan is the main active site responsible for moisture-retention ability [16, 15, 151].

Chitosan is used as anti-obesity agent, moisturizing agent, emollient and film former. Moisture-retention ability is also related to the MW, that is, higher MW helps improve moisture-retention ability. The results from human skin application tests showed that N-succinyl chitosan (N-SCS) and partially N-acylated chitosan pyrrolidine carboxylate increased the skin elasticity. These effects were not found in the case of hyaluronic acid. Stevens et al. studied the effect of shampoos containing 0.2% and 0.5% CMC on hair [143]. It was found that the shampoos significantly reduced the combing forces of hair. The results of customer tests showed that 0.2% CMC shampoo provided smoother and more manageable effects and was more desirable than shampoo without CMC. Partially deacetylated chitin and CMC–sodium salt were also evaluated [160]. N-CMCS as a 1.0% solution at pH 4.8 is a valuable functional ingredient of cosmetic hydrating creams in view of its durable moisturizing effect on the skin [80]. Similarly, the bacteriostatic activity of N-carboxymethyl chitosan (N-CMCS) and N-carboxybutyl chitosan (N-CBCS), together with other favorable properties, such as viscosifying action, the enhanced film forming ability, the moisturizing effect and the stabilization of emulsions, make these novel modified chitosans most suitable as functional cosmetic ingredients [114].

10.2.3   AGRICULTURAL APPLICATIONS

Attack by various fungi to certain seeds may result in decreased germination [35, 155]. Chitosan-based coatings exerting antifungal activity help preserve the quality of the stored seeds. The treatment consists in seed immersion in a chitosan suspension (up to 4%) followed by drying.

Chitin contributes to their cycling of nutrients such as nitrogen. When chitin decomposes, it produces ammonia, which takes part in the nitrogen cycle. Furthermore, chitin is a main constituent in geochemical recycling of both carbon and nitrogen. Fungi, arthropods, and nematodes are the major contributor of chitin in the soil. Among these, the fungi provide the largest amount of chitin in the soil (6–12% of the chitin biomass, which is in the range 500–5000 kg/ha). In another study, Kokalis-Burelle (2001) reported that chitin contributes significantly to soil enrichment. It was found that chitin could control plant pathogens and pathogenic nematodes and provoke the development of host plant resistance against these pathogens. Chitin led to an increase in microorganism population; this sharp increase could shift and prompt their action as anti-plant pathogens in two ways. First, the microorganism may act as parasite for plant pathogens. Second, they can kill or inhibit these pathogens through production of toxins or metabolites or enzymes. Furthermore, the increase in microorganism numbers increases the number of nonparasitic nematodes, which results in a decline in the number of pathogenic nematodes.

Chitosan forms a coating on fruit and vegetable, and thus the respiration rate of fruit and vegetable was reduced by adjusting the permeability of carbon dioxide and oxygen [37]. Chitosan is also used in many postharvest fruits and vegetables, such as grape, berry, jujube and fresh-cut lotus root [129, 203, 191]. Chitin or chitosan is used to control postharvest diseases of many fruits such as pear [202], strawberry [42, 9], table grape [106] tomato [6], citrus [17], and longan [66].

Though chitosan coating has many advantages for the preservation of postharvest fruit and vegetable, as for specific fruit or vegetable, single chitosan coating sometimes demonstrates a certain defect, which includes limited inhibition to especial microorganism that leads fruit to decay, and poor coating structure to adjust the permeability of carbon dioxide and oxygen [138]. To overcome this deficiency of single chitosan coating, there are two main methods to improve the property of chitosan coating. One method is that the chitosan were combined with organic compounds such as essential oil, organic acid, or inorganic compound including metal ions and inorganic nano-material, as well as biological control agents. Zeng and Shi (2009) obtained another safer, cheaper and more environmentally friendly seed coating agent using chitosan combined with plant growth regulators and other additives. Such a novel seed coating agent significantly enhanced sprout growth in regard of the traditional agents. It stimulated the seedling growth of rice, advanced the growth of root, improved root activity and increased the crop yield in the germination test and field trial.

10.2.4 TEXTILE INDUSTRIES

Textiles are among the most widely used materials in everyday use. The end-use of a textile material dictates its desirable properties. A lot of research was done on the application of chitosan in textile for antibacterial purpose, and the studies were focused on the effects of degree of deacetylation, molecular weight and other characters of chitosan to the antibacterial activity in textile. The need for antimicrobial textiles goes hand-in-hand with the rise in resistant strains of micro-organisms. The use of antimicrobial agents for textiles has also become indispensible to avoid cross-infection by pathogenic microorganisms, to control the infestation by microbes, and arrest metabolism in microbes in order to reduce odor formation. Antimicrobial treated fabric protects garments from staining, discoloration, and quality deterioration [183]. A number of chemicals were used to impart antibacterial activity to textiles [84, 102–104]. Many of these chemicals, however, are toxic and not biodegradable. These facts facilitate chitosan as a new antibacterial agent for textiles.

The biological properties of chitosan includes bacteriostatic and fungistatic properties [126, 152, 153, 144, 83, 52, 154, 163]. Mechanism studies suggest that the positively charged chitosan interacts with negatively charged residues at the cell wall of fungi or bacteria. The interaction changes cell permeability and causes the leakage of intracellular substances [94, 201]. Other studies suggest that the formation of the polymeric substance around the bacterial cell prevents the nutrients from entering the cell [51].

The applications of chitosan for improving dyeability of cotton fabric has been widely studied [73, 177, 48]. In the textile area, the higher the active site of chitosan favors the higher the dye adsorption (including natural dye) as well as film formation on fiber surface [92]. Chitosan can easily adsorb anionic dyes, such as direct, acid and reactive dyes, by electrostatic attraction due to its cationic nature in an acidic condition. It is postulated that the affinity of chitosan to cotton would be by Van der Waals forces between them because of the similar structures of chitosan and cotton.

Gupta et al. showed that, chitosan treated cotton has better dyeability with direct and reactive dyes and treatment with modified chitosan makes it possible to dye cotton in bright shades with cationic dyes having high wash fastness. Treated samples showed good antimicrobial activity against *Escherichia coli* and *Staphylococcus aureus* at 0.1% concentration as well as improved wrinkle recovery [36].

Hasebe et al. (2001) synthesized two different chitosan hybrids to use as deodorant agents for textiles. An aqueous solution of chitosan and methacrylic acid (MAA) was mixed with organic solvents, emulsified, and polymerized to give polymer A, which is a porous polymer particle (8–20 μm) and a composite between chitosan and polymethacrylate (PMAA). The unique property of this particle was its possession of numerous basic groups inside the porous structure. Polymer B was synthesized by the polymerization of the emulsion containing an aqueous solution of chitosan and MAA, and lauryl methacrylate (LMA). The resulting polymer formed a particle (0.1–3.0 μm, suitable for application to fabrics) consisting of a hydrophobic core (polylaurylmethacrylate (PLMA)) covered with a hydrophilic shell, and which has the same composite structure as polymer A. The polymers A and B were simply composites where chitosan and PMAA or chitosan and PMAA-LMA are physically entangled. Polymer A showed high deodorizing performance due to its amphoteric property, that is, an acidic substance is absorbed on to an amino group of chitosan and a basic substance is absorbed on the carboxylic acid groups of PMAA. The cotton fabric treated (pad (100% WPU) – cure (120°C for 1 min)) with polymer B containing a binder effectively absorbed acetic acid, isovaleric acid, and ammonia compared with the untreated fabric.

10.2.5   ENVIRONMENTAL APPLICATIONS

Recently, numerous approaches have been studied for the development of cheaper and most effective adsorbents containing biopolymers. The most widespread biopolymers are polysaccharides [190], chitin [167, 139, 7] and cyclodextrin [157, 26, 32]. These biopolymers reach the increasing demand for treatment of industrial wastewater before their use or disposal. Because the pollutants creates environmental and health difficulties, associated with heavy metals and pesticides and their deposit through the food chain [39]. Traditional methods for the elimination of heavy metals from industrial wastewater may be inefficient or costly, particularly when metals are present at low concentrations [30, 184]. Chitin, chitosan and oligosaccharides represent interesting and attractive alternative adsorbents because of their particular structure, physico-chemical characteristics, chemical stability, high reactivity and excellent selectivity towards metals. Moreover, they are abundant, renewable and biodegradable resources and have a capacity to associate by physical and chemical interactions with a wide variety of molecules [22, 131].

Numerous studies have demonstrated the effectiveness of chitin and its derived products in the uptake of metal cations such as lead, chromium, copper and nickel and the uptake of oxyanions as well as complex metal ions [125, 145]. Their structure allows excellent complexation capacity with metal ions, particularly transition and post-transition metals [116]. It was supported that the chelation of a single metal ion by several -NH or NHCOCH$_3$ groups effectively isolates each metal ion from its neighbors

[88]. Smither-kopperl (2001) found that chitin exhibits several functions, including retention of nutrients, in the soil. Chitin contributes to their cycling of nutrients such as nitrogen. When chitin decomposes, it produces ammonia, which takes part in the nitrogen cycle. Furthermore, chitin is a main constituent in geochemical recycling of both carbon and nitrogen. Fungi, arthropods, and nematodes are the major contributor of chitin in the soil. Among these, the fungi provide the largest amount of chitin in the soil (6–12% of the chitin biomass, which is in the range 500–5000 kg/ha).

Chitosan's functional groups and natural chelating properties make chitosan useful in wastewater treatment by allowing for the binding and removal of metal ions such as copper, lead, mercury, and uranium from wastewater. Chitosan chelates five to six times greater amount of metals than chitin [197]. Indeed, nitrogen atoms hold free electron doublets that can react with metal cations. Amine groups are thus responsible for the uptake of metal cations by a chelation mechanism. It chelates strongly with the metal ions hence forms the co – ordinate complex [169]. Jha et al. (1988) reported Cd (II) ions from wastewater were efficiently removed by chitosan. Adsorption of Cu (II) and Cr (VI) ions by chitosan was documented by Schmuhl et al. (2001), Taboada et al. (2003) used chitosan to adsorb Cu (II) and Hg (II) ions. Micera et al. (1986) showed that chitosan has a high binding capacity with metals such as copper and vanadium. Chitosan is a well-known solid sorbent for transition metals because the amino groups on chitosan chain can serve as coordination sites [125, 197]. However, the amine groups are easily protonated in acidic solutions. Hence the protonation of these amine groups may cause electrostatic attraction of anionic compounds including metal anions (or) anionic dyes [44, 45]. It can also be used to remove dyes and other negatively charged solids from wastewater streams and processing outlets. Chitosan due to its high content of amine and hydroxyl functional groups has an extremely high affinity for many classes of dyes including disperse, direct, anionic, vat, sulfur and naphthol [25].

Chitosan grafted with poly (acrylonitrile) has been further modified to yield amidoximated chitosan [67] a derivative having a higher adsorption for Cu^{2+}, Mn^{2+}, and Pb^{2+}, compared to crosslinked chitosan. The adsorption capacity had a linear dependence on pH in cases of Cu^{2+} and $Pb2+$. However, a slight decrease in the adsorption capacity was observed in case of Zn^{2+} and Cd^{2+} [68].

Chitosan has been modified with different mono as well as disaccharides. (Yang et al., 2003) have also reported the metal uptake abilities of macrocyclic diamine derivative of chitosan. The polymer has high metal uptake abilities, and the selectivity property for the metal ions was improved by the incorporation of aza crown ether groups in the chitosan. Mc Kay et al. (1989) used chitosan for the removal of Cu^{2+}, Hg^{2+}, Ni^{2+}, and Zn^{2+} within the temperature range of 25–60°c at neutral pH. Further adsorption parameters for the removal of these metal ions were reported by Yang et al. (1984).

The most characteristics feature of cyclodextrin (CD) is the ability to form inclusion compounds with various molecules, especially aromatics. The interior cavity of the molecule provides a relatively hydrophobic environment onto which a polar pollutant can be trapped [32].

10.3   RECENT RESEARCH IN THE APPLICATIONS OF CHITIN, CHITOSAN AND OLIGOSACCHARIDES

The intriguing properties of chitin, chitosan and oligosaccharides have been known for many years and this polycationic polymer (in acidic environments) has been used in the fields of industrial and biomedical.

The use of chitosan-based edible films is also used to preserve the microbial quality of pork meat hamburger. Tripathi et al. (2009) developed a novel antimicrobial coating based on chitosan and PVA and evaluated its effect on minimally processed tomato. The results indicated the film may be a promising material for food packaging applications. Chitosan is used as a preservative in low-pH foods, either alone or in combination with other preservative systems. The constituents of the food matrix appear to have an important effect on the antimicrobial efficacy of chitosan [140]. Several workers [87, 8, 12] have reported that chitosan coating is effective in preserving the internal quality of eggs. In the field of ophthalmology, chitosan is used in the making of therapeutic contact lenses and eye dressing. Recent studies have demonstrated that chewing the chitosan oligomer containing gum effectively inhibited the growth of carcinogenic bacteria in the saliva [50] and also the growth of periodontic bacteria (*P. gingivalis*) in saliva. Bio-inspired bi-layered physical hydrogels that are only constituted of chitosan and water were processed and applied in the treatment of full-thickness burn injuries [10]. While Aly et al. (2010), showed innovative multi finishing using O-PEG-g-chitosan/citric acid aqueous system for preparation of medical textiles. The cotton treated with the copolymer has been evaluated as healthcare worker uniforms and medical products, acquiring antimicrobial and anti-crease properties.

In addition to oral vaccination, another attractive application is the oral delivery of DNA for therapeutic gene expression as a so-called "gene pill." The benefits of such a delivery system have been delineated by Sheu et al. (2003) and include safety, patient compliance, and dose regulation. It is worth noting, however, that one of the arguments proposed for increased safety from an oral nonviral DNA pill is targeting to short-lived gut epithelial cells and lack of systemic cell transfection. However, plasmid DNA can be detected in systemic tissues after oral delivery, albeit at very low copy numbers [11] and oral delivery of DNA vaccines can produce detectable systemic immune responses, indicating that the effects of an orally delivered formulation may not be locally confined. The use of chitosan-based edible films is also used to preserve the microbial quality of pork meat hamburger. Tripathi et al. (2009) developed a novel antimicrobial coating based on chitosan and PVA and evaluated its effect on minimally processed tomato. The results indicated the film may be a promising material for food packaging applications. Chitosan is used as a preservative in low-pH foods, either alone or in combination with other preservative systems. The constituents of the food matrix appear to have an important effect on the antimicrobial efficacy of chitosan [140]. Several workers [87, 8, 12] have reported that chitosan coating is effective in preserving the internal quality of eggs.

In the field of ophthalmology, chitosan is used in the making of therapeutic contact lenses and eye dressing. In cartilage engineering, chitosan and HA can be blended together to create a three-dimensional scaffold that has a subchondral (bone layer) and cartilage layer. By creating a bilayered scaffold, it eliminates donor site morbidity associated with traditional autografts. Oliveira et al. (2006) created a bilayer scaffold using the freeze-dry method and pouring a 3% chitosan solution onto a sintered HA scaffold. The scaffolds demonstrated a high connectivity, adequate water uptake and porosity, good mechanical properties, and cellular adhesion.

Yamane et al. (2005) also indicated that chitosan-based hyaluronan hybrid polymer fibers show a great potential as a desirable biomaterial.

Chitosan/coral sponge with platelet-derived growth factor B (PDGFB) encoding pDNA was prepared to construct periodontal tissue. Increased expression of PDGFB and significant cell proliferation was observed in vitro, and increased expression of PDGFB and new vascular tissue growth were observed in vivo [209].

Novel methods have been recently devised for the preparation of chitin threads for the fabrication of absorbable suture materials, dressings, and biodegradable substrates for the growth of human skin cells (keratinocytes and fibroblasts) [168].

Several studies have focused on the use of chitosan as a component in calcium-based cements in the development of bone substitutes. Yokoyama et al. (2002) used chitosan as a component of the liquid phase that included citric acid and glucose in combination with a-TCP (tricalcium phosphate) and tetra-calcium phosphate to produce easily moldable cement. Novel poly (L-lactic acid) (PLLA)-chitosan hybrid scaffolds were prepared as tissue engineering scaffolds and simultaneously as drug release carriers [132].

Yuan et al. (2014) reported that the chitosan derivatives can be used as NO-releasing scaffolds has also been investigated since these materials contain large concentrations of primary amines, necessary for N-diazeniumdiolate NO (nitric oxide)donor formation.

Like chitosan, the antibacterial effects of COS are influenced by a number of factors such as DP, DD, type of microorganism, and certain other physical–chemical properties. Three kinds of COSs with different molecular weights were produced by employing a dual reactor system and tested against Gram-positive (*Escherichia coli, E. coli* O-157, *Salmonella typhi*, and *Pseudomonas aeruginosa*) and Gram-negative (*Streptococcus mutans, Micrococcus luteus, Staphylococcus aureus, Staphylococcus epidermidis*, and *Bacillus subtilis*) bacteria [64].

It was demonstrated that COSs can inhibit the growth of tumor cells by exerting immune-enhancing effects. Suzuki et al. (1986) demonstrated that COSs inhibited tumor growth through an increase in immune effects. The COSs with DP4–7showed strong inhibition of ascites cancer in BALB/c mice, while N-acetylchitohexaose and chitohexaose exhibited strong inhibiting effects for Sarcoma 180 (S180) and MM156solid tumor growth in syngenic mice. The same results were also observed by Tokoro et al. (1998).

COSs also have been reported to have protective effects against oxidative stress in various cell lines. In human umbilical vein endothelial cells, hydrogen peroxide-induced stress injuries were effectively protected against COSs by means of inhibiting intracellular ROS formation, suppressing the production of lipid peroxidation compounds such as malondialdehyde, and restoring the activities of antioxidant enzymes including superoxide dismutase and GSH peroxidase [96].

10.4 FUTURE DIRECTIONS FOR RESEARCH

In future, efforts are made to improve these novel bio materials for further enhancement in their application results. Moreover, designing of new scaffolds from other natural polymer such as alginate, gelatin would be possible for soft tissue engineering such as skin. Chitin/Chitosan has a great potential in a variety of biomedical, industrial applications and chitosan physicochemical and mechanical properties used in fabricating particles and films can be modulated for specific purposes. Efforts should be made to prepare nanofibrous scaffolds from other natural polymers including silk for hard and soft tissue engineering. And the best use of these marine sources in the field of Food, Cosmetics Industries, in Effluent treatment and in medical field should be made

10.5 CONCLUSION

Progress over these 30 years can be considered in various steps in chitin, chitosan and oligosaccharides. This review summarizes the industrial and biomedical applications of marine carbohydrates such as (Chitin, Chitosan and oligosaccharides) based nanomaterials in tissue engineering, wound dressing, drug delivery and cancer diagnosis. In addition, this review also opens up the novel applications for which these natural biopolymers can be put to use in a variety of nanostructural forms and sizes. Nanostructured composite scaffolds can be developed as promising tissue engineered constructs or for wound healing. Multifunctional use of chitin and chitosan based nanomaterials have been proved to aid simultaneous cancer targeting and drug delivery. We expect that this chapter will provide insights on the use of these marine carbohydrates for researchers working to discover new materials with new properties for the valuable applications of these materials.

ACKNOWLEDGMENTS

The authors are grateful to authorities of DKM College for Women and Thiruvalluvar University, Vellore, Tamil Nadu, India for the support.

KEYWORDS

- **biomedical applications**
- **biopolymers**
- **chitin**
- **chitosan**

REFERENCES

1. Aider, M. Chitosan application for active bio based films production and potential in the food industry, Review. *Food Sci Technol-LEB.* 2010, 43, 837–842.
2. Aly, A. S., Abdel-Mohsen, A. M., Hebeish, A. Innovative multifinishing using chitosan-O-PEG graft copolymer citric acid aqueous system for preparation of medical textiles. *J Text Inst.* 2010, 101, 76–90.
3. Amit Bhatnagar., Mika Sillanpää. Applications of chitin and chitosan derivatives for the detoxification of water and wastewater. A short review. *Advances in Colloid and Interface Science.* 2009, 152, 26–38.
4. Arnold, J. N., Dwek, R. A., Rudd, P. M., Sim, R. B. Mannan binding lectin and its interaction with immunoglobulins in health and in disease. *Immunology Letters.* 2006, 106, 103–110.
5. Atiba, A., Nishimura, M., Kakinuma, S., Hiraoka, T., Goryo, M., Shimada, Y., Ueno, H., Uzuka, Y. Aloe vera oral administration accelerates acute radiation-delayed wound healing by stimulating transforming growth factor β and fibroblast growth factor production. *American Journal of Surgery.* 2011, 201, 809–818.

6. Badawya, M. E. I., Rabeab, E. I. Potential of the biopolymer chitosan with different molecular weights to control postharvest gray mold of tomato fruit. *Postharvest Biol. Technol.* 2009, 51, 110–117.

7. Bailey, S. E., Olin, T. J., Bricka, R. M., Adrian, D. D. A review of potentially low-cost sorbents for heavy metals. *Water Resource.* 1999, 33, 2469 – 2479.

8. Bhale, S., No, H. K., Prinyawiwatkul, W., Farr, A. J., Nadarajah, K., Meyers, S. P. Chitosan coating improves shelf life of eggs. *J Food Sci.* 2003, 68, 2378–83.

9. Bhaskara, R. M. V., Belkacemi, K., Corcuff, R., Castaigne, F., Arul, J. Effect of pre-harvest chitosan sprays on post-harvest infection by Botrytis cinerea and quality of strawberry fruit. *Postharvest Biol. Technol.* 2000, 20, 39–51.

10. Boucard, N., Vitona, C., Agayb, D et al. The use of physical hydrogels of chitosan for skin regeneration following third-degree burns. *Biomaterials.* 2007, 28, 3478–3488.

11. Bowman, K., Sarkar, R., Raut, S., et al. Oral delivery of non viral DNA nanoparticles for hemophilia A. American Society of Gene Therapy, 8th Annual Meeting. *Mol Ther.* 2005, 11, 779.

12. Caner, C. The effect of edible eggshell coatings on egg quality and consumer perception. *J Sci Food Agric.* 2005, 85, 1897–1902.

13. Chao Wang., Alan R. Esker. Nanocrystalline chitin thin films. *Carbohydrate Polymers.* 2014, 102, 151–158.

14. Chen, J. P., Wang, Z. Z., Xiong, Z. Y., Li, Q. X. Morphological and structural characterization of a polysaccharide from Gynostemma pentaphyllum Makino and its anti exercise fatigue activity. *Carbohydrate Polymers.* 2008, 74, 868–874.

15. Chen, L., Du, Y., Wu, H., Xiao, L. *J Appl Polym Sci.* 2002, 83, 1233–1241.

16. Chen, R. H. *Adv Chitin Sci.* 2002, 5, 593–598.

17. Chien, P. J., Sheu, F., Lin, H. R. Coating citrus (Murcott tangor) fruit with low molecular weight chitosan increases postharvest quality and shelf life. *Food Chem.* 2007, 100, 1160–1164.

18. Cho, Y. W., Cho, Y. N., Chung, S. H., Yoo, G., Ko, S. W. Water soluble chitin as a wound healing accelerator. *Biomaterials.* 1999, 20, 2139–2145.

19. Chowdary, K. P. R., Rao, Y. S. Mucoadhesive microspheres for controlled drug delivery. *Biol Pharm Bull.* 2004, 27, 1717–1724.

20. Christman, K. L., Lee, R. J. Biomaterials for the treatment of myocardial infarction, *J. Am. Coll. Cardiol.* 2006, 5, 907–913.

21. Chung, M.H., Choi, S. A review on the relationship between aloe vera components and their biologic effects. *Seminars in Integrative Medicine.* 2003, 1, 53–62.

22. Ciesielski, W., Lii, C.Y., Yen, M. T., Tomasi, K. P. Interactions of starch with salts of metals from the transition groups. *Carbohydrate Polymer.* 2003, 51, 47–56.

23. Costa-Pinto, A. R., Reis, R. L., Neves, N. M. Scaffolds based bone tissue engineering: the role of chitosan. Tissue Eng Part B. 2011, 17, 331–47.

24. Crini, G. Recent developments in polysaccharide based materials used as adsorbentsin wastewater treatment, *Progress in Polymer Science.* 2005, 30, 38–70.

25. Crini, G., Badot, P. M. *Prog Polym Sci.* 2008, 33, 399.

26. Crini, G., Morcellet, M. Synthesis and applications of adsorbents containing cyclodextrins. *Journal of Separation Science.* 2002, 25, 789–813.

27. Cummings, R. D. *Mol. BioSyst.* 2009, 5, 1087–1104.

28. Dang, J. M., Liyang, K. W. *Adv Drug Deliv Rev.* 2006, 58, 487–499.
29. Deacon, M. P., McGurk, S., Roberts, C. J., et al. Atomic force microscopy of gastric mucin and chitosan mucoadhesive systems. *J. Biochem.* 2000, 348, 557–563.
30. Deans, J. R., Dixon, B. G. Bioabsorbents for waste water treatment. In Advances in Chitin and Chitosan, *Elsevier Applied Science*, Oxford, UK. 1992, 648–656.
31. Defaye, J., Gadelle, A., Pedersen, C. Chitin and chitosan oligosaccharides. In "Chitin and Chitosan," (Skjak-Braeak, G., Anthonsen, T., Sandford, P., Eds), *Elsevier Applied Science*, London. 1989, 415.
32. Del Valle, E. M. Cyclodextrins and their uses, a review. *Process Biochemistry.* 2004, 39, 1033 – 1046.
33. Delattre, C., Michaud, P., Courtois, B., Courtois, J. *Minerva Biotechnol.* 2005, 17, 107.
34. Delipo Taboada., Gustavo Cabrera., Galo Cardenas. Rentention capacity of chitosan for copper and mercury ions. *J. Chil. Chem. Soc.* 2003.
35. Donald, W. W., Mirocha, C. J. Chitin as a measure of fungal growth in stored corn and soybean seed. *Cereal Chemistry.* 1977, 54, 466–474.
36. Dragan Jocic., Susana Vı'lchez., Tatjana Topalovic., Antonio Navarro., Petar Jovancic., Maria Rosa Julia., Pilar Erra. Chitosan acid dye interactions in wool dyeing system, *Carbohydrate Polymers.* 2005, 60, 51–59.
37. Elsabee, M. Z., Abdou, E. S. Chitosan based edible films and coatings: A review. Mater Sci Eng C Mater *Biol Appl.* 2013, 33, 1819–1841.
38. Feng, Z. Q., Chu, X., Huang, N. P et al. The effect of nanofibrous galactosylated chitosan scaffolds on the formation of rat primary hepatocyte aggregates and the maintenance of liver function. *Biomaterials.* 2009, 30, 2753–2763.
39. Fereidoon Shahidi., Janak Kamil., Vidana Arachchi., You-Jin Jeon. Food applications of chitin and chitosans. *Trends in Food Science & Technology.* 1999, 10, 37–51.
40. Fujita, M., Ishihara, M., Shimizu, M., Obara, K., Ishizuka, T., Saito Y., et al. Vascularization in vivo caused by the controlled release of fibroblast growth factor-2 from an injectable chitosan non-anticoagulant heparin hydrogel. *Biomaterials.* 2004, 25, 699–706.
41. Gavini, E., Hegge, A. B., Rassu, G., Sanna, V., Testa, C., Pirisino, G. et al. Nasal administration of carbamazepine using chitosan microspheres, in vitro in vivo studies. *Int J Pharm.* 2006, 307, 9–15.
42. Ge, L. L., Zhang, H. Y., Chen, K. P., Ma, L. C., Xu, Z. L. Effect of chitin on the antagonistic activity of Rhodotorula glutinis against Botrytis cinerea in strawberries and the possible mechanisms involved. *Food Chem.* 2010, 120, 490–495.
43. Gerrit, B. Chitosans for gene delivery. *Advanced Drug Delivery Reviews.* 2001, 2, 145–150.
44. Gibbs, G., Tobin, J. M., Guibal, E. Sorption of Acid Green 25 on chitosan. influence of experimental parameters on uptake kinetics and sorption isotherms. *Journal of Applied Polymer Science.* 2003, 90, 1073–1080.
45. Guibal, E. Interactions of metal ions with chitosan based sorbents A review. *Separation and Purification Technology.* 2004, 38, 43–74.
46. Habraken, W. J. E. M., Wolke, J. G. C., Jansen, J. A. *Adv Drug Deliv Rev.* 2007, 59, 234–238.

47. Haipeng, G., Zhong, Y., Li, J., Gong, Y., Zhao, N., Zhang, X. Studies on nerve cell affinity of chitosan derived materials. *J. Biomed. Mater. Res.* 2000, 52, 285–295.
48. Hakeim, O., Abou-Okeil, A., Abdou, A. W., Waly, A. The Influence of Chitosan and Some of Its Depolymerized Grades on Natural Color Printing, *J. App. Polym. Sci.* 2005, 97, 559–563.
49. Hasebe, Y., Kuwahara, K., Tokynaga, S. Chitosan hybrid deodorant agent for finishing textiles, *AATCC Rev.* 2001, 1, 23–27.
50. Hayashi, Y., Ohara, N., Ganno, T. Chewing chitosan containing chewing gum effectively inhibits the growth of carcinogenic bacteria. *Archs. Oral Biol.* 2007, 52, 290–94.
51. Helander, I. M., Nurmiaho Lassila, E. L., Ahvenainen, R., Rhoades, J., Roller, S. Chitosan disrupts the barrier properties of the outer membrane of Gram-negative bacteria. *International Journal of Food Microbiology.* 2001, 71, 235–244.
52. Hima Bindu, T. V. L., Vidyavathi, M., Kavitha, K., Sastry, T. P., Suresh Kumar, R. V. Preparation and evaluation of chitosan gelatin composite films for wound healing activity. *Trends Biomater. Artif. Organs.* 2010, 24, 123–130.
53. Hirano, S., Nagano, N. *Agri Biol Chem.* 1989, 53, 3065–3066.
54. Hirano, S., Noishiki, Y. The blood compatibility of chitosan and N-acylchitosans. *J. Biomed. Mater. Res.* 1985, 19, 413–417.
55. Holemann, A., Seeberger, P. H. Curr. Opin. *Biotechnol.* 2004, 15, 622.
56. Horowitz, S. T., Roseman, S., Blumenthal, H. J. The preparation of glucosamine oligosaccharides. I. Separation. *J. Am. Chem. Soc.* 1957, 709, 5046–5049.
57. Illum, L., Watts, P., Fisher, A. N., Hinchcliffe, M., Norbury, H., Jabbal Gill, I., et al. Intranasal delivery of morphine. *J Pharma col Exp Ther.* 2002, 301, 391–400.
58. Jain, S., Singh, P., Mishra, V., Vyas, S. P. Mannosylated niosomes as adjuvant carrier system for oral genetic immunization against Hepatitis B. *Immunol Lett.* 2005, 101, 41–49.
59. Jayakumar, R., Chennazhi, K. P., Muzzarelli, R. A. A., Tamura, H., Nair, S. V., Selvamurugan, N. Chitosan conjugated DNA nanoparticles in gene therapy. *Carbohydrate Polymers.*2010, 79, 1–8.
60. Jayakumar, R., Divya Rani, V. V., Shalumon, K. T., Sudheesh Kumar, P. T., Nair, S. V., Furuike, T., et al. Bioactive and osteoblast cell attachment studies of novel α and β-chitin membranes for tissue engineering applications. *Int J Biol Macromol.* 2009, 45, 260–4.
61. Jayakumar, R., Nwe, N., Tokura, S., Tamura, H. Sulfated chitin and chitosan as novel biomaterials. *International Journal of Biological Macromolecules.* 2007, 40, 175–181.
62. Jayakumar, R., Prabaharan, M., Nair, S. V., Tokura, S., Tamura, H., Selvamurugan, N. Novel carboxymethyl derivatives of chitin and chitosan materials and their biomedical applications. *Progress in Materials Science.* 2010, 55, 675–709.
63. Jayakumar, R., Prabaharan, M., Reis, R. L., Mano, J. F. Graft copolymerized Chitosan Present status and applications. *Carbohydrate Polymers.* 2005, 62, 142–158.
64. Jeon, Y. J., Park, P. J., Kim, S. K. Antimicrobial effect of chitooligo saccharides produced by bioreactor. *Carbohydr. Polym.* 2001, 44, 71–76.
65. Jha, N., Leela, I., Prabhakar Rao, A. V. S. Removal of cadmium using chitosan, *J. Environ. Eng.* 1988, 114, 962.

66. Jiang, Y. M., Li, Y. B. Effect of chitosan coating on postharvest life and quality on longan fruit. *Food Chem.* 2001, 73, 139–143.
67. Kang, D. W., Choi, H. R., Kweon, D. K. Selective adsorption capacity for metal ions of amidocimated chitosan bead-g-PAN copolymer. *Polymer.* 1996, 20, 989–995.
68. Kang, D. W., Choi, H. R., Kweon, D. K. Selective adsorption capacity for metal ions of amidoximated chitosan bead-g-PAN copolymer. *Polymer* (Korea). 1996, 20, 989–95.
69. Kang, M. L., Kang, S. G., Jiang, H. L., Shin, S. W., Lee, D. Y., Ahn J. M, et al. In vivo induction of mucosal immune responses by intranasal administration of chitosan microspheres containing Bordetella bronchiseptica DNT. *Eur J Pharm Biopharm.* 2006, 63, 215–220.
70. Khanafari, A., Marandi, R., Sanatei, Sh. Recovery of chitin and chitosan from shrimp waste by chemical and microbial methods. *Iranian Journal of Environmental Health Science and Engineering.* 2008, 5(1), 19–24.
71. Khor, E. Chitin: a biomaterial in waiting. *Curr Opin Solid State Matter Sci.* 2002, 6, 313–317.
72. Kim, I.Y., Seo, S. J., Moon, H.S et al. Chitosan and its derivatives for tissue engineering applications. *Biotechnol. Adv.* 2008, 26, 1–21.
73. Kittinaovarat, S. Using Chitosan for Improving the Dye ability of Cotton Fabrics with Mangosteen Rind Dye, *J. Sci. Res. Chula. Univ.* 2004, 29, 155–164.
74. Koeller, K. M., Wong, C. H. *Glycobiology.* 2000, 10, 1169.
75. Kokalis Burelle, N. Chitin amendments for suppression of plant nematodes and fungal pathogens, *Phytopathology.* 2001, 91, 5168–5175.
76. Kong, H. J., Alsberg, E., Kaigler, D., Lee, K. Y., Mooney, D. L. Controlling degradation of hydrogels via the size of crosslinked junctions. *Adv. Mater.* 2004, 16, 1917–1921.
77. Kuen, Y. L., Wan, S. H., Won, H. P. Blood compatibility and biodegradability of partially N acylated chitosan derivatives. *Biomaterials.* 1995, 16, 1211–1216.
78. Kurita, K. Chitin and chitosan, Functional biopolymers from marine crustaceans. *Marine Biotechnology.* 2006, 8, 203–226.
79. Kweon, D. K., Song, S. B., Park, Y. Y. Preparation of water soluble Chitosan heparin complex and its application as wound healing accelerator. *Biomaterials.* 2003, 24, 1595–601.
80. Lapasin, R., Stefancic, S., Delben, F. *Agro-Food Ind Hi-Tech.* 1998, 7, 12–18.
81. Laurienzo, P. Marine polysaccharides in pharmaceutical applications: an overview. *Mar Drugs,* 2010, 8, 2435–2465.
82. Lee, D. W., Lim, H., Chong, H. N., Shim, W. S. Advances in chitosan material and its hybrid derivatives, A review. *Open Biomater. J.* 2009, 1, 10–20.
83. Lee, E. J., Shin, D. S., Kim, H. E., Kim, H. W., Koh, Y. H., Jang, J. H. Membrane of hybrid chitosan silica xerogel for guided bone regeneration. *Biomaterials.* 2009, 30, 743–750.
84. Lee, J. K., Kim, S. U., Kim, J. H. Modification of chitosan to improve its hypocholesterolemic capacity. *Biosci. Biotechnol. Biochem.* 1999, 63, 833–839.
85. Lee, K. H., Shin, S. J., Kim, C. B et al. Microfluidic synthesis of pure chitosan microfibers for bio artificial liver chip. *Lab. Chip.* 2010, 10, 1328–1334.

86. Lee, K.Y., Ha, W.S., Park, W. H. Blood compatibility and biodegradability of partially N-acylated chitosan derivatives. *Biomaterials.* 1995, 16, 1211–1216.
87. Lee, S. H. Effect of chitosan on emulsifying capacity of egg yolk. *J Korean Soc Food Nutr.* 1996, 25, 118–122.
88. Lepri, L., Desideri, P. G., Tanturli, G. Chromatographic behavior of inorganic ions on chitosan thin layers and columns. *J. Chromatogr.* 1978, 147, 375–381.
89. Li, K., Hwang, Y., Tsai, T., Chi, S. *Food Sci.* 1996, 23, 608–12.
90. Li, Q., Dunn, E. J., Grandmaison, E. W., Goosen, M. F. A. Applications and properties of chitosan. *Journal of Bioactive and Compatible Polymers.* 1992, 7(4), 370–397.
91. Li, X., Kong, X., Shi, S., Gu, Y., Yang, L., Guo, G., Luo, F., Zhao, X., Wei, Y. Q., Qian, Z. Y. Biodegradable MPEG-g-chitosan and methoxypoly (ethylene glycol)-b-poly ([epsilon]-caprolactone) composite films: Part 1. Preparation and characterization. *Carbohydr Polym.* 2009, 79, 429–436.
92. Lim, S. H. Synthesis of a Fiber-Reactive Chitosan derivative and its Application to Cotton Fabric as an Antimicrobial Finish and a Dyeing-Improving Agent, PhD Dissertation, North Carolina State University, USA. 2002.
93. Lim, S. H., Hudson, S. M. Review of chitosan and its derivatives as antimicrobial agents and their uses as textile chemicals. *J Macromol Sci Polym Rev.* 2003, 43, 223–269.
94. Lim, S. H., Hudson, M. Review of chitosan and its derivatives as antimicrobial agents and their uses as textile chemicals. *Journal of Macromolecular Science.* 2004, 43, 223–269.
95. Limam, Z., Selmi, S., Sadok, S., El-abed, A. Extraction and characterization.
96. Liu, H. T., Li, W. M., Xu, G., Li, X. Y., Bai, X. F., Wei, P., Yu, C., Du, Y. G. Chitosan oligosaccharides attenuate hydrogen peroxide-induced stress injury in human umbilical vein endothelial cells. *Pharmacol. Res.* 2009, 59, 167–175.
97. Liu, X., Guan, Y., Yand, D., Li, Z., Yao, K. Antibacterial action of chitosan and carboxymethyl chitosan. *J. Appl. Polym. Sci.* 2001, 79, 1324–1335.
98. Lloyd, L. L., Kennedy, J. F., Methacanon, P., Paterson, M., Knill, C. J. *Carbohydr Polym.* 1998, 37, 315–322.
99. Macquarrie, D. J. Modified mesoporous materials as acid and base catalysts, Nano porous Materials, *Science and Engineering.* 2005.
100. Madhumathi, K., Binulal, N. S., Nagahama, H., Tamura, H., Shalumon, K. T., Selvamurugan, N., et al. Preparation and characterization of novel-chitin-hydroxyapatite composite membranes for tissue engineering applications. *International Journal of Biological Macromolecules.* 2009, 44, 1–5.
101. Madhumathi, K., Sudheesh Kumar, P. T., Abilash, S., Sreeja, V., Tamura, H., Manzoor, K., et al. Development of novel chitin nanosilver composite scaffolds for wound dressing applications. *Journal of Material Science*: Materials in Medicine. 2010, 21, 807–813.
102. Mahltig, B., Bottcher, H. Modified silica sol coatings for water repellent textiles. *Journal of Sol-Gel Science and Technology.* 2003, 27, 43–52.
103. Mahltig, B., Fiedler, D., Bottcher, H. Antimicrobial sol-gel coatings. *Journal of Sol-Gel Science and Technology.* 2004, 32, 219–222.

104. Mahltig, B., Haufe Helfried., Bottcher, H. Functionalization of textiles by inorganic sol gel coatings. *Journal of Materials Chemistry.* 2005, 15, 4385–4398.

105. Mao, H. Q., Roy, K., Troung-Le, V. L., et al. Chitosan DNA nanoparticles as gene carriers: synthesis, characterization and transfection efficiency. *J Control Release.* 2001, 70, 399–421.

106. Mark, H. F., Bikales, N. M., over Berger, C. G., Menges, G. *Encyclopedia of polymer Science and engineering.* 1985, 1, 20.

107. Martino, A. D., Sittinger, M., Risbud, M. V. *Biomaterials.* 2005, 26, 5983–5990.

108. Matsuyama, H., Kitamura, Y., Naramura, Y. Diffusive permeability of ionic solutes in charged chitosan membrane. *Journal of Applied Polymer Science.* 1999, 72, 397–404.

109. McKay, G., Blair, H. S., Findon, A. Equilibrium studies for the sorption of metal ions onto chitosan. *Ind. J. Chem.* 1989, 28, 356–360.

110. Micera, G., Deiana, S., Dessi, A., Decock, P., Dubois, B., Kozlowski, H. Copper and vanadium complexes of chitosan. *In Chitin in Nature and Technology.* 1986, 565–567.

111. Miller, R. F. Chitin paleoecology, *Biochem. Syst. Ecol.* 1991, 19, 401–411.

112. Mincheva, R., Manolova, N., Rashkov, I. Biocomponent aligned nanofibers of N-carboxyethyl chitosan and poly (vinyl alcohol). *Eur. Polym. J.* 2007, 43, 2809–2818.

113. Mori, T., Okumura, M., Matsuura, M., Ueno, K., Tokura, S., Okumoto, Y et al. *Biomaterials.* 1997, 18, 947–51.

114. Muzzarelli, R. A. A. *Carbohydr Polym.* 1988, 8, 1–21.

115. Muzzarelli, R. A. A. Carboxy methylated chitins and chitosan. *Carbohydrate Polymers.* 1989, 8, 1–21.

116. Muzzarelli, R. A. A., Tubertini, O. Chitin and chitosan as chromatographic supports and adsorbents for from organic and aqueous solutions and seawater. *Talanta.* 1969, 16, 1571–1577.

117. Muzzarelli, R. A. A. *Chitin.* New York, Pergamon Press. 1977.

118. Neamnark, A., Rujiravanit, R., Supaphol, P. Electrospinning of hexanoyl chitosan. *Carbohydr. Polym.* 2006, 66, 298–305.

119. Nezlin, R., Ghetie, V. Interactions of immunoglobulins outside the antigen combining site. *Advances in Immunology.* 2004, 82, 155–215.

120. Ngimhuang, J., Furukawa, J., Satoh, T., Furuike, T., Sakairi, N. Synthesis of a novel polymeric surfactant by reductive N-alkylation of chitosan with 3-O-dodecyl-glucose. *Polymer.* 2004, 45, 837–841.

121. Okamoto, Y., Kawakami, K., Miyatake, K., Morimoto, M., Shigemasa, Y., Minami, S. Analgesic effects of chitin and chitosan. *Carbohydrate Polymers.* 2002, 49, 249–252.

122. Oliveira, J. M., Rodrigues, M. T., Silva, S. S et al. Novel hydroxyapatite chitosan bilayered scaffold for Osteochondral tissue engineering applications: Scaffold Design and its performance when seeded with goat bone marrow stromal cells. *Biomaterials.* 2006, 276, 123–137.

123. Omura, H., Uehara, K., Tanaka, Y. Manufacture of low-molecular weight chitosan. *Japan patent.* 1991, 3, 2203.

124. Onishi, H., Nagai, T., Machida, Y. Applications of chitin, chitosan, and their derivatives to drug carriers for microparticulated or conjugated drug delivery systems. *Application of chitin and chitosan.* 1997, 407–409.

125. Onsoyen, E., Skaugrud. Metal recovery using chitosan. *Journal of Chemical Technology and Biotechnology.* 1990, 49, 395–404.
126. Paul, W., Sharma, C. P. Chitosan and alginate wound dressings, A short review. *Trends Biomater. Artif. Organs.* 2004, 18, 18–23.
127. Pellizzoni, M., Ruzickova, G., Kalhotka, L., Lucini, L. Antimicrobial activity of different Aloe barbadensis Mill. And Aloe arborescens Mill. leaf fractions. *Journal of Medicinal Plants Research.* 2012, 6, 1975–1981.
128. Peniston, Q. P., Johnson, E. L. Process for depolymerization of chitosan. Chitooligosaccharides as Potential Nutraceuticals. 1975, US patent No. 3922260, 335.
129. Perdones, A., Sánchez-González, L., Chiralt, A., Vargas, M. Effect of chitosan lemon essential oil coatings on storage keeping quality of strawberry. *Postharvest Biol Tec.* 2012, 70, 32–41.
130. Pillai, C. K. S., Willi Paul., Chandra., Sharma, P. Chitin and chitosan polymers. Chemistry, solubility and fiber formation Progress in Polymer Science. 2009, 34, 641–678.
131. Polaczek, E., Staarzyk, F., Malenki, K., Tomasik, P. Inclusion complexes of starches with hydrocarbons. *Carbohydrate Polymer.* 2000, 43, 291–297.
132. Prabaharan, M., Rodriguez-Perez, M.A., de Saja, J.A., Mano, J. F. Preparation and characterization of poly (l-lactic acid)-chitosan hybrid scaffolds with drug release capability. *J. Biomed. Mater. Res. Part B: Appl.* Biomater. 2007, 81, 427–434.
133. Prashanth, K. V. H., Tharanathan, R. N. Chitin chitosan modifications and their unlimited application potential an overview. *Trends Food Sci Technol.* 2007, 18, 117–131.
134. Pukin, A. V., Weijers, C. A. G. M., van Lagen, B., Wechselberger, R., Sun, B., Gilbert, M. *Carbohydr. Res.* 2008, 343, 650.
135. Radhakumary, C., Prabha, D. N., Nair, C. P. R., Suresh, M. Chitosan comb graftpolyethylene glycol monomethacrylate synthesis, characterization, and evaluation as a biomaterial for hemodialysis applications. *J Appl Polym Sci.* 2009, 114, 2873–2886.
136. Ramírez-Coutiño, L., Marín-Cervantes, M. D. C., Huerta, S., Revah, S., Shirai, K. Enzymatic hydrolysis of chitin in the production of oligosaccharides using Lecanicillium fungicola chitinases. *Process Biochemistry.* 2006, 41, 1106–1110, 1359–5113.
137. Ravi Kumar, M. N. V., Muzzarelli, R. A. A., Muzzarelli., C., Sashiwa, H., Domb, A. J. Chitosan chemistry and pharmaceutical perspectives. *Chemical Reviews.* 2004, 104, 6017–6084.
138. Ravi, K. M. N. V. A review of chitin and chitosan applications. *React funct Polym.* 2000, 46, 1–27.
139. Ravikumar, M. N. V. A review of chitin and chitosan applications. *Reactive and Functional Polymers.* 2000, 46, 1–27.
140. Rhoades, J., Roller, S. Antimicrobial Actions of Degraded and Native Chitosan against Spoilage Organisms in Laboratory Media and Foods. *Applied and environmental microbiology.* 2000, 66, 80–86.
141. Ribeiro, M. P., Espiga, A., Silva, D., Baptista, P., Henriques, J., Ferreira, C., Silva, J. C., Borges, J. P., Pires, E., Chaves, P., Correia, I. *J. Development of a new chitosan hydrogel for wound dressing. Wound Rep Reg.* 2009, 17, 817–824.

142. Rinaudo, M. Chitin and chitosan properties and applications. *Prog Polym Sci.* 2006, 31, 603–632.
143. Ruksiriphong, K., Chandrkrachang, S., Stevans, W. F. *Adv Chitin Sci.* 2002, 5, 599–604.
144. Sajeev, U. S., Anoop Anand, K., Deepthy Menon., Shanti Nair. Control of nanostructures in PVA, PVA chitosan blends and PCL through electrospinning. *B. Mater. Sci.* 2008, 31, 343–351.
145. Sakaguchi, T., Hirokoshi, T., Nakajima, A. Adsorption of uranium by chitin phosphate and chitosan phosphate. *Agricultural and Biological Chemistry.* 1981, 45, 2191–2195.
146. Sandford, P. A., Sjak-Braek, G., Anthonsen, T., Sandford, P. A. editors. Chitin and chitosan. London, *Elsevier Applied Science.* 1989, 51–69.
147. Sashiwa, H., Aiba, S. Chemically modified chitin and chitosan as Biomaterials. *Progress in Polymer Science.* 2004, 29, 887–908.
148. Schipper, N. G., Vårum, K. M., Artursson, P. Chitosans as absorption enhancers for poorly absorbable drugs. Influence of molecular weight and degree of acetylation on drug transport across human intestinal epithelial (Caco-2) cells. *Pharm Res.* 1996, 13, 1686–1692.
149. Schmuhl, R., Krieg, H. M., Keizer, K. Adsorption of Cu(II) and Cr(VI) ions by chitosan Kinetics and equilibrium studies. *Water SA.* 2001.
150. Seeberger, P. H. Synthesis and medical applications of oligosaccharides. *Nature.* 2007, 446, 1046–1051.
151. Seino, H., Akamatsu, E., Kamimura, Y., Hamada, K., Yoshioka, H. Kichin, Kitosan Kenkyu. 1999, 5, 162–164.
152. Shanmugasundaram, O. L. Chitosan coated cotton yarn and its effect on antimicrobial activity. *J. Text. Apparel Technol. Management.* 2006, 5, 1–6.
153. Shanmugasundaram, O. L., Giridev, V. R. Neelakandan, R., Madhusoothanan, M., Suseela Rajkumar, G. Drug release and antimicrobial studies on chitosan coated cotton yarns. *Ind. J. Fiber. Text.* 2006, 31, 543–547.
154. Shanmugasundaram, O. L., Gowda, R. V. M. Development and characterization of bamboo gauze fabric coated with polymer and drug for wound healing. *Fiber Polym.* 2011, 12, 15–20.
155. Sharma, P. D., Fisher, P. J., Webster, J. Critique of the chitin assay technique for estimation of fungal biomass. *Transactions of the British Mycological Society.* 1977, 69, 479–483.
156. Sheu, E., Rothman, S., German, M., et al. The "Gene Pill" and its therapeutic applications. *Curr Opin Mol Ther.* 2003, 5, 420–427.
157. Singh, M., Sharma, R., Banerjee, U.G. Biotechnology applications of cyclo dextrin. *Biotechnology Advances.* 2002, 20, 341–359.
158. Smither Kopperl, M. L. Chitin as bio mass its origin and role in nutrient cycling. Phytopathology. 2001, 91, 167–168.
159. Somashekar, D., Joseph, R. Chitosanases properties and applications, A review. *Bioresources Technology.* 1996, 55, 35–45.
160. Srikumlaithong, S., Laixuthai, P., Aiba, S., Shinagawa, S. *Adv Chitin Sci.* 1998, 3, 418–423.
161. Stone, C. A., Wright, H., Clarke, T., Powell, R., Devaraj, V. S. *Brit J Plast Surg.* 2000, 53, 601–608.
162. Sudha, P. N., Aisverya, S., Maximas, H. Rose., Jayachandran Venkatesan., Se-Kwon Kim. Bionanocomposites of Chitosan for Multi-tissue Engineering Applications In

Se-Kwon Kim. (Eds). Chitin and Chitosan Derivatives, *Advances in Drug Discovery and Developments.* 2013, 24, 451–462.

163. Sun, Z. H., Li, K. Preparations, properties and applications of chitosan based nanofibers fabricated by electrospinning. *Express Polym. Lett.* 2011, 5, 342–361

164. Surenjav, U., Zhang, L. N., Xu, X. J., Zhang, X. F., Zeng, F. B. Effects of molecular structure on antitumor activities of -b-D-glucans from different Lentinus edodes. *Carbohydrate Polymers.* 2006, 63, 97–104.

165. Suzuki, K., Mikami, T., Okawa, Y., Tokoro, A., Suzuki, S., Suzuki, M. Antitumor effect of hexa-N-acetylchitohexaose and chitohexaose. *Carbohydr. Polym.* 1986, 151, 403–408.

166. Suzuki, S., Watanabe, T., Mikami, T., Suzuki, M. Proceedings of the 5th international conference on chitin and chitosan USA. 1990, 96–105.

167. Synowiecki, J., Al-Khateeb, N. A. Production, properties, and some new applications of chitin and its derivatives. Critical Reviews in *Food Science and Nutrition.* 2003, 43, 145–171.

168. Tamura, H., Hamaguchi, T., Tokura, S. Destruction of rigid crystalline structure to prepare chitin solution. In: Boucher, I., Jamieson, K., Retnakaran, A. editors. Advances in chitin science, proceedings of the ninth international conference on chitin and chitosan. Montreal, *European Chitin Society.* 2004, 84–87.

169. Tirmizi, S. A., Iqbal, J., Isa, M. Collection metal ions present in water samples of different sites of Pakistan using biopolymers chitosan. *J. Chem. Soc. Pakistan.* 1996, 18, 312–315.

170. Tokura, S., Baba, S., Uraki, Y., Miura, Y., Nishi, N., Hasegawa, O. Carboxymethylchitin as a drug carrier of sustained release. *Carbohydrate Polymers.* 1990, 13, 273–281.

171. Tokura, S., Tamura, H. Bio *macromolecules.* 2001, 2, 417–421.

172. Tokura, S., Tamura, H., Azuma, I., Jolles, P., Muzzarelli, R. A. A. editors. Chitin and chitinases. Basel Switzerland. 1999, 279–292.

173. Tripathi, S., Mehrotra, G. K., Dutta, P. K. Physicochemical and bioactivity of cross-linked chitosan PVA film for food packaging applications. *International Journal of Biological Macromolecules.* 2009, 4, 372–376.

174. Tuzlakoglu, K., Alves, C. M., Mano, J. F., Reis, R. L. *Macromol Biosci.* 2004, 4, 811–819.

175. Ueno, H., Nakamura, F., Murakami, M., Okumura, M., Kadosawa, T., Fujinaga, T. *Biomaterials.* 2001, 22, 2125–2130.

176. Uneno, Y., Haruta, S., Ishii, M., Igarashi, Y. Microbial community in anaerobic hydron-producing microflora enriched from sludge compost. *Appl. Microbiol. Biotechnology.* 2001, 57, 555–562.

177. Vakhito, N. A., Safonov, V. V. Effect of chitosan on the efficiency of dyeing textiles with active dyes, *Fiber Chem.* 2003, 35, 27–28.

178. van der Lubben, I. M., Kersten, G., Fretz, M. M., Beuvery, C., Coos Verhoef, J., Junginger, H. E. Chitosan microparticles for mucosal vaccination against diphtheria. oral and nasal efficacy studies in mice. *Vaccine.* 2003, 21, 1400–1408.

179. van Kooyk, Y., Rabinovich, G. A. Nat. *Immunol.* 2008, 9, 593–601.

180. Varki, A., *Glycobiology.* 1993, 2, 97.

181. Varki, A., Lowe, J. B. In Essentials of Glycobiology. Cold Spring Harbor Laboratory Press, Cold Spring Harbor, New York. 2009, 80.

182. Varshosaz, J., Sadrai, H., Alinagari, R. Nasal delivery of insulin using chitosan microspheres. *J Microencapsulate.* 2004, 21, 761–774.
183. Vigo, T. L. *Textile processing and properties.* Elsevier Sci. B.V. Netherlands. 1994, 252–254.
184. Volesky, B. Biosorbents for metal recovery, *Trends Biotechnol.* 1987, 5, 96–99.
185. Vondran, J. L. Fabrication, optimization, and characterization of carboxymethylated chitosan nanofiber mats for cartilage regeneration applications. Materials science and engineering Philadelphia: Drexel University. 2007, 123.
186. Wang, A. J., Xu, J. J., Chen, H. Y. In-situ grafting hydrophilic polymer on chitosan modified poly (dimethylsiloxane) microchip for separation of biomolecules. *J Chromatogr.* 2007, 1147, 120–126.
187. Wang, J., Tauchi, Y., Deguchi, Y., et al. Positively charged gelatin microspheres as gastric mucoadhesive drug delivery system for eradication of H. pylori. *Drug Deliv.* 2000, 7, 237–243.
188. Wang, X., Yan, Y., Lin, F., Xiong, Z., Wu, R., Zhang, R., et al. Preparation and characterization of a collagen chitosan heparin matrix for an implantable bioartificial liver. *J Biomater Sci Polym Ed.* 2005, 16, 1063–1080.
189. Wongpanit, P., Sanchavanakit, N., Pavasant, P., Supaphol, P., Tokura, S., Rujiravanit, R. *Macromol Biosci.* 2005, 5, 1001–112.
190. Wurzburg, O. B. Modified starches, properties and uses, CRC Press, Boca Raton. 1986.
191. Xing, Y., Li, X., Xu, Q., Jiang, Y. H., Yun, J., et al. Effects of chitosan based coating and modified atmosphere packaging (MAP) on browning and shelf life of fresh-cut lotus root (Nelumbo nucifera Gaerth). *Innov Food Sci Emerg.* 2010, 11, 684–689.
192. Xu, G., Huang, X., Qin, L., Wu, J., Hu, Y. Mechanism study of chitosan on lipid metabolism in hyperlipidemic rats. *Asia Pac. J. Clin. Nutr.* 2007, 16, 313–317.
193. Yamane, S., Iwasaki, N., Majima, T et al. Feasibility of chitosan-based hyaluronic acid hybrid biomaterial for a novel scaffold in cartilage tissue engineering. *Biomaterials.* 2005, 26, 611–619.
194. Yang, J. H., Du, Y. M., Huang, R. H., Sun, L. P., Liu, H., Gao, X. H., et al. Chemical modification and antitumor activity of Chinese lacquer polysaccharide from lac tree Rhus vernicifera. *Carbohydrate Polymers.* 2005, 59, 101–107.
195. Yang, J. M., Lin, H. T., Wu, T. H., Chen, C. C. Wettability and antibacterial assessment of chitosan containing radiation-induced graft nonwoven fabric of polypropylene-g-acrylic acid. *J Appl Polym Sci.* 2003, 90, 1331–1336.
196. Yang, M. C., Wang, S. S., Chou, N. K et al. The cardiomyogenic differentiation of rat mesenchymal stem cells on silk fibroin polysaccharide cardiac patches in vitro. *Biomaterials.* 2009, 30, 3757–3765.
197. Yang, T. C., Zall, R. R. Absorption of metals by natural polymers generated from seafood processing. *Industrial and Engineering Chemistry Product Research and Development.* 1984, 23, 168–172.
198. Yang, T. C., Zall, R. R. I & EC Product R & D. 1984, 23, 168.
199. Yokoyama, A., Yamamoto, S., Kawasaki, T., Kohgo, T., Nakasu, M. Development of calcium phosphate cement using chitosan and citric acid for bone substitute materials. *Biomaterials.* 2002, 23, 1091–1101.

200. Yong-Chang Shi., Yong-Ming Jiang., De-Xin Sui., Yan-Li Li., Tian Chem., Lin, Ma., Zhong-Tian Ding. Affinity chromatography of trypsin using chitosan as ligand support. *J. Chromatogr.* 1996, 742, 107–112.

201. Young, D. H., Kohle, H., Kauss, H. Effect of Chitosan on Membrane Permeability of Suspension Cultured Glycine max and Phaseolus vulgaris Cells. *Plant Physiology.* 1982, 70, 1449–1454.

202. Yu, T., Wang, L. P., Yin, Y., Wang, Y. X., Zheng, X. D. Effect of chitin on the antagonistic activity of Cryptococcus laurentii against Penicillium expansum in pear fruit. *Int. J. Food Microbiol.* 2008, 122, 44–48.

203. Yu, Y. W., Zhang, S. Y., Ren, Y. Z., Li, H., Zhang, X. N., et al. Jujube preservation using chitosan film with nano silicon dioxide. *J Food Eng.* 2012, 113, 408–414.

204. Yuan, Lu., Danielle Slomberg, L., Mark, H. Schoenfisch Nitric oxide-releasing chitosan oligosaccharides as antibacterial agents *Biomaterials.* 2014, 35, 1716–1724.

205. Yuan, Lu., Slomberg, L., Mark, H. Schoenfisch Nitric oxide releasing chitosan oligosaccharides as antibacterial agents. *Biomaterials.* 2014, 35, 1716–1724.

206. Yuan, Y., Zhang, P., Yang, Y., Wang, X., Gu, X. The interaction of Schwann cells with chitosan membranes and fibers in vitro. *Biomaterials.* 2004, 25, 4273–4278.

207. Zeng, D. F., Shi, Y. F. Preparation and application of a novel environmentally friendly organic seed coating for rice. *Journal of the Science of Food and Agriculture.* 2009, 89, 2181–2185.

208. Zhang, C. X., Huang, K. X. Characteristic immunostimulation by MAP, a polysaccharide isolated from the mucus of the loach, Misgurnus anguillicaudatus. *Carbohydrate Polymers.* 2005, 59, 75–82.

209. Zhang, Y.Z., Su, B., Venugopal, J., Ramakrishna, S., Lim, C.T. Biomimetic and bioactive nanofibrous scaffolds from electrospun composite nanofibers. *Int. J. Nanomedicine.* 2007, 2, 623–638.

210. Zhu, X. L., Lin, Z. B. Modulation of cytokines production, granzyme B and perforin in murine CIK cells by Ganoderma lucidum polysaccharides. *Carbohydrate Polymers.* 2006, 63, 188–197.

211. Zimmermann, W. H., Melnychenko, I., Eschenhagen, T. Engineered heart tissue for regeneration of diseased hearts. *Biomaterials.* 2004, 25, 1639–1647.

212. Zobell, C. E., Rittenberg, S. C. The occurrence and characteristics of chitonoclastic sediments. *J. Bacteriology.* 1983, 35, 275.

CHAPTER 11

NATURAL FIBER COMPOSITES AND APPLICATIONS

P. N. SUDHA, K. NASREEN, and P. ANGELIN VINODHINI

Department of Chemistry, D.K.M. College for Women, Thiruvalluvar University, Vellore, Tamilnadu, India, Tel: (+91) 98429 10157; E-mail: drparsu8@gmail.com

CONTENTS

ABSTRACT

This chapter is a review on the natural fiber reinforced polymer composites. Natural fibers have recently become attractive to researchers, engineers and scientists as an alternative reinforcement for fiber reinforced polymer (FRP) composites. Natural fibers can be used as reinforcement in polymers. The greatest challenge in working with natural fiber reinforced polymer composites is their large variation in properties and characteristics due to their low cost, good mechanical properties, high specific strength, non-abrasive, eco-friendly and bio-degradability characteristics. Several chemical modifications are employed to improve the interfacial matrix-fiber bonding resulting in the enhancement of tensile properties of the composites. In general, the tensile strengths of the natural FRP composites increase with fiber content, are discussed. This review deals with the universal review report on natural fiber composites. The applications of natural fiber composites and their potential development of different kinds of engineering and domestic products are also discussed in detail.

11.1   INTRODUCTION

Since, 1990s, natural fiber composites are emerging as realistic reinforced composites in many applications. In recent years, polymer composites containing natural fibers have obtained considerable attention. The interest in the natural fiber reinforced polymer composite arises rapidly due to high performance in mechanical properties and significant processing advantages.

Natural fiber reinforced polymer composites have attracted the attention of the research community [88] and extended to almost all the fields. Much work is done in the application of natural fiber as reinforcement in polymer composite [114]. Natural fibers are an attractive research area because they are eco-friendly, inexpensive, abundant and renewable, lightweight, have low density, high toughness, high specific properties, biodegradability and non-abrasive to processing characteristics, and lack of residues upon incineration [120, 119]. Natural fiber composites such as hemp fiber-epoxy, flax fiber-polypropylene (PP), and china reed fiber-PP are particularly attractive in automotive applications because of lower cost and lower density.

The efforts to produce economically attractive composite components have resulted in several innovative manufacturing techniques currently being used in the composites industry [111]. These fiber composites have many properties which make them an attractive alternative to traditional materials due to high specific properties such as stiffness [125], impact strength [133], flexibility [85] and modulus [35]. Several authors have reported the chemical composition, properties of sisal fibers and their composites by incorporating the fiber in different matrices before and after treatment by different methods [78].

In addition, they are available in large amounts [83] which are renewable and biodegradable and other desired properties include low cost, low density, less equipment abrasion [85, 139], less skin and respiratory irritation [70]. The composites industry has begun to recognize the commercial applications of composites which promises to offer much larger business opportunities to the aerospace sector due to the sheer size of transportation industry [23]. Due to the need for more environmental friendly materials, natural fiber composites are regaining the attention that once has been shifted to synthetic products. The first known utilization of natural fiber composites was straw reinforced clay for bricks and pottery [14]. Many of the early research and development in fiber composites are dominated by the use of synthetic fibers. Although synthetic fiber composite materials such as glass fibers, carbon fibers are high performance materials, they are less biodegradable and are sourced from non-renewable resources. Therefore, the use of natural fibers may bring environmental benefits as well as cost benefits.

The modification of natural fiber composites continued development, and improvement of these materials could bring on a paradigm shift in the world of composite reinforcements and green materials, and could have a significant positive impact on the automotive industry [140]. Carbon-based research has gained boost with the introduction of carbon fibers in carbon-based composites. Polymeric nanocomposites (PNCs) or polymer nanostructured materials represent a radical alternative to conventional- filled polymers. The reinforcement of polymers is done by fillers, which play a major role in strengthening the composite. In contrast to the conventional systems where the reinforcement is on the order of microns, discrete constituents on the order of a few nanometers (~10,000 times finer than a human hair) exemplify PNCs. Uniform dispersion of these nanoscopically sized filler particles produces ultra-large interfacial area per volume between the nanoelement and host polymer. Over the past decades, there have been endless efforts to find suitable application for CNTs. Their extraordinary electrical, mechanical, thermal, and electrochemical properties [38] seem to have several possible applications: field emission devices [40], electronic circuits, devices, and interconnects [103], super capacitors and batteries [37], separation membranes [24], nanoscale sensors [74], drug delivery systems [106], and composite materials—both polymeric and metal matrix filled [3]. However, the biggest obstacle is industrial or large-scale manufacturing of the material.

Accordingly, manufacturing of high-performance engineering materials from renewable resources has been pursued by researchers across the world owning to the facts that they are renewable raw materials, environmentally sound, and do not cause health problem. The prominent advantages of natural fibers include acceptable specific strength properties, low cost, low density, high toughness, and good thermal properties. The Journal of Reinforced Plastics and Composites (2014) presents research studies on a broad range of today's reinforced plastics and composites and reports on new materials. R&D are often related to the service requirements of specific application areas, such as automotive, marine, construction and aviation.

11.1.1 COMPOSITES: A DEFINITION

A composite material is made by combining two or more materials to give a unique combination of properties, one of which is made up of stiff, long fibers and the other, a binder or 'matrix' which holds the fibers in place.

Kelly (1967) stated as the composites should not be regarded simple as a combination of two materials. In the broader significance; the combination has its own distinctive properties. In terms of strength to resistance to heat or some other desirable quality, it is better than either of the components alone or radically different from of them.

Van Suchetclan (1972) explained composite materials as heterogeneous materials consisting of two or more solid phases, which are in intimate contact with each other on a microscopic scale. They can be also considered as homogeneous materials on a microscopic scale in the sense that any portion of it will have the same physical property.

Beghezan (1966) defined as "The composites are compound materials which differ from alloys by the fact that the individual components retain their characteristics but are so incorporated into the composite as to take advantage only of their attributes and not of their short comings," in order to obtain improved materials.

11.1.2 PROPERTIES OF COMPOSITES

A Composite material consists of one or more discontinuous phase. The discontinuous phase is usually harder and stronger than the continuous phase and is called the "reinforcement" or "reinforcing" materials. Whereas the matrix is the monolithic material into which the reinforcement is embedded, and is completely continuous. This means that there is a path through the matrix to any point in the material, unlike two materials sandwiched together. In structural applications, the matrix is usually a lighter metal such as aluminum, magnesium, or titanium, and provides a complete support for the reinforcement. In high temperature applications, cobalt and cobalt-nickel alloy matrices are common. It serves to strengthen the composites and improves the overall mechanical properties of the matrix.

The reinforcement material is embedded into the matrix. The reinforcement does not always serve a purely structural task (reinforcing the compound), but is also used to change physical properties such as wear resistance, friction coefficient, or thermal conductivity. The reinforcement can be either continuous, or discontinuous. Discontinuous metal matrix composites can be isotropic, and can be worked with standard metalworking techniques, such as extrusion, forging or rolling. In addition, they may be machined using conventional techniques, but commonly would need the use of polycrystalline diamond tooling (PCD).

Properties of composites are strongly dependent on the properties of their constituent materials, their distribution and the interaction among them. The composite properties may be the volume fraction sum of the properties of the constituents or better properties. Apart from the nature of the constituent materials, the geometry of the reinforcement (shape, size and size distribution) influences the properties of the composite to a great extent. The concentration distribution and orientation of the reinforcement also affect the properties.

The shape of the discontinuous phase (which may be spherical, cylindrical or rectangular cross-sectioned prisms or platelets), the size and size distribution (which controls the texture of the materials), and the volume fraction determine the interfacial area, which plays an important role in determining the extent of the interaction between the reinforcement and the matrix.

Concentration, usually measured as volume or weight fraction determines the contribution of a single constituent to the overall properties of the composites. It is not only the single most important parameter influencing the properties of the composites, but also an easily controllable manufacturing variable used to alter its properties. Further, the need of composite for high strength to weight ratio, corrosion resistance, lighter construction materials and more seismic resistant structures has placed high emphasis on the use of new and advanced materials that not only decreases weight but also absorbs the shock and vibrations through tailored microstructures.

11.2 NATURAL FIBER COMPOSITES

Natural fibers, as reinforcement, have recently attracted the attention of researchers because of their advantages over other established materials [45]. They are environmentally friendly, fully biodegradable, abundantly available, renewable, cheap and have low density. The various advantages of natural fibers over man-made synthetic and carbon fibers are low cost, low density, competitive specific mechanical properties, reduced energy and biodegradability. Thermoplastic materials that currently dominate as matrices for natural fibers are polypropylene (PP), polyethylene, and poly(vinyl chloride) while thermosets, such as phenolics and polyesters, are common matrices.

11.2.1 CLASSIFICATION OF NATURAL FIBERS AND THEIR PROPERTIES

11.2.1.1 Classification of Natural Fibers

Natural fibers are subdivided into three categories based on their origin i.e., whether they are derived from Plants, Animals and Minerals. Classification of natural fibers which can be used as reinforcement for polymers [96] is shown in the Figure 11.1.

Some of the important natural fibers used as reinforcement in composites are listed in Table 11.1 and its origin is also explained [35, 127, 54, 56, 141].

11.2.1.2   Properties of Natural Fibers

Table 11.2 shows the mechanical properties of natural (plant) and synthetic fibers [18]. Natural fibers are non abrasive towards mixing and molding equipment. This can contribute to significant equipment maintenance cost reductions. The elementary fibers are bound together by a pectin interphase. This interphase is much stronger than the interphase between the technical fibers [39]. Natural fibers are safe to handle and the working conditions are better when compared to synthetic reinforces, such as glass fibers.

Their processing is environmental friendly, offering better working conditions and therefore, there is a reduction in the risk of dermal or respiratory problems. The most interesting aspects of natural fibers are their positive environmental impact. These fibers are renewable resources, which are biodegradable and their production requires little

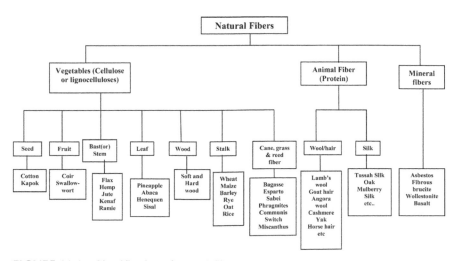

FIGURE 11.1   Classification of natural fibers.

TABLE 11.1   List of Important Natural Fibers.

Fiber Source	Species	Origin
Abaca	*Musa textiles*	Leaf
Alfa	*Stippa tenacissima*	Grass
Bagasse	–	Grass
Bamboo	*(>1,250 species)*	Grass
Banana	*Musa indica*	Leaf
Broom root	*Muhlenbergia macroura*	Root
Cantala	*Agave cantala*	Leaf
Coir	*Cocos nucifera*	Fruit
China jute	*Abutilon theophrasti*	Stem
Cotton	*Gossypium sp.*	Seed
Flax	*Linum usitatissimum*	Stem
Hemp	*Cannabis sativa*	Stem
Henequen	*Agave fourcroydes*	Leaf
Isora	*Helicteres isora*	Stem
Istle	*Samuela carnerosana*	Leaf
Jute	*Corchorus capsularis*	Stem
Kapok	*Ceiba pentranda*	Fruit
Kenaf	*Hibiscus cannabinus*	Stem
Nettle	*Urtica dioica*	Stem
Sisal	*Agave sisilana*	Leaf
Straw (Cereal)	–	Stalk
Wood	*(>10,000 species)*	Stem

energy. Therefore, the major task to be solved, in order to boost the acceptance of natural fibers as a quality alternative to conventional reinforcing fibers, is that to develop a high performance natural fiber reinforced composites [63, 105]. Georgopoulos et al. (2005) have investigated that the loading of LDPE with natural fibers leads to a decrease in tensile strength of the pure polymer. Joly et al. (1996) have optimized the fiber treatments for PP/cellulosic-fiber composites. Bledzki and Faruk (2006) proved that the Maleic anhydride-Polypropylene has improved the physico-mechanical properties up to 80%. Composites of an aliphatic polyester with natural flax fibers are prepared by batch mixing and the effect of

TABLE 11.2 Mechanical Properties of (Natural) Plant and Synthetic Fibers [115].

Fibers	Tensile Strength (mpa)	Young's modulus (spa)	Elongation at break (%)
Flax	345–1500	27.6	2.7–3.2
Hemp	690	70	1.6
Jute	393–800	1.3–26.5	1.16–1.5
Kenaf	930	53	1.6
Ramie	400–938	61.4–128	1.2–3.8
Nettle	650	38	1.7
Sisal	468–700	90–4.22	3–7
Henequen	–	–	–
PALF	413–1627	34.5–82.5	1.6
Abaca	430–760	–	–
Oil palm EFB	287–800	5.5–12.6	7–8
cotton	131–220	4–6	15–40

processing conditions on fiber length distribution and the dependence of the composite mechanical properties on fiber content were investigated by Baiardo et al. (2004).

Costa et al. (2000) have investigated the tensile and flexural performance of PP–wood fiber composites. The effect of these variables on tensile strength, Young's modulus, elongation at yield and flexural strength was determined. Arbelaiz et al. (2006) have determined the effect of fiber treatments with maleic anhydride, maleic anhydride PP copolymer, vinyltrimethoxy silane, and alkalization on thermal stability of flax fiber and crystallization of flax fiber/PP composites. The tensile strength of hemp strand/PP composites can be as high as 80% of the mechanical properties of glass fiber/PP composites [101]. Doan et al. (2006) investigated the effect of maleic anhydride grafted PP (MAPP) coupling agents on the properties of jute fiber/PP composites.

Joseph et al. (1999, 2003) prepared sisal fiber reinforced PP composites by melt-mixing and solution-mixing methods. The methods enhanced the tensile properties of the composites. Natural fibers generally contain large amounts of the hydroxyl group, which makes them polar and hydrophilic in nature. However most plastics are hydrophobic in nature. The addition of hydrophilic nature fibers to hydrophobic plastic will result in a composite with poor mechanical properties due to non-uniform fiber dispersion in the matrix and an interior fiber matrix interphase [89]. This polar nature also results in high moisture sorption in natural fiber based composites, leading to fiber swelling and voids in the fiber matrix interphase. Moisture if not removed from natural fiber prior to compounding by drying, will result in a porous product. High moisture absorption could also cause deterioration in mechanical properties and loss of dimensional stability [13, 6]. Another major limitation, in exploiting the use of natural fiber, is the limited thermal stability, possessed by natural fibers. They undergo degradation when processed beyond 200°C, this further limits the choice of plastic material to be used as matrix [46]. Girisha et al. (2012) have tested the composites and found that the tensile strength of the composite increased for approximately 25% of weight fraction of the fibers and further increase in the weight fraction of fiber, decreased the strength. Also it is found that for the hybrid combination of ridge guard and sisal fibers, there is 65% increase in the tensile strength.

11.2.2   PLANT FIBER VS. ANIMAL FIBER COMPOSITES

Much attention is paid on the natural fiber reinforced composites by researchers. Some of the natural (plant) fibers are sisal, coir, jute, hemp, ramie, kenaf, pineapple leaf, banana fiber etc. Banana fiber is a natural (plant) bast fiber. It can be explored as a potential reinforcement and it has its own physical and chemical characteristics and many other properties that make it as fine quality fiber. Appearance of banana fiber is similar to that of bamboo fiber and ramie fiber, but its fineness and spinnability is better than the two. It is highly strong fiber, bio-degradable, eco-friendly fiber, smaller elongation, light weight and it has somewhat shiny appearance depending upon the extraction and spinning process. It has strong moisture absorption quality and as well as release moisture very fast. The chemical composition of banana fiber is cellulose, hemicellulose and lignin (Table 11.3) [87].

TABLE 11.3   Composition of Banana Fiber [87].

Composition	Percentage
Cellulose	63–64
Hemicellulose	19
Lignin	5
Moisture content	10–11

Kenaf is one of the natural (plant) fibers used as reinforced in polymer matrix composites. It belongs to hibiscus family (Hibiscus cannabinus L) which is 4000 year old annual crop, native to Africa. Kenaf fibers are stronger, whiter and more lustrous. Kenaf grows quickly and will achieve 5 to 6 m in height and 2.5 to 3.5 cm in diameter within 5 to 6 months [128]. The properties of Kenaf fiber is tabulated as shown below (Table 11.4)

The Kenaf plant is composed of many useful components (e.g., stalks, leaves and seeds) and within each of these, there are various usable portions (e.g., Fibers and fiber strands, protein, oils and allelopathic chemicals) [146]. Kenaf filaments consists of discrete individual fibers of generally 2–6 mm. Filaments and individual fiber properties can vary depending on sources, age, separating technique and history of the fibers. The barks constituent 30–40% of the stem dry weight and shows a rather dense structure. The core reveals an isotropic and almost amorphous pattern. However, the bark shows an oriented high crystalline fiber pattern. Yusoff et al. (2010) studied the mechanical properties of short random oil palm fiber reinforced epoxy (OPF/epoxy) composites.

TABLE 11.4   Physical and Mechanical Properties of Kenaf Fiber [102, 21].

Properties	Value
Density	1200–1400 kg/m^3
Tensile strength	295–1191 MPa
Young's modulus	22–60
Specific strength	246–993 MPa
Specific Modulus	18–50 GPa

11.2.2.1 Animal Fiber Composite

Fibers that are taken from animals or hairy mammals such as Sheep, Wool, Goat (hair), Alpaca (hair), Horse (hair), etc., are animal fibers. Animal fibers generally comprise proteins, for example wool, silk, human hair, and feathers, etc. [99].

11.2.2.2 Silk Fiber

Silks are fibrous protein, some forms of which can be woven into textiles. The best known type of silk have a range of functions, including cocoons of the larvae of the mulberry silkworm. Bombyx mori are reared in captivity [41, 143, 145]. They have repetitive protein sequence [29] with a predominance of alanine, glycine and serine (which is high in silkworm silks but low in spider silks). Silk proteins are comprised of four different structural components: (1) elastic β-spirals, (2) crystalline β-sheets rich in alanine, (3) tight amino acid repeats forming α-helices, and (4) spacer regions [47, 126].

Silks are produced by several other insects, but only the silk of moth caterpillars has been used for textile manufacturing. Silks are mainly produced by the larvae of insects.

Silk emitted by the silkworm consists of two main proteins, sericin and fibroin. Fibroin is the structural center of the silk and sericin is the sticky material surrounding it. Some of the properties of silk are given below:

- silk has a smooth, soft texture that is not slippery, unlike many synthetic fibers.
- it is one of the strongest natural fibers but loses up to 20% of its strength when wet.
- it has a good moisture regain of 11%.
- its elasticity is moderate to poor.
- it can be weakened if exposed to too much sunlight.
- silk is poor conductor of electricity and thus susceptible to static ling.
- it is a measurement of linear density in fibers.
- silk has high proportion (50%) of glycine, which is a small amino acid.
- silk is resistant to most mineral acids, except for sulfuric acid, which

dissolves it.

The shimmering appearance of silk is due to the triangular prism-like structure of the silk fibers, which allows silk cloth to refract incoming light at different angles, thus producing different colors. The woven natural silk was chosen considering its environmental and mechanical properties, which is given Table 11.5 [20, 30, 107]. The bombyx mori silk produced by silkworm is among the strongest fibers produced in nature. It is also extremely elastic and resilient.

Silkworm is often processed via the production of fibroin solution where degummed silk is dissolved, and thus the inherent mechanical properties of the silk fibers are lost and any damage caused by the removal of sericin becomes insignificant. In its simplest form, fibroin can be regenerated into a film or coated onto other materials [90, 51, 62, 9]. The manufacture of nano-Hydroxyapatite-silk sheets [135, 42] has also been reported.

Oxygen and water vapor permeability are high for methanol-treated fibroin membranes [25, 100, 94, 93]. Another study suggested that B.mori silk braided into yarns elicited a mild inflammatory response after seven days in vivo, whereas sericin-coated silk yarns and Polyglycolide (PGA) caused an acute inflammatory response [121]. Gelatin coated B.mori silk has also been reported to initiate a minimal inflammatory response [79]. Sericin coatings on synthetic polymer fibers are reported to have antibacterial and antifungal properties [118]. Recent reviews concerning the use of silk-based biomaterials are available [5, 52, 145]. Fibroin films [132] and fibroin-alginate sponges have been found to enhance skin wound healing in vivo compared to clinically used materials. The majority of the research

TABLE 11.5   Properties of Woven Silk Fiber [20, 30, 107]

Properties of woven natural silk fiber	
Density	$1.47/cm^3$
Elongation	15%
Modulus of elasticity	22GNm
Thickness	0.42 mm
Ultimate strength	11GNm

used silk-based biomaterials for bone tissue engineering and fibroin solution to form films [129].

11.2.3 MODIFICATION OF NATURAL FIBERS

The application of natural fibers as reinforcement in composite materials is constantly in development. It has some limitations such as poor moisture resistance, inferior fire resistance, limited processing temperatures, lower durability, variation in quality and price, and difficulty in using established manufacturing process [10, 32]. These limitations are overcome by modification of natural fiber composites. Pothen et al. (2004) researched on reinforced polyester composites with banana fiber. It shows that 30 mm fiber length gave maximum tensile strength and 40 mm fiber length shows maximum impact strength. Sherely Annie Paul et al. (2008) reported that the NaOH concentration has an influence on the thermophysical properties of the composites. A 10% NaOH treated banana fiber composites showed better thermophysical properties than 2% NaOH treated banana fiber composites. Pothen et al. (2006) investigated the influence of chemical modification on dynamic mechanical properties of banana fiber-reinforced polyester composites. Lally et al. (2003) reported that the optimum content of banana fiber in polyester composite to be 40%. Rajesh Ghosh et al. (2011) researched on the tensile properties of the banana fiber with vinyl ester resin composites. Joseph et al. (2006) studied the environmental durability of chemically modified banana fiber- reinforced phenol formaldehyde (PF) composites. Indicula et al. (2006) investigated the thermophysical properties of Banana-sisal hybrid-reinforced composites as function of chemical modification.

Sreekumar et al. (2008) have reported the effect of fiber content in polyester composites and have reported 40% volume fraction to show maximum tensile strength. Sapuan et al. (2006) investigated the mechanical properties of woven banana fiber reinforced with epoxy composites. Tensile strength, Flexure, impact and fracture surface study of woven pseudo stem banana fiber reinforced with epoxy composites was reported and the chemical modification of kenaf fiber was carried out by Edeerozey et al. (2007). Anuar et al. (2011) studied the thermal properties of injection molded kenaf fiber/PLA biocomposite. They have found that the glass crystallization temperature and cold-crystallization enthalpy increases with fiber content, cold-crystallization and melting temperature slightly decreased while the storage modulus increases at higher kenaf fiber content. Ben et al. (2007) studied the properties of PLA/kenaf fiber composites and was found that the value of unidirectional fiber orientation (UD) composite of 0° shows double the tensile strength of PLA alone. Ma et al. (2010) reported the interfacial shear strength of jute fiber/PLA composite. It generally increased with increase in the treating time and the concentration of NaOH solution, and the maximum value was obtained at 8 h treatment with 12% NaOH solution. Khondker et al. (2005) studied the processing conditions of unidirectional jute yarn reinforced polypropylene composites fabricated by film stacking methods.

11.2.4 NATURAL FIBER AS A SUBSTITUTE TO SYNTHETIC FIBER IN POLYMER COMPOSITES

Natural fibers traditionally have been used to fill and reinforce thermosets, natural fiber reinforced thermoplastics, especially polypropylene composites, have attracted greater attention due to their added advantage of recyclability [95]. Investigation done by Alavudeen et al. (2011) reveals that the woven natural fibers, especially Banana and Kenaf as reinforcing agent in polymer based composites, were used for non-loading bearing automotives components and the factors which affects the mechanical properties of composites. Various concentration of NaOH was used and the morphological changes were determined by SEM. The authors observed that the NaOH treated kenaf fibers exhibited better mechanical properties than untreated fibers. A trunk model "Manaca" was developed and tested by using banana fiber mixed with epoxy resin and hardner by Maleque et al. (2007) and Al-Qureshi (1999). However, some special and critical panels of a hybrid composite of fiber glass/banana chopped fiber/epoxy was laminated.

Rajesh Ghosh et al. (2011) researched on the tensile properties of the Banana fiber-Vinyl ester resin composites. At 35% of fiber volume fraction, the tensile strength is increased by 38.6%. At lower volume fraction of banana fiber, the strength of the composite specimen is reduced when compared with Virgin resin. Banana fiber having high specific strength makes a light weight composite material and can be used to make light weight automobile interior parts.

Raw kenaf bast fiber have to be mechanically processed because of the coarseness, brittleness and low cohesive of the fiber bundles. Carding of kenaf is used to further back up the fiber bundles after the chemical processing of kenaf fibers. Generally optimum tensile properties and Young's modulus are dictated by the volume of reinforcing fiber used for the composites.

11.3   CARBON NANOTUBE FIBER NANOCOMPOSITES

Carbon nanotubes (CNTs) discovered by Iijima (1991), are seamless cylinders made of rolled up hexagonal network of carbon atoms. Single-wall nanotubes (SWNTs) consist of a single cylindrical layer of carbon atoms. The details of the structure of multi-wall carbon nanotubes (MWNTs) are still being resolved, but can be envisioned as a tubular structure consisting of multiple walls with an inter-layer separation of 0.34 nm. The diameter for inner most tube is on the scale of nanometer. In one model, the MWNTs are considered as a single sheet of graphene that is rolled up like a scroll to form multiple walls [33].

CNTs have received an enormous degree of attention in recent years, due to the remarkable physical and mechanical properties of individual perfect CNTs, which are considered to be one of the most promising new reinforcements for structural composites. The researchers have particularly focused on CNTs as toughening elements to overcome the intrinsic brittleness of the ceramic or glass material. Although there are now a number of studies published in the literature, these inorganic systems have received much less attention than CNT/polymer matrix composites.

The properties of the various composite systems are discussed, with an emphasis on toughness; a comprehensive comparative summary is provided, together with a discussion of the possible toughening mechanism that may operate. The extraordinary stiffness and specific tensile strength of CNTs makes them well-suited for use as reinforcing elements in polymer composites. The incorporation of carbon nanotubes can greatly increase the strength and stiffness of a polymer matrix with minimal increase in weight. There is a growing body of evidence that CNTs are ideal candidates for reinforcement in composite materials due to their nanoscale structure, outstanding mechanical, thermal and electrical properties [122, 27, 53]. A number of studies have reported the fabrication of CNT-grafted fibers using both direct growth of CNTs onto fibers [138, 149, 22, 142, 148] and chemical reactions between modified fibers and CNTs.

Meng and Hu (2008) incorporated MWNT into the shape memory polyurethane fiber by in-situ polymerization with treatment of MWNT in concentrated nitric acid and sulfuric acid. Shen et al. (2008) premixed CNT with poly(ethylene terephthalate) (PET) in a solvent followed by melt spinning after drying the mixture. They demonstrated that the tensile strength of the composite fibers increased by 36.9% (from 4.45 to 6.09 cN/dtex), and the tensile modulus increased by 41.2% (from 80.7 to 113.9 cN/dtex) by adding 0.02 wt.% of acid treated MWNT.

Kim et al. (2005) prepared electro-spun composite fibers based on polycarbonate with MWNT. They reported that the membrane composed of the composite fibers exhibited strong and tough properties. They suggested that the results may provide a feasible consideration of such electro-spun composite fibers for use as the reinforcing elements in a polymer based composite of a new kind.

11.4 APPLICATIONS OF NATURAL FIBER REINFORCED COMPOSITES

11.4.1 APPLICATIONS OF PLANT FIBER REINFORCED COMPOSITES

In the recent past considerable research and development have been expanded in natural fibers as reinforcement in thermoplastic resin matrix. These reinforced plastics serve as an inexpensive, biodegradable, renewable, and nontoxic alternative to glass or carbon fibers. The various advantages of natural fibers over man-made glass and carbon fibers are low cost, low density, competitive specific mechanical properties, reduced energy consumption and biodegradability. Natural fiber reinforced composites with thermoplastic matrices have successfully proven their qualities in various fields of application [15]. Thermoplastic materials that currently dominate as matrices for natural fibers are polypropylene (PP), polyethylene, and poly(vinyl chloride) while thermosets, such as phenolics and polyesters, are common matrices. With a view to replace the wooden fittings, fixtures and furniture, organic matrix resin reinforced with natural fibers such as jute, kenaf, sisal, coir, straw, hemp, banana, pineapple, rice husk, bamboo, etc., have been explored in the past two decades. In recent time plant fibers have been receiving considerable attention as substitutes for synthetic fiber reinforcements. Natural fibers such as hemp, flax, jute, kenaf etc. have been introduced as reinforcement in both thermoplastic and thermoset polymer based composites and have found extensive applications in transportation (automobile and railway coach interior, boat, etc.), construction as well as in packaging industries worldwide [95, 82, 50, 113]. Holbery and Houston (2006) reviewed different aspects of the application of Natural fiber reinforced polymer composites (NFRPCs) in automotive applications.

Unlike the traditional synthetic fibers like glass and carbon these lignocellulosic fibers are able to impart certain benefits to the composites such as low density, high stiffness, low cost, renewability, biodegradability and high degree of flexibility during processing [45]. The biodegradability of the natural plant fibers may present a healthy ecosystem while the low costs and good performance of these fibers are able to fulfill the economic interest of industry [69]. The use of plant fiber based automotive parts such as various panels, shelves, trim parts and brake shoes are attractive for automotive industries worldwide because of its reduction in weight about 10%, energy production of 80% and cost reduction of 5% [59]. Natural fiber composites are being used for manufacturing many components in the automotive sector [134, 99].

The fibers could be used as an effective reinforcement for making composites, which have an added advantage of being lightweight. Sapuan et al. (2011) have studied the tensile and flexural strengths of coconut spathe and coconut spathe-fiber reinforced epoxy composites and evaluated the possibility of using it as a new material in engineering applications. Jute is a strong, coarse and rigid fiber with very low extensibility which makes it suitable to act as reinforcing material in a composite. Jute is relatively inexpensive and process friendly [43]. Bast fibers commonly show the highest

DP among all the plant fibers (~10,000). However, Natural fiber composite has been used also in biomedical applications for bone and tissues repair and reconstruction [26]. In the area of structural rehabilitation, jute mats reinforced composites have been used for trenchless rehabilitation of underground drain pipes and water pipes [147]. Abaca fiber reinforced polypropylene composite has got remarkable and outstanding interest in the automobile industries owing to low cost availability, high flexural and tensile strength, good abrasion and acoustic resistance, relatively better resistance to mold and very good resistance to UV rays [112]. The combination of sugarcane fiber cellulose (SCFC) as reinforcement fiber, tapioca as the matrix material and glycerol as plasticizer can be used to produce a composite which is an alternative material for food packaging application [61].

11.4.2 APPLICATIONS OF ANIMAL FIBER REINFORCED COMPOSITES

Applications of animal fibers in composites have not yet been exploited fully. Animal based natural fibers can also be used as alternatives for producing composite materials which may have great scope in value added application including bio-engineering and medical applications. The contents of these fibers like wool, spider and silkworm silk, are mainly made by proteins. Wool is the most popular natural material. In the textile industries, a lot of waste wool fibers and their products induce actions which lead to the regeneration of wool keratin materials. It has good fire-resistance and noise-absorbing properties. The silk fibers are environmentally stable as compared to the proteins because of their extensive hydrogen bonding. Silk fiber composites is expected to be light weight and very tough with good impact strength bearing materials. It can be shaped into complex shapes with suitable matrix. Fibroin films [132] and fibroin-alginate sponges have been found to enhance skin wound healing in vivo compared to clinically used materials. Majority of the research was carried out into the use of silk-based biomaterials for bone tissue engineering. Fibroin solution is used to form films [129]. Spider silk has been investigated for bone tissue engineering. It has been found to have similar visco-elastic properties to human bone [12]. Silk fiber have higher tensile strength than glass fiber or synthetic organic fiber, good elasticity, and excellent resilience [108]. Silk's knot strength, handling characteristics and ability to lay low to the tissue surface make it popular suture in cardiovascular applications where bland tissue reactions are desirable for the coherence of the sutured structures [109]. A recent patent reported on a PVA/sericin crosslinked hydrogel membrane produced by using dimethyl urea as the crosslinking agent, had a high strength, high moisture content and durability for usage as a functional film [104]. Meinel et al. (2004) concentrated on cartilage tissue engineering with the use of silk protein scaffold and the authors identified and reported that silk scaffolds are particularly suitable for tissue engineering of cartilage starting from human mesenchymal stem cells (hMSC), which are derived from bone marrow, mainly due to their high porosity, slow degradation, and structural integrity.

Animal fiber reinforced composite materials have found applications in the automotive, aerospace and sports equipment industries. Advantage of animal fiber composite materials is that they can be tailored to meet the specific structure. Composites may expect to be cost competitive and a very attractive alternative to conventional materials. Further the feathers take up a lot of space in landfills and take a long time to decay because of the keratin proteins that make up the feathers. There is also the fear of bird flu, which makes converting feathers into animal feeds undesirable. The cement-bonded feather board developed is more resistant to decay and termite attack due to the keratin. Feather boards could be used for paneling, ceilings and as insulation but not for weight-bearing building components like walls or pillars. Thus use of feather in composites would be an attractive and better alternative for safe management especially in value added engineering.

11.4.3 APPLICATIONS OF CARBON NANOTUBE FIBER REINFORCED COMPOSITES

The first report on the preparation of a CNT/polymer nanocomposite in 1994 [2], a myriad of research efforts have been made to understand their structure–property relationship and find useful applications. The effective utilization of carbon nanotubes in composite applications depends strongly on the ability to homogeneously disperse them throughout the matrix without destroying their integrity. Furthermore, good interfacial bonding is required to achieve load transfer across the CNT–matrix interface, a necessary condition for improving the mechanical properties of the composite [7]. Lynam et al. (2007) suggested that combining the conductivity of CNT with biomolecules such as chitosan or heparin involved in tissue repair should produce novel platforms, that is, scaffolds, with the properties of expediting cell growth. Wang et al. (2007) presented new battery materials that consist of a solid polyaniline (PANI)/CNT composite fibers which exhibited a discharge capacity of 12.1 mAh/g with a CNT content of 0.25 wt.%. They have the capacity to protect sensitive bioactive materials from enzymatic and chemical degradation in vivo and during storage, and to facilitate the transport of charged molecules across the absorptive epithelial cells [86].

Carbon nanotubes could replace conventional conductive fillers for a range of applications like electrostatic discharge (ESD) and electromagnetic interference (EMI) shielding, and a much lower loading of carbon nanotubes can be used to achieve desired conductivity levels [49, 48]. The fabrication of CNT-epoxy composites has been studied concerning their screen printing performance [55]. Tehrani et al. (2013) suggested the Hybrid carbon fiber with carbon nanotube composites for structural damping applications. De Valve et al. (2014) has investigated the operational modal analysis of rotating CNT infused composite beams and the application of carbon nano tube composite involving helicopter rotors and wind turbine blades.

11.5   CONCLUSION

The potential use of natural fiber composites is discussed in this chapter. One of the most important reasons for the increasing trend in the production of natural fiber composites is due to easy and economical disposal of the wastes. There is a lot of scope in the future for researchers and industrialists in the field of natural fiber reinforced composites. A significant amount of research work has been carried out in the field of natural fiber composites and yet more investigations have to be made with polymer composites. To wider the applications of these fibers in solving environmental problems, more studies have to be continued in the future.

ACKNOWLEDGMENT

The authors are grateful to management of DKM College for Women and Thiruvalluvar University, Vellore, Tamil Nadu, India for the support.

KEYWORDS

- applications
- carbon nanotube fiber reinforced composites
- natural fiber reinforced composites
- polymer composites

REFERENCES

1. A1-Qureshi, H. A. In *2nd International Wood and Natural Fiber Composite Symposium Germany*, 1999.
2. Ajayan, P. M., Stephan, O., Colliex, C., Trauth, D. Aligned carbon nanotube arrays formed by cutting a polymer resin-nanotube composite. *Science*.1994, *265,* 1212–1214.
3. Ajayan, P. M., Stephan, O., Redlich, P., Colliex, C. Carbon nanotubes as removable templates for metal oxide nanocomposites and nanostructures. *Nature,* 1995, *375(6532),* 564–567.
4. Alavudeen, A., Thiruchitrambalam, M., Venkateshwaran, N., Athijayamani, A. Review of natural fiber Reinforced woven composite. *Rev. Adv. Materials. 2011, 27,* 146–150.
5. Altman, G. H., Diaz, F., Jakuba, C., Calabro, T., Horan, R. L., Chen, J. S., Lu, H,; Richmond, J., Kaplan, D. L. Silk-based biomaterials. *Journal of Biomaterials,* 2003, *24(3),* 401–416.
6. Alvarez, V., Fraga, A., Vazquez, A. Effect of the moisture and fiber content on the mechanical properties of biodegradable polymer sisal fiber biocomposites. *Journal of Applied Polymer science,* 2004, *91,* 4007–16.
7. Andrew Rodney; Jacques David; Qian Dali; Rantell Terry. *Acc. Chem. Res.* 2002, *35,* 1008.
8. Anuar, H., Zuraid, A., Malays. *Polym. J.* 2011, *6,* 51.
9. Arai, T., Freddi, G., Innocenti, R., Tsukada, M. Biodegradation of bombyx mori silk fibroin fibers and films. *Journal of Applied Polymer Science,* 2004, *91(4),* 2383–2390.
10. Araújo, J. R., Waldman, W. R., De Paoli, M. A. Thermal properties of high density polyethylene composites with natural fibers: coupling agent effect. *Polym Degrad Stabil.* 2008, *93,* 1770–5.

11. Arbelaiz, A., Ferna' ndez, B., Ramos, J. A., Mondragon, I. Thermal and Crystallization Studies of Short Flax Fiber Reinforced Polypropylene Matrix Composites: Effect of Treatments. *Thermochimica Acta*, 2006, *440*, 111–121.
12. Bai, J., Ma, T., Chu, W., Wang, R., Silva, L., Michal, C., Chiao, J. C., Chiao, M. Regenerated spider silk as a new biomaterial for MEMS. *Biomedical Microdevices*, 2006, *8(4)*, 317–323.
13. Baiardo, M., Zini, E., Scandola, M. Flax fiber–polyester composites. *Comp.: Part A* 2004, *35*, 703–710.
14. Barbero, E. J. *Introduction to composite materials design.* Department of Mechanical & Aerospace Engineering, West Virginia University, Taylor & Francis: USA, 1999.
15. Begum, K., Islam, M. A. Natural Fiber as a substitute to Synthetic Fiber in Polymer Composites: A Review. *Research Journal of Engineering Sciences*, 2013, *2(3)*, 46–53.
16. Ben, G., Kihara, Y., Nakamori, K., Aoki; Yoshio. *Adv. Compos. Mater.* 2007, *16*, 361.
17. Berghezan, A. Nucleus, 8(5), *Nucleus A Editeur,* rhe, Chalgrin, Paris, 16(e), (1966).
18. Bismark, A., Misha, S., Lampke, T. *Natural fibers Biopolymers and biocomposites.* Edited by Mohanty, A. K., Mista, M., Drazal, L.T. *CRT Press,* 2005, 37.
19. Bledzki, A. K., Faruk, O. Injection Molded Microcellular Wood Fiber–Polypropylene Composites. *Composites: Part A,* 2006, *37,* 1358–1367.
20. Bledzki, A., Gassan, J. Composites reinforced with cellulose – based fibers. *Progress in Poly. Sci.* 1999, *24,* 221–274.
21. Bodros, E., Baley, Ch. Study of the tensile properties of stinging nettle fibers (Urtica dioica). *Materials Letters,* 2008, *62(14),* 2143–2145.
22. Boskovic, B. O., Golovko, V. B., Cantoro, M., Kleinsorge, B., Chuang, A. T. H., Ducati, C., et al. Low temperature synthesis of carbon nanofibers on carbon fiber matrices. *Carbon,* 2005, *43,* 2643–8.
23. Bullions, T., Hoffman, D., Gillespie, R., Prince –O'Brien, J., Loos, A. Contributions of feather fibers and various cellulose fibers to the mechanical properties of polypropylene matrix composites. *Journal of Composites Sci and Tech.* 2006, *66,* 102–1.
24. Che, G., Lakshmi, B. B., Fisher, E. R., Martin, C. R. Carbon nanotube membranes for electrochemical energy storage and production. *Nature, 1998, 393(6683),* 346–349.
25. Chen, C., Chuanbao, C., Xilan, M., Yin, T., Hesun, Z. Preparation of non-woven mats from all-aqueous silk fibroin solution with electro spinning method. *Journal of Polymer,* 2006, *47(18),* 6322–6327.
26. Cheung, H.-y., Ho, M.-p., Lau, K.-t., Cardona, F., Hui, D. Natural fiber-reinforced composites for bioengineering and environmental engineering applications. *Composites: Part B* 2009, *40,* 655–663.
27. Coleman, J. N., Khan, U., Blau, W. J., Gun'ko, Y. K. Small but strong: A review of the mechanical properties of carbon nanotube-polymer composites. *Carbon,* 2006, *44,* 1624–1652.
28. Costa, T. H. S., Carvalho, D. L., Souza, D. C. S., Coutinho, F. M. B., Pinto, J. C., Kokta, B. V. Statistical Experimental Design and Modeling of Polypropylene–Wood Fiber Composites. *Polymer Testing,* 2000, *19,* 419–428.
29. Craig, C.L., Riekel, C. Comparative architecture of silks, fibrous proteins and their encoding genes in insects and spiders. *Comparative Biochemistry and Physiology B Biochemistry& Molecular Biology,* 2002, *133(4),* 493–507.

30. Craven, J. P., Cripps, R., Viney, C. Evaluating the silk/epoxy interface by means of the Microbond Test. *Composites Part A: Applied Science and Manufacturing,* 2000, *31,* 653–660.
31. De Valve, C., Ameri, N., Tarazaga, P., Pitchumani, R. In *Modal Analysis of Rotating Carbon Nanotube Infused Composite Beams,* Conference Proceedings of the Society for Experimental Mechanics Series, 45(7), pp. 533–541, 2014.
32. Dittenber, D. B., GangaRao, H. V. S. Critical review of recent publications on use of natural composites in infrastructure. *Compos A Appl Sci Manuf.* 2012, *43,* 1419–29.
33. Dravid, V. P., Lin, X., Wang, Y., Wang, X. K., Yee, A., Ketterson, J. B., et al. Buckytubes and Derivatives: Their Growth and Implications for Buckyball Formation. Science, 1993, *259(5101),* 1601–1604.
34. Eddrozay, A. M. M., Ahil, H. M., Azhar, A. B., Arrifin, M. I. Z. *Mater Lett.* 2007, *61,* 2023–2025.
35. Eicchorn, S. J., Baillie, C. A., Zafeiropoulos, N., Mwai Kambo, Ly; Ansell, M. P., Dufresne, A. Current International research into cellulosic fibers and composites. *J. Mater Sci.* 2001, *39(9),* 2107–2131.
36. Eicchorn, S. J. C. A., Baillie, N., Zafeiropoulous, L. Y., Mwaikambo, M. P., Ansell, A., Dufresne, K. M., Entwistle, P. J., Herrare-Franco, G. C., Escamilla, L., Groom, M., Hugnes, C., Hill, T. G., Rials; Wild, P. M. *Journal of Mater. Sci.* 2001, *36,* 2107.
37. Endo, M., Kim, C., Nishimura, K., Fujino, T., Miyashita, K. Recent development of carbon materials for Li ion batteries. *Carbon,* 2000, *38(2),* 183–197.
38. Endo, M., Strano, M. S., Ajayan, P. M. Potential applications of carbon nanotubes. *Topics in Applied Physics,* 2008, *111,* 13–62.
39. Ever, O. M. J. A. V., Bos, H. C., Venkemenade, M. J. J. M. Influence of the physical structure of flax fiber on the mechanical properties of flax fiber. *Applied composites materials,* 2000, *7,* 387–402.
40. Fan, S., Chapline, M. G., Franklin, N. R., Tombler, T. W., Cassell, A. M., Dai, H. Self-oriented regular arrays of carbon nanotubes and their field emission properties. *Science,* 1999, *283(5401),* 512–514.
41. Foelix, R. F. *Biology of spiders.* Cambridge, MA, Harvard University Press: USA, 1992.
42. Furuzono, T., Kishida, A., Tanaka, J. Nano-scaled hydroxyapatite/polymer composite I. Coating of sintered hydroxyapatite particles on poly (γ-methacryloxypropylt rimethoxysilane)-grafted silk fibroin fibers through chemical bonding. *Journal of Materials Science: Materials in Medicine,* 2004, *15(1),* 19–23.
43. Ganguly, P. K., Samajpati, S. Book of papers, seminar on technology today – transfer tomorrow, NIRJAFT, Kolkata, India, 1996.
44. Georgopoulos, S. Th., Tarantili, P. A., Avgerinos, E., Andreopoulos, A. G., Koukios, E. G. Thermoplastic Polymers Reinforced with Fibrous Agricultural Residues. *Polymer Degradation and Stability,* 2005, *90,* 303–312.
45. Girisha, C., Sanjeevamurthy; Gunti Rangasrinivas. Tensile properties of natural fiber-reinforced epoxy-hybrid composites. *International Journal of Modern Engineering Research (IJMER)* 2012, *2(2),* 471–474.
46. Glasser, W. G., Taib, R., Jain, R. K., Kander, R. Fiber-reinforced cellulosic thermoplastic composites. *Journal of Applied polymer science,* 1999, *73,* 1329–1340.

47. Gosline, J. M., Guerette, P. A., Ortlepp, C. S., Savage, K.N. The mechanical design of spider silks: From fibroin sequence to mechanical function. *Journal of Experimental Biology,* 1999, *02(23),* 3295–3303.
48. Grimes, C. A., Dickey, E. C., Mungle; Ong, K. G., Qian, D., *J. Appl. Phys.* 2001, *90,* 4134.
49. Grimes, C. A., Mungle, C., Kouzoudis, D., Fang, S., Eklund, P. C. *Chem. Phys. Lett.* 2000, 319–460.
50. Gross, R. A., Karla, B. Biodegradable polymers for the environment, *Science,* 2002, *297,* 1803–1807.
51. Gupta, M. K., Khokhar, S. K., Phillips, D. M., Sowards, L. A., Drummy, L. F., Kadakia, M. P., Naik, R. R. Patterned silk films cast from ionic liquid solubilized fibroin as for cell growth. *Langmuir,* 2007, *23(3),* 1315–131.
52. Hakimi, O., Knight, D. P., Vollrath, F., Vadgama, P. Spider and mulberry silkworm silks as compatible biomaterials. *Journal of Composites Part B: Engineering,* 2007, *38(3),* 324–337.
53. Harris, P. Carbon nanotube composites. *International Materials Reviews,* 2004, *49(1),* 31–43.
54. Hattallia, S., Benabioura, A., Ham-Pichavant, F., Nourmamode, A., Castellan, A. *Journal of Polym Degrad Stab.* 2002, *75,* 259.
55. Heimann, M., Rieske, R., Telychkina, O., Boehme, B., Wolter, K. Application of carbon nanotubes as conductive filler of epoxy adhesives in Microsystems. *GMM* Workshop Mikro-Nano-Integration, Seeheim, Germany, 2007.
56. Hoareau, W., Trindade, W. G., Siegmund, B., Alam Castellan, A., Frollini, E. *Journal of Poly. Degrad. Stab.* 2004, *86,* 567.
57. Idicula, M., Boudence, A., Umadevi, L., Ibos, L., Candau, Y., Thomas, S. *Journal of Composite. Sci. Technology,* 2006, *6,* 2719.
58. Iijima, S. Helical microtubules of graphitic carbon. *Nature,* 1991, *354,* 56–58.
59. Ing. Eva Aková, Development of natural fiber reinforced polymer composites. *Transfer inovácií,* 2013, 25.
60. James, H., Dan, H. Natural-Fiber-Reinforced Polymer Composites in Automotive Applications. *JOM: J. Min. Met. Mat. S.,* 2006, *58(1),* 80–86.
61. Jeeferie, A. R., Nurul Fariha, O., Mohd, A. R., Warikh; Yuhazri, M. Y., Haeryip Sihombing., Ramli, J. Preliminary Study on the Physical and Mechanical Properties of Tapioca Starch/Sugarcane Fiber Cellulose Composite. *ARPN Journal of Engineering and Applied Sciences,* 2011, *6,* 7–15.
62. Jin, H. J., Park, J., Valluzzi, R., Cebe, P., Kaplan, D. L. Biomaterial films of Bombyx mori silk fibroin with poly(ethylene oxide). *Journal of Biomacromolecules,* 2004, *5(3),* 711–717.
63. John, M.J., Thomas, S. Biofibers and biocomposites. *Journal of Carbohydrate polymers,* 2008, *71,* 343–364.
64. Joly, C., Gauthier, R., Chabert, B. Physical Chemistry of the Interface in Polypropylene/Cellulosic-Fiber Composites. *Composites Science and Technology,* 1996, *56,* 761–765.
65. Joseph, P. V., Joseph, K., Thomas, S., Pillai, C. K. S., Prasad, V. S., Groeninckx, G., Sarkissova, M. The Thermal and Crystallization Studies of Short Sisal Fiber Reinforced Polypropylene Composites. *Composites: Part A* 2003, *34,* 253–266.

66. Joseph, P. V., Kuruvilla, J., Sabu, T. Effect of Processing Variables on the Mechanical Properties of Sisal Fiber-reinforced Polypropylene Composites. *Composites Science and Technology,* 1999, *59,* 1625–1640.

67. Joseph, S., Oommen, Z., Thomas, S. *Journal of. Appl. Polym. Sci.* 2006, *100,* 2421.

68. Journal of reinforced plastics and composites. Publisher SAGE Publications. [Online] 2014, http://www.researchgate.net/journal/07316844_Journal_of_Reinforced_Plastics_and_Composites.

69. Karmaker, A. C., Hoffmann, A., Hinrichsen, G. J. Influence of water uptake on the mechanical properties of jute fiber-reinforced polypropylene. *J. Appl. Polym. Sci.* 1994, *54,* 1803–1807.

70. Karnani, R., Krishnan, M., Narayan, R. Biofiber – reinforced polypropylene composites. *Journal of Poly. Eng Sci.* 1997, *37(2),* 476–82.

71. Kelly, A. The nature of Composite Material, *Sci. Amer. Mag.* 1967, *217,* 161.

72. Khondker, O. A., Ishiaku, U. S., Nakai, A., Hamada, H. Fabrication and Mechanical Properties of Unidirectional Jute/PP Composites Using Jute Yarns by Film Stacking Method. *Journal of Polymers and Environment.* 2005, *13,* 115–126.

73. Kim, G. M., Michler, G. H., Potschke, P. Deformation processes of ultrahigh porous multi-walled carbon nanotubes/polycarbonate composite fibers prepared by electrospinning. *Polymer,* 2005, *46(18),* 7346–7351.

74. Kong, J., Franklin, N. R., Zhou, C. et al., Nanotube molecular wires as chemical sensors. *Science,* 2000, *287(5453),* 622–625.

75. Kuruvilla Joseph; Romildo Dias Tolêdo Filho; Beena James; Sabu Thomas; Laura Hecker de Carvalho. A review on sisal fiber reinforced polymer composites. *Revista Brasileira de Engenharia Agrícola e Ambiental,* 1999, *3(3),* 367–379.

76. Laly Pothan, A., Zachariah Oomenb; Sabu Thomas. Dynamic Mechanical Analysis of Banana fiber reinforced composites. *Journal of composites science and technology,* 2003, *63(2),* 283–293.

77. Laly, A., Pothan, Nn., Neelakantan, R., Bhaskar Rao; Sabu Thomas. Stress relaxation behavior of banana fiber reinforced polyester composter. *Journal Reinforced Plastics and Composites,* 2004, *23(2).*

78. Li, Y., Mai, Y. W., Lin, Y. Sisal fiber and its composites: a review of recent developments. *Comp. Sci. Tech.* 2000, *60,* 2037–2055.

79. Liu, H., Ge, Z., Wang, Y., Toh, S. L., Sutthi Khum, V., Goh, J. C. H. Modification of sericin-free silk fibers for ligament tissue engineering application. *Journal of Biomedical Materials Research – Part B Applied Biomaterials,* 2001, *82(1),* 129–138.

80. Lynam, C., Moulton, S. E., Wallace, G. G. Carbon-nanotube biofibers. *Advanced Materials,* 2007, *19(9),* 1244–1248.

81. Ma, H., Joo, C.W. *J. Compos. Mater.* 2010, *45,* 1451.

82. Maguro, A. Vegetable fibers in automotive interior components. *Angew Makromol Chem.* 1999, *272,* 99–107.

83. Maldas, D., Kokta, Br., Raj, R. G., Daneault, C. Improvement of the mechanical properties of sawdust wood fiber – polystyrene composites by chemical treatment. *Journal of Polymer,* 1998, *29(7),* 1255–1265.

84. Maleque, M. A., Belal, F.Y., Sapuan, S. M. *The Arabian journal for science and Engineering,* 2007, *32,* 359.

85. Manikandan, K. C., Diwan, S. M., Thomas, S. Tensile properties of short sisal fiber reinforced polystyrene composites. *J. Appl. Poly. Sci.* 1997, *60*, 1483–1497.

86. Mao, H-Q., Roy, K., Troung-Le, V. L., et al. Chitosan-DNA nanoparticles as gene carriers: synthesis, characterization and transfection efficiency. *J. Control Release,* 2001, *70*, 399–421.

87. Maries Idicula, S. K., Malhotra, Kuruvilla Joseph; Sabu Thomas. Dynamic mechanical analysis of randomly oriented intimately mixed short banana/sisal hybrid fiber reinforced polyester composites. *Composites Science and Technology,* 2005, *65*, 1077–1087.

88. Maya Jacob John; Rajesh, D., Anandjiwala. Recent Development in chemical modification and characterization of Natural Fiber – Reinforced composites. *J. Poly. Composite,* 2008.

89. Mehta, G., Mohanty, A., Missa, M., Drazal, L. Effect of novel sizing on the mechanical and morphological characteristics of natural fiber reinforced unsaturated polyester resin based bio-composites. *Journal. of materials. Science,* 2004, *39*, 2961–4.

90. Meine, L., Hofmann, S., Karageorgiou, V., Kirker-Head, C., McCool, J., Gronowicz, G., Zichner, L., Langer, R., Vunjak-Novakovic, G., Kaplan, D.L. The inflammatory responses to silk films in vitro and in vivo. *Journal of Biomaterials,* 2005, *26(2),* 147–155.

91. Meinel, L., Hofmann, S., Karageorgiou, V., Zichner, L., Langer, R., Kaplan, D. Engineering cartilage-like tissue using human mesenchymal stem cells and silk protein scaffolds. *Biotechnol Bioeng.* 2004, *88,* 379–391.

92. Meng, Q. H., Hu, J. F. Self-organizing alignment of carbon nanotube in shape memory segmented fiber prepared by in situ polymerization and melt spinning. *Composites Part a-Applied Science and Manufacturing,* 2008, *39(2),* 314–321.

93. Min, B. M., Jeong, L., Lee, K. Y., Park, W. H. Regenerated silk fibroin nanofibers: Water Vapor-induced structural changes and their effects on the behavior of normal human cells. *Journal of Macromolecular Bioscience,* 2006, *6(4),* 285–292.

94. Minoura, N., Tsukada, M., Nagura, M. Fine structure and oxygen permeability of silk fibroin membrane treated with methanol. *Journal of Polymer,* 1990, *31(2),* 265–269.

95. Mohanty, A. K., Drazl, L. T., Misra, M. Engineered natural fiber reinforced polypropylene composites: influence of surface modifications and novel powder impregnation processing. *J Adhes Sci Technol.* 2002, *16(8),* 999–1015.

96. Mohanty, A. K., Misra, M., Drzal, L. T. Surface Modification Cations of natural fibers and performance of the resulting biocomposites – an overview. *Journal of Comp interfaces,* 2001, *815,* 313 –43.

97. Mohanty, A. K., Misra, M., Drzal, L. T. Sustainable biocomposites from renewable resources: Opportunity and challenges in the green materials world. *J. Polym. Environ.* 2002, *10,* 19–26.

98. Mohd Zuhri Mohamed Yusoff; Mohd Sapuan Salit; Napsiah Ismail; Riza Wirawan. Mechanical Properties of Short Random Oil Palm Fiber Reinforced Epoxy Composites. *Sains Malaysiana,* 2010, *39(1),* 87–92.

99. Mohini Saxena; Ashokan Pappu; Anusha Sharma; Ruhi Haque; Sonal Wankhede. Composite Materials from Natural resources Recent Trends and future potentials, 2011.

100. Motta, A., Fambri, L., Migliaresi, C. Regenerated silk fibroin films: Thermal and dynamic mechanical analysis. *Journal of Macromolecular Chemistry and Physics,* 2002, *203(10–11),* 1658–1665.
101. Mutje, P., Lopez, A., Vallejos, M. E., Lopez, J. P., Vilaseca, F. Full Exploitation of Cannabis Sativa as Reinforcement/Filler of Thermoplastic Composite Materials. *Composites: Part A* 2007, *38,* 369–377.
102. Mwaikambo, L. Y., Review of the history, properties and applications of plant fibers. *African Journal of Science and Technology,* 2006, *7(2),* 120–133.
103. Naeemi, A., Sarvari, R., Meindl, J. D. Performance comparison between carbon nanotube and copper interconnects for gigascale integration (GSI). *IEEE Electron Device Letters,* 2005, *26(2),* 84–86.
104. Nakamura, K., Koga, Y. Sericin-containing polymeric hydrous gel and method for producing the same. *Japan Patent,* 2001, 106794A.
105. Nishino, T. Green composites: polymer composites and the environment, United Kingdom, 2004.
106. Pastorin, G., Wu, W., Wieckowski S. et al., Double functionalization of carbon nanotubes for multimodal drug delivery. *Chemical Communications,* 2006, *11,* 1182–1184.
107. Perez-Rigueiro, J., Viney, C., Llorca, J., Elices, M. Mechanical properties of singlebrin silkworm silk. *Journal of Applied Polymer Science,* 2000, *75,* 1270–1277.
108. Perez-rigueiro, J., Viney, C., Llorca, J., Elices, M. Silkworm silk as an engineering material. *J Appl Polym Sci.* 1998, *70,* 2439–47.
109. Postlethwait, R. W., Dumphy, J. E., Van Winkle, W. Tissue reaction to surgical sutures. In: *editors. Repair and regeneration.* New York: McGraw-Hill.1969, pp. 263–85.
110. Pothan, L. A., Thomas, S., Groenilox, G. *Journal of comp. A.* 2006, *37,* 1260.
111. Prakash Tudu. Processing and characterization of Natural fiber Reinforced polymer Composites, 2009.
112. Proemper, E. New automotive interior parts from natural fiber materials, *Proceedings, International AVK-TV Conference* (Baden-Baden, Germany), 2004.
113. Puglia, D., Biagiotti, J., Kenny, J. M. A review on natural fiber-based composites-part II: Application of natural reinforcements in composite materials for automotive industry. *Journal of Natural Fibers,* 2005, *1,* 23–65.
114. Rajesh Ghosh; Rama Krishna, A., Reena, G., Lakshmiathi Raju, Bh. Effect of fiber volume fraction on the tensile strength of banana fiber reinforced vinyl ester composites. *Journal of Advanced Engineering Science and Technology,* 2011, *4(1),* 089–091.
115. Rouison, D., Sain, M., Conturier, M. Resin transfer molding of natural fiber reinforced composites. *Journal of Cure simulation composites science & Technology,* 2004, *64,* 629–44.
116. Sapuan, S. M., Leeni, A., Harimi, M., Beng, G. K. *Journal. of materials and Design,* 2006, *27,* 689.
117. Sapuan, S. M., Zan, M. N. M., Zainudin, E. S., Prithvi Raj Arora. Tensile and flexural strengths of coconut spathe-fiber reinforced epoxy composites. *Journal of Tropical Agriculture,* 2005, *43 (1–2),* 63–65.
118. Sarovart, S., Sudatis, B., Meesilpa, P., Grady, B. P., Magaraphan, R. The use of sericin as an antioxidant and antimicrobial for polluted air treatment. *Reviews on Advanced Materials Science.* 2003, *5(3),* 193–198.

119. Satyanarayana, K. G., Pai, B. C., Sukumaran, K., Pillai, S. G. K. *Lingocellulosic fiber reinforced polymer composites* (New York, Marcel Decker, Vol.1, p 339, 1990.

120. Satyanarayana, K. G., Sukumaran, K., Mukherjee, P. S., Pavithran, C., Pillai, S. G. K. *International conference on low cost housing for developing countries*, Rookree, 1984; pp. 171–181.

121. Seo, Y. K., Choi, G. M., Kwon, S. Y., Lee, H. S., Park, Y. S., Song, K.Y., Kim, Y. J., Park, J. K. The biocompatibility of silk scaffold for tissue engineered ligaments. *Key Engineering Materials*, 2007, 73–76.

122. Shaffer, M., Sandler, J. Carbon Nanotube/Nanofiber Polymer Composites. In: *Processing and Properties of Nanocomposites*, Advani, S. G., Ed., World Scientific: New Jersey, 2007; 1–59.

123. Shen, L. M., Gao, X. S., Tong, Y., Yeh, A., Li, R. X., Wu, D. C. Influence of different functionalized multiwall carbon nanotubes on the mechanical properties of poly(ethylene terephthalate) fibers. *Journal of Applied Polymer Science*, 2008, *108(5)*, 2865–2871.

124. Sherely Annie Paul, et al. Effect of fiber loading and chemical treatment on thermophysical properties of banana fiber /propylene commingled composite materials. *Composites part A: Applied science and Manufacturing*, 2008, *39(9)*, 1582 –1588.

125. Sherman, L. M. Natural fibers, the new fashion in automotive plastics. *Journal of Plast. Technol.* 1999, *45(10)*, 62–8.

126. Sirichaisit, J., Brookes, V. L., Young, R. J., Vollrath, F. Analysis of structure/property relationships in silkworm (Bombyx mori) and spider dragline (Nephila edulis) silks using Raman Spectroscopy. *Journal of Biomacromolecules*, 2003, *4 (2)*, 387–394.

127. Sjorstrom, E. In wood Chemistry: *Fundamentals and Applications Academic Press London*, 169, 1981.

128. Smith, H. About the kenaf plant. Retrieved from: http://www.visionpaper.com/kenaf2.html, 1998.

129. Sofia, S., McCarthy, M. B., Gronowicz, G., Kaplan, D. L. Functionalized silk-based biomaterials for bone formation. *Journal of Biomedical Materials Research*, 2001, *54(1)*, 139–148.

130. Sreekumar, P. A., Pradessh Albert, G., Unnikrishnan Kuruvilla Joseph; Sabu Thomas. Mechanical and Water Sorption studies of ecofriendly banana fiber reinforced polyester composites. *J. App. poly Sci.* 2008, *109*, 1547–1555.

131. Suchetclan Van. *Philips Research Reports.* 1972, *27*, 28.

132. Sugihara, A., Sugiura, K., Morita, H., Ninagawa, T., Tubouchi, K., Tobe, R., Izumiya, M., Horio, T., Abraham, N. G., Ikehara, S. Promotive Effects of a Silk Film on Epidermal Recovery from Full-Thickness Skin Wounds. *Proceedings of the Society for Experimental Biology and Medicine*, 2000, *225(1)*, 58–64.

133. Sydenstricker, T. H., Mochnaz, S., Amico, S. C. Pull-out and other evaluations in sisal-reinforced polyester bio composites. *Journal of Poly. Test.* 2003, *22(4)*, 375–80.

134. Taj, S., Munawar, M. A., Khan, S.U. A Review: Natural fiber-reinforced polymer composites. *Proc Pakistan Acad Sci.* 2007, *44(2)*, 129–144.

135. Tanaka, T., Hirose, M., Kotobuki, N., Ohgushi, H., Furuzono, T., Sato, J. Nanoscaled hydroxyapatite/silk fibroin sheets support osteogenic differentiation of rat

bone marrow mesenchymal cells. *Journal of Materials Science and Engineering*: C 2007, *27(4)*, 817–823.

136. Tehrani, M., Safdari, M. A., Boroujeni, Y., Razavi, Z., Case, S. W., Dahmen, K., Garmestani, H., Al-Haik, M. S. Hybrid carbon fiber/carbon nanotube composites for structural damping applications. *Nanotechnology*, 2013, *24*.

137. Thi-Thu-Loan Doan; Shang-Lin Gao; Mader, E. Jute/Polypropylene Composites, I. Effect of Matrix Modification. *Composites Science and Technology*, 2006, *66*, 952–963.

138. Thostenson, E. T., Li, W. Z., Wang, D. Z., Ren, Z. F., Chou, T. W. Carbon nanotubes/ carbon fiber hybrid multiscale composites. *J Appl Phys*. 2002, *91*, 6034–6037.

139. Toriz, G., Denes, F., young, R.A. Lignin – polypropylene composites. Part 1: composites from unmodified lignin and polypropylene. *Journal of Polymer composite*, 2002, *23(5)*, 801–811.

140. Vaidya, U. *Composites for automotive, truck and mass transit: materials, design, manufacturing*, DEStech Publications: Lancaster (PA), 2011.

141. Valadez-Gonzalez, A., Cervantes, J. M., Olayo, U. C. R., Herrera-Franco, P. J. *Compos. B*. 1999, *30*, 30.

142. Veedu, V. P., Cao, A. Y., Li, X. S., Ma, K. G., Soldano, C; Kar, S., et al. Multifunctional composites using reinforced laminae with carbon-nanotube forests. Nat Mater. 2006, *5(6)*, 457–462.

143. Vollrath, F. Strength and structure of spiders' silks. *Journal of Biotechnology*, 2000, *74(2)*, 67–83.

144. Wang, C. Y., Mottaghitalab, V., Too, C. O., Spinks, G. M., Wallace, G. G. Polyaniline and polyaniline-carbon nanotube composite fibers as battery materials in ionic liquid electrolyte. *Journal of Power Sources*, 2007, *163(2)*, 1105–1109.

145. Wang, Y., Kim, H. J., Vunjak-Novakovic, G., Kaplan, D.L. Stem cell-based tissue Engineering with silk biomaterials. *Journal of Biomaterials*, 2006, *27 (36)*, 6064–6082.

146. Webber, C. L., III, Bledsoe, V. K. Kenaf yield composite and plant composition. *Trends in New Crops and New Uses*, ASHS press, Alexandria, VA, EVA, 2002; pp. 348–357.

147. Yu, H. N., Kim, S. S., Hwang, I. U., Lee, D. G. Application of natural fiber reinforced composites to trenchless rehabilitation of underground pipes. *Composite Structures, 2008*, 86, 285–290.

148. Zhao, J. O., Liu, L., Guo, Q. G., Shi, J. L., Zhai, G. T., Song, J. R., et al. Growth of carbon nanotubes on the surface of carbon fibers. *Carbon*, 2008, *46(2)*, 380–383.

149. Zhu, S., Su, C. H., Lehoczky, S. L., Muntele, I., Ila, D. Carbon nanotube growth on carbon fibers. *Diamond Relat Mater.* 2003, *12*, 1825–1828.

CHAPTER 12

REMOVAL AND RECOVERY OF HEAVY METALS USING NATURAL POLYMERIC MATERIALS

P. N. SUDHA, T. GOMATHI, K. VIJAYALAKSHMI, and R. NITHYA

Department of Chemistry, D.K.M. College for Women, Thiruvalluvar University, Vellore, Tamilnadu, India, Tel: (+91) 98429 10157; E-mail: drparsu8@gmail.com

CONTENTS

ABSTRACT

Water is the resource for sustenance of all the species in the earth. The potable water demand is ever increasing since there has been a continuous growth in world population. Continuous increase in the world population and development of industrial applications made environmental pollution problem more important. Removal of toxic metal contaminant from wastewater has been a cause of major concern. Much attention has recently been focused on various biosorbent materials such as fungal or bacterial biomass and biopolymers that can be obtained in large quantities and that are harmless to nature. Special attention has been given to polysaccharides such as chitin and chitosan, a natural amino polymer. Biosorption is an emerging technology, which uses natural materials as adsorbents for wastewater treatment. This chapter reviews the treatment of wastewater up to the present time using marine polysaccharides and its derivatives. Special attention is paid to advantages properties of the natural adsorbents, which is a wondering gift for the human survival.

12.1 INTRODUCTION

Water is one of the essential items needed for all forms of plants and animal life for the survival and growth [192]. For our planet, surface water is undoubtedly the most precious natural resource. Subsequently, we are slowly but surely harming our planet through pollution with unsustainable anthropogenic activities [99]. Water pollution is a worldwide problem and its potential to influence the health of human populations is great [52, 155]. Water pollution is mainly caused by the indiscriminate disposal of water after use in the form of waste which has become a major source of concern and a priority for most industrial sectors faced with more and more stringent regulations [66]. These wastes often contain a wide range of contaminants such as chlorinated hydrocarbons and heavy metals, petroleum hydrocarbons, various alkalis, acids, dyes and other chemicals which greatly change the pH of water [158]. In order to combat water pollution, we must understand the problems and become part of the solution. Pollution of the aquatic environment by inorganic chemicals has been considered as a major threat. In recent days, heavy metal concentrations, besides other pollutants, have increased to reach dangerous levels for living environment in many regions. The most anthropogenic sources of metals are industrial, petroleum contamination and sewage disposal [168].

12.1.1 PROBLEM OF HEAVY METAL POLLUTION

Heavy metals are an element which is having atomic weights between 63.546 and 200.590 and a specific gravity greater than 4.0, that is, at least 5 times that of water. Inorganic effluent from the industries contains heavy metals such as Cadmium, Zinc, Lead, Chromium, Nickel, Copper, Vanadium, Platinum, Silver, Aluminum, Arsenic, Cadmium, Mercury and titanium [176]. Heavy metals have high solubility in the aquatic environment, it can exist in water in colloidal, particulate and dissolved phases [2] with their occurrence in water bodies being either of natural origin (e.g., eroded minerals within sediments, leaching of ore deposits and volcanism extruded products) or of anthropogenic origin (i.e., solid waste disposal, industrial or domestic effluents, harbor channel dredging) [126].

Heavy metals enter into the human body via the food chain, drinking water, air or absorption through the skin. Trace amounts of some heavy metals, e.g., iron, copper and zinc are required by human body in order to maintain the metabolism, but these metals become poisoning or hazardous at high concentration because they do not have the tendency to degrade, so they tend to bioaccumulate [34]. Since the body does not metabolize these heavy metals, so they accumulate in the soft tissues and become toxic.

Mostly the heavy metals are toxic or carcinogenic [55], when it present beyond the acceptable limits. They produce their toxicity by forming complexes with proteins, in which carboxylic acid (–COOH), amine (–NH2), and thiol (–SH) groups are involved. These modified biological molecules lose their ability to function properly and result in the malfunction or death of the cells.

For example, Aluminium accumulation was associated with Alzheimer's and Parkinson's disease, senility and presenile dementia. Arsenic exposure cause cancer, abdominal pain and skin lesions. Cadmium exposure produces kidney damage and hypertension. Nickel exceeding its critical level might bring about serious lung and kidney problems aside from gastrointestinal distress, pulmonary fibrosis and skin dermatitis [23].

From many research activities and studies it is concluded that very high level of nickel in human hair are related to cardiovascular problems by altering the immunoglobulin levels [183].

Lead is a commutative poison and a possible human carcinogen [13] while for Mercury, toxicity results in mental disturbance and impairment of speech, hearing, vision and movement. Copper in high doses can cause anemia, liver, kidney damage and intestinal irritation. People with Wilson's disease are at greater risk for health effects from over exposure to copper.

While mammals are not as sensitive to copper toxicity as aquatic organisms, toxicity in mammals includes effects such as liver cirrhosis, necrosis in kidneys and the brain, gastrointestinal distress, lesions, low blood pressure, and fetal mortality [11, 93, 206, 200]. Since chromium exits in the aquatic environment mainly in two states: Cr(III) and Cr(VI). Cr(VI) is more toxic than Cr(III). Chromium (VI) is also considered as a highly carcinogenic and creating adverse effect on urinary system, dermatitists, lung cancer and nasopharynx cancer [51, 100].

Thus, faced with more and more stringent regulations, nowadays heavy metals are the environmental priority pollutants and are becoming one of the most serious environmental problems. So these toxic heavy metals should be removed from the wastewater to protect the people and the environment.

12.1.2 SOME COMMON METHODS USED TO REMOVE HEAVY METALS FROM WASTEWATER

Nowadays, numerous methods (physical and chemical processes) have been proposed for efficient heavy metal removal from waters, including but not limited to chemical precipitation, ion exchange, ultrafiltration, adsorption, ion-exchange, reverse osmosis, oxidation, ozonation, coagulation, flocculation, membrane filtration processes, sonication [175, 20] and electrochemical technologies [55, 204, 144, 109, 57].

Chemical precipitation is the most traditional common simple method. By this technique the acidic effluent is first neutralized and heavy metals are then precipitated as metal hydroxides [88]. Janson et al. (1982) used hydroxide precipitation technique for the removal of Zn^{2+}, Pb^{2+}, Cr^{3+} by adjusting the pH of the solution to a value greater than 10. Cu^{2+} can be removed as metal sulfide precipitate. This method produce large amount of sludge, in which an aquatic pollution problem transforms to a solid waste pollution problem. In addition, chemical precipitation is usually inefficient to deal with low concentration of heavy metals.

Ultrafiltration and reverse osmosis methods require high pressures and are thus fairly costly in terms of energy. Reverse osmosis systems operate at pressures ranging from 2 to 10 Mpa, while ultra filtration systems operate between 70–700 KPa [71]. One of the most widely used alternatives is ion exchange method. Ion exchange resins contain functional groups which are capable of complexing or ion exchanging with metal ions to treat heavy metals from industrial wastewater. This process is extensively used for metal finishing bath purification, effluent polishing after primary treatment, and recovering precious metal. The resin is contacted with the contaminated solution, located with metal ions, and stripped with an appropriate elute [15]. The disadvantage of ion exchange resins is slow kinetics and low selectivity of metal ions.

Because of its ability to remove minute particles such as fats, protein and pathogens [29, 216], membrane filtration is another widely used technology of choice for superior water and large-scale reclamation of wastewater. The factors such as pressure, chemical composition, temperature, feed flow and interactions between components in the feed flow and the membrane surface influence the separation performance of membranes [116]. Conventional techniques have their own inherent limitations such as less efficiency, sensitive operating conditions, production of secondary sludge and further the disposal is a costly affair [4, 41, 199]. Another powerful technology is adsorption of heavy metals from industrial wastewater [77, 79].

12.1.3   ADSORPTION A PROMISING TOOL FOR HEAVY METAL REMOVAL

There is an increasing demand for treatment of industrial wastewater before their use or disposal because of the environmental and health difficulties associated with heavy metals and pesticides and their deposit through the food chain [173]. Adsorption technology has good potential to treat wastewater and industrial residues because it is cost-effective, easy for application and efficient in various kinds of heavy metal removal. Adsorption is one of the most commonly used methods to remove heavy metal ions from various aqueous solutions with relatively low metal ion concentrations. In addition to this the adsorption process are reversible, the adsorbents can be regenerated by suitable desorption processes for multiple use [147]. Many desorption processes are of low maintenance cost, high efficiency, and ease of operation [133]. Therefore, the adsorption process has come to the forefront as one of the major techniques for heavy metal removal from water/wastewater. Efficient treatment of wastewater is gaining increasing importance to make wastewater treatment contribute more towards sustainability. The adsorbents may be of mineral, organic or biological origin, zeolites, industrial byproducts, agricultural wastes, biomass, and polymeric materials [108]. Various adsorption materials have been studied, such as activated carbons, clays, polymeric synthetic resins, metal oxides, and some natural materials.

Adsorption on activated carbon is a recognized method for the removal of heavy metals from wastewater. For example, adsorption with activated carbon has been reported for Cu^{2+}, Cd^{2+}, Zn^{2+}, Pb^{2+}, Hg^{2+}, CO_2^{+}, Mn^{2+}, Ca^{2+}, Ni^{2+}, Cr^{3+} [39]. The high cost of activated carbon limits its use in adsorption. A search for a low-cost and easily available adsorbent has led to the investigation of materials of agricultural and biological origin, along with industrial byproducts, as potential metal sorbents.

Biosorption is a sub branch of adsorption, aims to use cheaper materials of biological origin as adsorbents. Biosorption, a passive non-metabolically mediated process, removes metals or metalloid species, compounds and particulates from solution, by materials of living or dead biomass [56] and has served as an important means for purifying industrial wastewaters containing toxic heavy metal ions. The naturally occurring material including cellulose, alginates, carrageenan, lignins, proteins, chitosan and chitin derivatives are extensively used as a biosorbents [214, 49, 91, 148, 44]. The salient feature of biopolymers is that they possess a number of different functional groups, such as hydroxyls and amines to which the metal ions can bind either by chemisorption or by physisorption.

Adsorption carried out by these biopolymers has some advantages.

- Cheap: the cost of biosorbents is low since they are made from abundant or waste material.
- Metal selective: the metal adsorbing feature of different types of biosorbents will be almost selective on different metals.
- Regenerative: the biosorbents can be used again after recycling the metals adsorbed.
- No sludge formation: the secondary problems due to sludge formation will not arise here as in the case of precipitation.
- Metal recovery: the adsorbed metals can be recovered after the process.

Among the biopolymers worked with for adsorption of metal ions, cellulose, chitin and the derivatives of chitin have played significant role in their capacity as adsorbent and complexing agent by virtue of their hydroxyl, acetate, amido and amino groups.

12.2   CHITIN AND CHITOSAN AS ADSORBENTS

Chitosan is well known as an excellent biosorbent for metal cation removal in near-neutral solutions because the large number of NH_2 groups. The excellent adsorption characteristics of chitosan for heavy metals can be attributed to (1) high hydrophilicity due to large number of hydroxyl groups of glucose units, (2) presence of a large number of functional groups (acetamido, primary amino and/or hydroxyl groups) (3) high chemical reactivity of these groups and (4) flexible structure of the polymer chain [34]. The reactive amino group selectively binds to virtually all group III transition metal ions but does not bind to groups I and II (alkali and alkaline earth metal ions) [135]. Also, due to its cationic behavior, in acidic media, the protonation of amine groups leads to adsorption of metal anions by ion exchange [67, 106].

Chitin can also be used as blends with natural or synthetic polymers. It can be crosslinked by the agents used for cellulose (epichlorhydrin, glutaraldehyde, etc.) or grafted in the presence of ceric salt [162] or after selective modification [107]. Grafting of chitosan allows the formation of functional derivatives by covalent binding of a molecule, the graft, on to the chitosan backbone. Chitosan has two types of reactive groups such as free amino groups and hydroxyl groups that can be grafted [7].

Chitosan – g – maleic anhydride – g –(acrylonitrile) copolymer was prepared via free radical polymerization using ceric ammonium nitrate as the initiator. This graft copolymer was used as an adsorbent in dye effluent treatment [73].

A composite adsorbent was prepared by entrapping crosslinked chitosan and nano-magnetite (NMT) on heulandite (HE) surface to remove Cu(II) and As(V) in aqueous solution [47]. Recently, chitosan composites have been developed to adsorb heavy metals and dyes from wastewater. A series of polyurethane (PU)/chitosan composite foams were prepared with different chitosan content of 5~20 wt.% and investigated their adsorption performance of acid dye (Acid Violet 48) [74]. The adsorption of lead (II) from aqueous solutions onto chitosan was investigated [46]. Chitosan can be modeled in several shapes: gels, flakes, powders, beads, membranes and particles.

A new composite biosorbent has been prepared by coating chitosan onto acid treated oil palm shell charcoal (AOPSC) and is used for the removal of chromium [142]. Bromine pretreated chitosan was found to be promising adsorbent for lead (II) removal from water [157]. The ability of chitosan as an adsorbent for Cu (II) and Cr (VI) ions in aqueous solution was studied by Schmuhl et al. (2001). Polymer blend films of chitin and bentonite were prepared and this blended polymer was used as an adsorbent for the removal of copper and chromium from the dye effluent [169]. The alginate–chitosan hybrid gel beads were prepared and it is used to adsorb divalent metal ions [182].

Chitosan coated carbon was evaluated for the removal of chromium (VI) and cadmium (II) from its aqueous solution [177]. The application of chitosan in the form of beads to the removal of copper, zinc and chromium ions from water was reported by Katarzyna Jaros et al. (2005). The composite chitosan magnetite microparticles for the removal of Co^{2+} and Ni^{2+} ions in aqueous solution have been evaluated by Donia Hitchu et al. (2012). The adsorption of Cu^{2+} on the chitin surface has been studied by Melchor Gonzalez-Davila and Frank J. Millero (1989). The maximum adsorption of the chitosan beads for the removal of formaldehyde and ammonia were in the order of those derived from cuttlefish bone, white leg shrimp, horseshoe crab and mud crab, respectively [172]. The adsorption of chromium (VI) ions from aqueous solution by ethylenediamine-modified crosslinked magnetic chitosan resin (EMCMCR) was studied by Xin-Jiang Hua et al. (2011).

A magnetic composites was synthesized with chitosan, nanomagnetite, and heulandite to remove Cu(II) and As(V) from aqueous solution [47]. Adsorption of Reactive Black 5 dye onto chitosan was studied by Szygula et al. (2008). The amine groups on chitosan bind metal cations at pH close to neutral. At low pH, chitosan is more protonated and therefore it is able to bind anions by electrostatic attraction [67]. Chitosan has been used in a variety of forms, which include chitosan beads, flakes and membranes [53, 41, 146, 130]. The electrons present in the amino and N-acetylamino groups forms dative bonds with transition metal ions and some of the hydroxyl groups in these biopolymers may act as donors. Hence, deprotonated hydroxyl groups can be involved in the co-ordination with metal ions [114].

Lee et al. (1997) used crab shell particles as a biosorbent to remove lead from aqueous solutions [112]. The authors reported 20% lead removal levels at pH< 4. However, removal increased (>90%) at pH 5–8. Vijayaraghavan et al. (2006) investigated copper and cobalt biosorption onto crab shell particles, obtaining a maximum uptake of 243.9 mg/g for copper and 322.6 mg/g for cobalt at pH 6 with a biosorbent load of 5 g/L in batch tests. Zhou et al. (2004) investigated the adsorption of lead, cadmium, and copper onto cellulose/chitin beads; these metals were adsorbed at pH ranging from 3 to 6. Hawke et al. (1991) studied the uptake of iron and manganese from seawater onto chitin; results showed low Manganese(II) removal (<10%) at pH 6–8.7 and increased removal (>90%) at pH 9.5, while Fe(II) was removed at levels of 22–30% at pH 2–8.

Benguella and Benaissa (2002) studied the adsorption of cadmium from a 100 mg/L aqueous solution onto chitin at initial pH between 5.7 and 6.4. The maximum cadmium removal capacity of chitin was 12.5 mg/g at a load of 2 g/L. They followed this study by evaluating the effect of several ions (Na^+, Mg^{2+}, Ca^{2+}, Cl^-, SO_4^{2-}, and CO_3^{2-}) on the kinetics of cadmium sorption onto chitin. The authors found that Ca^{2+} and CO_3^{2-} had a large inhibitory effect over cadmium adsorption, while Mg^{2+} and SO_4^{2-} had a weak inhibitory effect. At the same time, Na^+ and Cl^- were found to have no effect on cadmium adsorption. Most of these studies focused on either single metal removal or on the effect of competitive ions on metal removal.

12.3   OTHER NATURAL POLYMERS AS PROMISING ADSORBENTS

In recent years, numerous approaches have been studied for the development of cheaper and more effective adsorbents. The specific properties which are required for the substance to act as adsorbent was the high ability to reduce the concentration of heavy metals below the acceptable limits, high adsorption capacity and the long lifetime. The adsorbents used may be of mineral, organic or biological origin: activated carbons [80, 28, 152, 164, 143, 92], minerals, fish bone charcoal, coconut shell carbon, rise husk carbon, and bio sorbent material, polymeric materials (organic polymeric resins) [216, 10] and macroporous hyper crosslinked polymers [68, 80, 104, 105, 121].

The various adsorbents such as fly ash [41], activated carbon [208] and bio adsorbents (adsorbents from plant- and animal-origin materials) for example peat moss, bark/tannin-rich materials, humus, modified cotton and wool, chitin, chitosan, seaweed, and biopolymers have been used for the heavy metal adsorption [87, 195, 16].

Using both mineral and organic matrices an important number of new hybrid materials [151] such as polyamine beads [140, 36], glass beads [119, 120], alginate beads [65], silica gel [50, 153, 154], sand [201], polyvinylalcohol [70, 27], polyethylene terephthalate [216], polyacrylic acid [110], polyurethane [89], polysiloxane [95], polyester [128], alumina [179], polypropylene [111, 127] etc. have been successfully prepared and described for the removal of pollutants from aqueous solutions.

Super absorbents, a three-dimensional hydrophilic polymers have the high capability to swell and absorb a large amount of water [48] and hence it was found to be valuable in some specialized applications, including controlled delivery of bioactive agents [103] and wastewater treatment [24]. The use of the naturally occurring cellulosic materials, their modified forms and their efficacy as adsorbents for the removal of heavy metals from waste streams was reported by many researchers [82]. Montmorillonite and kaolinite were used for removal of lead and cadmium [178]. In addition, carrageenan has shown that it is a good transport system and can facilitate in controlling pollution by heavy metals such as Pb, Cd, and Cr [14].

12.4   FACTORS AFFECTING ADSORPTION

There are many factors which are affecting the efficiency of adsorption. The factors are initial concentration of metal ion solution, adsorbent dose, pH of the metal solution, temperature, agitation speed and contact time. Also the nature of the adsorbent such as functional groups, porosity, active sites, size, and charges on the surface will participate in the adsorption process.

12.4.1   NATURE OF ADSORBENT

Performance of the adsorption dependents on the type of adsorbent used. The materials are versatile. This versatility allows the sorbent to be used under different forms, from insoluble beads, to gels, sponges, capsules, films, membranes or fibers. Materials are available in a variety of structures with a variety of properties. The surface of contact between any sorbent and the liquid phase plays an important role in the phenomena of sorption. The efficiency of adsorption depends on physicochemical characteristics such as porosity, surface specific area and particle size of sorbents. A fundamentally important characteristic of good adsorbents [188, 117] is their high porosity and consequent larger surface area with more specific adsorption sites.

The problem with polysaccharide-based materials is their poor physicochemical characteristics, in particular porosity. Polysaccharides are, in general, non-porous and their derivatives possess a low surface area. Chitosan has a very low specific area ranging between 2 and 30 $m^2 g^{-1}$. The tailoring of polymers generally converts the semi-crystalline polymers which are less porous into the amorphous one. Amorphous nature is mainly related to the surface area and porosity of the adsorbent. Also the increase in functionality increases the adsorption sites for the metal ions. Crosslinking reduces the amount of the crystalline domains in the polysaccharide and can then change the crystalline nature of the raw polymer. This parameter significantly influences the sorption properties because it may control the accessibility to sorption sites. Crosslinking drastically reduces segment mobility in the polymer and a number of chains are interconnected by the formation of new three dimensional network linkages.

When the crosslinking degree is high, the material is mostly amorphous [174]. Because they are covalently crosslinked networks, therefore called permanent or chemical gels [150]. The two most important factors controlling the extent of adsorption properties of polysaccharide-based materials are the hydrophilicity of the polymer and the crosslink density [69]. The adsorption properties depend on the extent of chemical activation and modifications.

Recently, Mi et al. (2002) proposed a new method for the preparation of porous chitosan beads via a wet phase-inversion method. Delval et al. (2003) prepared porous crosslinked starch by generating gas bubbles within the material during synthesis. The size of sorbent particles has also been shown to be a key parameter in the control of sorption performances [30, 43, 166].

Glutaraldehyde crosslinked chitosan beads [129], EPI–cyclodextrin gels [50] and EPI–starch beads [43] have a specific surface area around 60, 213 and 350 $m^2\,g^{-1}$ respectively. Most commercial activated carbons have a specific area of the order of 800–1500 m^2/g. Several crosslinked cyclodextrins gels were prepared by Crini and Morcellet [37, 35]. They proposed the use of these gels for the removal of various aromatic derivatives (chloro and nitro phenols, benzoic acid derivatives, dyes) from aqueous solutions. Results of adsorption experiments showed that these crosslinked polymers exhibited high sorption capacities.

Di Xu et al. (2008) identified that with increasing solid content, the functional groups at the montmorillonite surfaces increases and thereby more surface sites are available to form complexes with Ni^{2+} at solid surfaces, during the adsorption and desorption of Ni^{2+} on Na-montmorillonite.

12.4.2 THE FUNCTIONAL GROUPS

With repetitive functional groups, biopolymers provide excellent chelating and complexing materials for a wide variety of pollutants including dyes, heavy metals and aromatic compounds. The biopolymers chitin and chitosan are very much active towards the metal removal due to the presence of its functional groups. Chitin contains the hydroxyl and acetamido functional groups and chitosan contains hydroxyl and amine functions.

The other natural materials algae, fungi, nucleotides, oligosaccharides and other polysaccharides with its active groups act as a good adsorbents.

There are various chemical functional groups that would attract and sequester the metals in biomass: acetamido groups of chitin, structural polysaccharides of fungi, amido, amino, sulphydryl and carboxyl groups in proteins, hydroxyls in polysaccharide and mainly carboxyls and sulfates in polysaccharides of marine algae that belong to the divisions Phaeophyta, Rhodophyta and Chlorophyta.

The functional groups like O–H, C-N, N-H and C=O groups were found in all samples play a major role in heavy metal adsorption [54, 97]. Moreover, even though polysaccharides and their derivatives are highly sorptive in their natural state, their adsorption capacity can be improved selectively by the substitution of various functional groups onto the polymer backbone. They possess a high capacity and high rate of adsorption, high efficiency and selectivity in detoxifying either very dilute or concentrated solutions. Synthetic derivatives of chitosan substituted with other functional groups have specific complexing properties towards the effectiveness in removing heavy metals from water [193, 105, 160].

Lo et al. (2001) reported that the Pb^{2+} adsorption capacity of charcoal was mainly affected by the functional groups on the charcoal surface. Chao et al. (2004) proposed enzymatic grafting of carboxyl groups onto chitosan as a means to confer the ability to adsorb basic (cationic) dyes on beads. The reactive groups grafted may serve as electron donors in an alkaline environment. The oxygen-containing functional groups have higher adsorption capacity of carbonaceous surfaces for polar organic molecules (acetone) than non-polar (propane), because carbonaceous surfaces exposed to ambient conditions typically contain the kind of functionalities [22].

For grafted polymer materials, pollutant sorption results were found to be a function of the degree of attachment of the polyfunctional groups. In the case of chitosan, the best method of achieving selective extraction is to use a metal specific ligand, but it has proved impossible to find specific ligands for each metal ion.

Crini et al., 2002 proposed the synthesis of cyclodextrin–carboxymethyl cellulose gels. Results obtained with these gels showed that effective and efficient extraction of beta-naphtol is achieved. The presence of carboxyl groups in the polymer networks permit to increase significantly the sorption properties. The carboxylic acid groups present in the adsorbents plays an important role in the adsorption of Cu^{2+}, Pb^{2+} and Ni^{2+} [94].

The positive influence of acid functional groups and surface oxidation on the adsorption capacity for heavy metal ions was also reported by various researchers [159].

12.4.3 THE FORCES OF ATTRACTION

Weber and his co-workers summarized the possible interactions between solute and sorbent in adsorption as physical, chemical, and electrostatic [207]. Due to the complexity of materials used and their specific characteristics (such as the presence of complexing chemical groups, small surface area, poor porosity), the sorption mechanism of polysaccharide-based materials is different from those of other conventional adsorbents. The adsorption of metal ions on adsorbent materials can be attributed to the coulombic interaction [58]. The coulombic term was obtained from the electrostatic energy of interactions between the adsorbents and adsorbate. The charges on substrates as well as softness or hardness of charge on both sides are mostly responsible for the intensity of the interaction. Coulombic interaction can be observed from adsorption of cationic species versus anionic species on adsorbents.

These mechanisms are, in general, complicated because they implicate the presence of different interactions [161, 113]. In addition, a wide range of chemical structures, pH, salt concentrations and the presence of ligands often add to the complication. Some of the reported interactions include: Ion-exchange, Complexation, Coordination/Chelation, Electrostatic interactions, Acid–base interactions, Hydrogen bonding, hydrophobic interactions, Physical adsorption, Precipitation. An examination of the data in the literature indicates that it is quite possible that at least some of these mechanisms are to varying degrees acting simultaneously depending on the chemical composition of the sorbent, the nature of the pollutant and the solution environment.

Since chitosan with its main reactive groups amines and hydroxyls may contribute to adsorption, which is responsible for metal ion binding through chelation mechanisms. However, chitosan is also a cationic polymer and its pKa ranges from 6.2 to 7 (depending on the deacetylation degree and the ionization extent of the polymer) [9]. In acidic solutions it sorb metal ions through anion exchange mechanisms [166].

Physical adsorption plays little role in the interaction between cross-linked chitosan beads and pollutants because beads have a small surface area. The pH may also affect the speciation of metal ions, and changing the speciation of the metal may result in turning the chelation mechanism into the electrostatic attraction mechanism. Another parameter can play an important role in the mechanism: the presence of ligands grafted on the chitosan chains.

For crosslinked starch materials, physical adsorption in the polymer structure and chemisorption of the pollutant via hydrogen bonding, acid–base interactions, complexation and/or ion exchange are both involved in the sorption process [42, 43]. In most cases, though a combination of these interactions was proposed to explain adsorption mechanisms, the efficiency and the selectivity of these adsorbents are mainly attributed to their chemical network.

When the materials contain cyclodextrin molecules, the mechanism is due to the formation of an inclusion complex between the CD molecule and the pollutant through host–guest interactions. It has also been reported that the presence of pollutant–pollutant hydrophobic interactions can explain the adsorption properties [35, 43].

12.4.4 THE SIZE OF ADSORBENT/ADSORBATE

Macura et al. (1982), Jha et al. (1998) and Rorrer et al. (1993) indicate that the adsorption capacity was influenced by the particle's size and they also showed that the adsorption capacity increases with the reduction in external superficial area. According to Weber and Morris, the breaking of larger particles tends to open tiny cracks and channels on the particle surface of the material resulting in more accessibility to better diffusion, owing to the smaller particle size [171]. The intrinsic adsorption of the materials is determined by their surface areas which can be observed by the effect of different sizes of adsorbents on adsorption capacity [83].

Smaller particle sizes reduce internal diffusion and mass transfer limitation to penetrate the adsorbate inside the adsorbent (i.e., equilibrium is more easily achieved and nearly full adsorption capability can be attained). Investigation was done on the removal efficiency of Fe^{3+} ions by natural zeolite through three different particle sizes (45, 125 and 250 μm) by

Mohammed Al Anber et al. (2010). It can be observed that the maximum adsorption efficiency is achieved with particle size of 45 µm. This may be due to the fact that most of the internal surface of such particles might be used for the adsorption. The smaller particle size gives higher adsorption rates, in which the Fe^{3+} ion has short path to transfer inside zeolite pores structure of the small particle size. Large molecules may be too large to enter small pores and so this may reduce adsorption independently of other causes.

In general, adsorption capacity varies randomly due to variation in the particle size [43]. The sorption increases with a decrease in the size of the particle since the effective surface area is higher for the same mass of smaller particles, and the time required to reach the equilibrium significantly increases with the size of sorbent particles.

When compared to the spheres, the microspheres presented higher adsorption capacity values indicating that the adsorption is possibly limited by external mass transfer resistance (boundary layer effect), reducing intra particle diffusion [141]. In principle, a more finely ground sample of a given cellulose-based material is expected to adsorb more metal ions from solution, under specified conditions, compared to coarser particles. This statement has been proven in a few studies [5, 21]. When the particle size of sawdust was decreased from 500 µm to 100 µm an increase in metal sorption capacity by only a factor of about 2 was observed [5].

The kinetic and equilibrium studies were done on the removal of cadmium from aqueous solutions using chitin as an adsorbent by Benguella and Benaissa (2002). The effect of particle size of chitin on the cadmium removal was studied using six particle size groups such as 0- 0.20; 0.20–0.63; 0.63–1.25; 1.25– 2.50; 2.50–4.10; 4.10–6.30 mm. The increase in cadmium sorption capacity at the equilibrium with the decrease of chitin particle sizes indicates that cadmium ion sorption occurs by a surface mechanism. Similar results have been reported for the sorption of metal ions by natural polymers and their derivatives [211].

The variation in particle size appears to have an influence on the time required for equilibrium. The observed results reported that the time required to reach equilibrium is about 1h for particle sizes of 0.2 mm, with a quantity of cadmium removed at an equilibrium of 13.57 mg/g chitin; while for particle sizes 4.1–6.3 mm, the time necessary is about 4 h with a weak capacity of cadmium removal of about 7.5 mg/g chitin.

Consequently, the time needed to reach equilibrium gets increased with increasing particle size. These observations suggest that the cadmium sorption kinetic by chitin is largely determined by the particle size.

12.4.5   PH

In particular, uptake is strongly pH-dependent. The wastewater pH may be an important factor in the sorption of pollutants onto sorbents. The effect of pH on heavy metal adsorption from aqueous solutions has been reported [137, 145, 221]. Nomanbhay and Palanisamy (2005) reported that the metal ion adsorption effect was greatly influenced by the solution pH value which mainly concerns the solubility of metal ions, ion concentration of adsorbent functional groups, and degree of ionization of the adsorbate in the reaction. In addition, most pollutants are weak electrolytes, for which the adsorption equilibrium depends on the solution pH. In acidic solutions, amine groups present in the chitosan beads easily form protonation, which induces an electrostatic repulsion of metal ions. This induces a competition between protons and metal ions for adsorption sites. By changing the solution pH value the surface charge of the adsorbent can be improved. Rangel-Mendez and Streat (2002) also indicated the influence of surface oxidation and solution pH value on the adsorption capacity for heavy metal ions. The optimum pH values of tannic acid immobilized activated carbons for adsorption of Cu^{2+}, Cd^{2+}, Zn^{2+}, Mn^{2+}, and Fe^{3+} were 5.4, 5.7, 5.6, 5.4, and 4.0 was reported by Ucer et al. (2006).

Thilagan et al. (2013) carried out the adsorption of Copper (II) ions on varying the pH using crosslinked Chitosan-Cellulose and Chitosan-Red Soil beads. The optimum was reached maximum at pH of 5 and it shows a slow decrease from pH 6 in the case of crosslinked Chitosan-Cellulose and Chitosan-Red Soil beads. The maximum adsorption was found to be at pH 6 for the Chitosan-Banana stem fiber beads, and the adsorption slowly decreased from pH of 7. The crosslinked Chitosan-Cellulose beads has a good chemical stability in the pH range of 2 to 10 whereas the Chitosan-Red Soil and Chitosan-Banana stem fiber beads gets slightly dissolved at pH of 2 and they showed chemical stability in the pH range of 3 to 10.

Delval et al. (2003) showed that starch based materials containing tertiary amino groups have a low efficiency in dye uptake in the low pH range present in acidic wastewaters because of the protonation of the amino groups. Teker et al. (1999) identified that the optimum pH value for the adsorption of copper and cadmium ions was found to be in the range of 5–8 using activated carbon from rice hulls as an adsorbent. Kononova et al. (2001) indicated that the optimum pH value for adsorption of Zn^{2+}, Cu^{2+}, and Fe^{3+} using carbonaceous adsorbent. The decrease in the percentage removal of metal ions at lower pH was apparently due to the higher concentration of H^+ ions present in the reaction mixture, which compete with the M^{2+} ions for the adsorption sites. The influence of pH and adsorbent concentration on adsorption of lead and zinc on a natural goethite was studied by Abdus Salam and Adekola (2005).

When the pH of the adsorbing medium is increased, there was a corresponding increase in deprotonation of the geothite surface leading to a decrease in H^+ ion. This creates more negative charges on the adsorbent surface, which favors adsorption of positively charge species as a result of less repulsion between the positively charge species and the positive sites on the goethite surface [62, 94]. Removal of metal ions at higher pH values could be attributed to the formation of their hydroxides, which results in precipitates, this is consistent with the observation of Lisa et al. (2004); Xiao and Ju-Chang (2009). Therefore, removal of metal ions at higher pH values is due to the formation of precipitates rather than adsorption.

12.4.6   TEMPERATURE

The adsorption capacity also depends on the temperature. In general, an increase in temperature decreases the saturation adsorption capacity of chitosan. The temperature had a dramatic effect on the adsorption metal ions on peanut husk. The adsorption capacity rose at the beginning with increase in temperature but thereafter it shows a decrease. The process is controlled by the adsorbate-adsorbent interactive forces. The results indicate that chemical adsorption becomes stronger in comparison to physical adsorption as the temperature increases [12]. Aksu et al. (1992) reported that the temperature seems not to influence the biosorption performances in the range of 20–35°C. The kinetic energy of the metal ions increased with increase in temperature and as a result of this the forces of attraction between the metal ions and the adsorbent gets weakened. Increase in temperature resulted in decreased adsorption for Cu(II) and Fe(III) and an increased adsorption for Pb (II) ions, indicating that the processes were exothermic and endothermic respectively [86].

Naseem Zahra et al. (2012) reviewed that the uptake of Pb(II) ion using low cost adsorbnets. The adsorption of Pb(II) ions may involves chemical bond formation and ion exchange since the temperature is the main parameter affecting the above two processes. The Pb(II) adsorption increases with increase in temperature, indicating better adsorption at higher temperatures. The amount of Pb(II) adsorbed at equilibrium increases with increase in temperature. This may be due to the acceleration of some originally slow adsorption steps or to the creation of some active sites on the adsorbent surface [136]. The enhanced mobility of Pb(II) ions from the bulk solution towards the adsorbent surface should also be taken into account [139]. A similar type of result have been obtained in studies of the adsorption of Pb(II) on rice husk [213].

12.4.7   AGITATION

Surface diffusion will be rate-limiting when little agitation occurs. Pore diffusion will be rate-limiting in a highly agitated system Many researchers have found that the amount of metals adsorbed increased with increasing agitation during batch testing [3]. Agitation facilitate convective transport of metal ions to sorbent surfaces, and several studies indeed showed positive effects of agitation on the rate or extent of metal sorption [61]. Alternatively, the associated pressure pulses acting on suspended particles in a mixture might also create a pump-like action, creating intermittent flow into and out of pore spaces within cellulosic materials in suspension.

A study about adsorption and desorption characteristics of mercury(II) ions using aminated chitosan bead carried out by Choong Jeon and Kwang Ha Park (2005). The uptake capacity of mercury ion by aminated chitosan bead increased with agitation rate and attained a constant value at about 150 rpm. This result is similar to that reported on copper ion sorption using various chitosan sorbents by Wan Ngah et al. (2002). The kinetics of cadmium removal by chitin was studied using agitation speeds ranging from 0 rpm to, 1250 rpm by Benguella and Benaisa (2002) in his work. The obtained results indicate that the time necessary to achieve equilibrium is about 2 h for the very elevated agitation speeds (1000–1250 rpm) and 4–5 h for the middle agitation speeds (100–400 rpm).

The effect of agitation time on turbidity, BOD and COD removals from wastewater using chitosan was investigated by Thirugnanasambandham et al. (2013b). From the observed results it was found that the removal of turbidity, BOD and COD were increased with increasing agitation time from 1 to 2 min. This may be due to the fact that the mechanical agitation increased the specific surface area of chitosan for effective removal efficiency. Further increase in the agitation time, decreases removal efficiencies of turbidity, BOD and COD due to the creation of disturbance for the bond formation between chitosan and organic matters present in the wastewater.

12.4.8   IONIC NATURE OF METALS –CATIONIC/ANIONIC

The anion effects on Cd sorption by montmorillonite were reported by Garcia-Miragaya and Page (1976). They indicated that Cl- and SO_4^{2-} had a greater effect than other anions in altering the distribution of Cd. Cadmium behaves mainly as a neutral species ($CdCl_2^0$) and an anion ($CdCl^{3-}$ and $CdCl_4^{2-}$), rather than a cation (Cd^{2+}) in soil with high Cl^- concentrations.

Stefan Dultz et al., 2012 reported the competitive adsorption of the oxyanions of Cr(VI), including the species $Cr_2O_7^{2-}$, $HCrO^{4-}$ and CrO_4^{2-}, on 6 organo-clay samples of HDPy-, HDTMA- and BE-montmorillonite and vermiculite in the presence of the competitive anions Cl^-, NO_3^- and SO_4^{2-} was determined in the batch mode. For the description of competitive adsorption, selectivity (S) of anion A relative to another one B for an adsorbent was defined by the ratio of qA/qB in a binary system [84, 122]. In determinations of anion selectivity for adsorption on organo-bentonites by Behnsen and Riebe (2008) a higher affinity of different monovalent anions for HDPy-, HDTMA and BE-bentonites than of divalent anions SO_4^{2-} and SeO_3^{2-} corresponding to the sequence of increasing hydration energy of anions, was observed. Consequently the sequence of selectivity is for all organo-bentonites Cr(VI)>SO_4^{2-}>NO_3^-, whereas it is Cr(VI)>NO_3^- >SO_4^{2-} for HDPy- and BE-vermiculites.

12.5   DESORPTION

Recovery and regeneration of consumed adsorbents and metals respectively is very important as it will reduce the cost of remediation of heavy metals and other contaminants in our environment [204]. Adsorption/ desorption studies are useful for generating essential information on the mobility of chemicals and their distribution in the soil, water and air compartments of our biosphere [163, 167]. Indeed the energetic barriers for desorption are much higher in a number of situations than the corresponding barriers for adsorption. Particles desorb regardless of their local environment. The information on desorption of adsorbed metals from biomass is scanty in the literature [78, 220, 33].

For metal desorption from the biomass certain dilute solutions of mineral acids like hydrochloric acid, sulfuric acid, acetic acid and nitric acid were used [219, 76]. Batch system was carried out to study the desorption of the adsorbed Hg (II) from the biosorbent – immobilized and heat inactivated *Trametes versicolor* and *Pleurotus sajur-caju* [8]. Hg (II) ions adsorbed onto the biosorbents were eluted with 10 mmol dm^{-3} HCl and the results showed that more than 97% of the adsorbed Hg (II) ions were desorbed from the biosorbents. In order to evaluate the feasibility of applying the prepared biosorbents in the heavy metals removal processes, the metal desorption efficiency from loaded biosorbents, and the reusability of the biosorbent in repeated adsorption-desorption operations were determined. The charged species exhibited desorption-resistance fraction whereas the desorption of the neutral form was completely reversible. The difference in sorption and desorption between the neutral and charged species is attributed to the fact that the anionic species sorbs by a more specific exothermic adsorption reaction whereas the neutral form partition by the hydrophobic binding to the soil [206]. Desorption of soil-associated metal ions and possible mechanisms have received considerable attention in literature [148].

Desorption rates of metal ions can be characterized by three types of processes, rapid desorption, rate-limited desorption, and a fraction that does not desorbed over experimental time scale. Many factors affect the adsorption-desorption of metal ion type; soil properties, organic matter, clay content and environmental conditions [124]. A wide range of experimental parameters can affect the desorption of metal ions with a soluble polymer, for example: the residence time t in the fixed bed, the pH of the polymer solution, the type of metal adsorbed, the type and the molecular weight of soluble polymer.

12.6 FACTORS AFFECTING DESORPTION

There are many factors which affects the desorption process. The desorption rate depends mainly on the nature of the adsorbed material, pH, temperature, agitation speed and the nature of interaction between the adsorbent and the adsorbate.

12.6.1 NATURE OF DESORBING AGENTS

Several mechanisms were proposed for the recovery of metals on the basis of desorption process. Basically, there are three desorption mechanisms [81, 96, 60]:

(i) The adsorbed metal gets precipitated due to the formation of insoluble compounds with the desorbing agent (e.g., H_2S).

(ii) The reaction of adsorbed metal with compounds having the available pair of electrons to share with the metal cation gives rise to the formation of complex. Desorbent agents such as $NaHCO_3$, Na_2CO_3 and EDTA follow this mechanism.

(iii) Ion exchange. The exchange of cations from the desorbent agent for the metal adsorbed gives rise to good yields in ion exchange mechanism.

The most common desorbent agents in this category are: HCl, H_2SO_4 and NaOH. There are a large number of desorbent agents that can be used but acids, inorganic or organic, have the highest metal desorption capacity [40]. The four important basic requirements for the desorbing agents are: (i) the high elution efficiency; (ii) low damage of the biomass in order to be reused in subsequent cycles; (iii) low degree of contamination; and (iv) low cost [31, 131].

HCl was markedly efficient as desorbent agent (Desorption I) with metal recoveries of 85.7% for Cd, 66.7% for Cu and 63.2% for Pb. The efficiency of the reused biomass in a new biosorption test (Biosorption II) gets significantly decreased with respect to the first sorption cycle (Biosorption I). This could be due to several factors, such as: structural damages of the active centers provoked by the desorbent agent, or blockage of those sites due to the inefficiency of the eluant leaving less active sites available for the a new sorption cycle [156, 170, 115, 195, 197]. Yan reported similar effects of $NaHCO_3$ on Mucor rouxii after a second cycle under identical conditions [210].

12.6.2 THE FORCES OF ATTRACTION

Recent studies have indicated that, in general, a given pollutant-binding mechanism plays a dominant role in the sorption. Similarly the type of interaction between the metal ion and the adsorbent is the important factor for desorption. Crini (2003) demonstrated the reproducibility of the sorption properties of cyclodextrin polymers used as sorbents for the removal of various dyes and the regeneration of the sorbents after saturation. Since the interactions between the pollutant and cyclodextrin are driven mainly by hydrophobic interactions, organic solvents are good candidates for the regeneration of the material. As a result, the polymers were easily regenerated using ethanol as the washing solvent. The sorption capacity value remained unchanged after this treatment. This showed the chemical stability of crosslinked gels and reproducibility of the values.

In case of polysaccharides the complexation of the pollutant by the ligand can displaced using acidic solution. The interaction between free electron doublet of nitrogen on amine groups and metal cations on chitosan derivatives by change in the pH of the solution reverses the sorption. Because the chelation mechanism is very sensible to pH.

Modified silica beads were reported to keep a constant capacity towards phenolic compounds after several cycles of sorption and regeneration with methanol [63, 154, 153]. However, the question of the long-term stability of these sorbents is posed in different ways depending on whether they are prepared by grafting or coating. In the former case, the main factor is the stability of the covalent bond between the polysaccharide and the matrix. In spite of rather good results in terms of adsorption capacity, some stability problems were encountered with the grafted silicas [154]. When the coating method is used, the stability depends on the strength of the interactions between the polymer and the silica surface. In order to increase the stability and to avoid desorption of the polymer, several researchers proposed a crosslinking reaction, after the coating of the polymer. Phan et al. (2000) showed that for equal CD content, the coated supports are better than grafted supports.

12.6.3   PH

A study about the desorption of cadmium from porous chitosan beads was carried out by Tzu-Yang Hsien and Yu-Ling Liu (2012). At a final pH value of 2.0, 94% of cadmium desorption was achieved and 8.3 mmole H^+ per gram of beads were consumed and also it is interesting to note that even at a final pH of 4.7, 80% of cadmium desorption was accomplished and after 6stages of desorption in series, it was observed that 95.7% of the cadmium desorption was achieved at a final pH of 3.0.

The effect of pH on heavy metal ion adsorption capacity was studied by previous researchers using the shake flask experiments. Eric and Roux used the shake flask experiment to study the influence of pH on the heavy metal ion binding onto a fungus-derived bio-sorbent in the year, 1992. Also the evaluation of the effect of the hydrochloric acid concentration on the adsorption of platinum group metal ions onto chemically modified chitosan was done by Inoue et al., using the shake flask experiment [85]. Depending upon the type of P complexation with the surface such as monodentate, bidentate mononuclear, and bidentate binuclear the phosphorus desorption is potentially controlled. These complexes can be either non protonated or protonated depending on the suspension pH [184].

Bhattacharyya and Gupta (2008) covered that the adsorption of Cr(III) reduced with pH at low pH, but it increased with pH at higher pH, which could be due to metal hydroxide precipitation helping Cr(III) retention by the clay. The desorption mechanism of Cr(III) from GMZ bentonite at low pH can be explained as follows: (i) in acidic region, both the adsorbent and adsorbate are positively charged and the net interaction is the electrostatic repulsion; (ii) the positively charged metal ions face a good competition with the higher concentration of H^+ ions present in the reaction mixture.

Wang Min et al. (2009) reported that the desorption rate of Sr in the gels was 100% when acid solution was used as desorbent, but only 81.4% when KCl solution was used. It can be caused by the fact that besides the adsorption mechanism of ionic exchange on the carboxyl groups, the metal ion is captured by chelation interaction.

12.6.4   TEMPERATURE

Temperature also influence the desorption process. Ghulam Mustafa et al. (2006) observed that the cumulative Cd desorption (%) significantly decreased with increase in temperature from 20 to 70°C for all aging treatments and at both equilibrium pH values. Yong-gui Chen et al. (2013) conducted a study on the desorption of Cr (III) adsorbed onto GMZ bentonite. Desorption experiments of Cr (III) indicate that the desorption rate of Cr (III) from GMZ bentonite increases gradually from 293 to 303 K. According to the 2nd order model, the rate limiting step may be a chemical desorption involving valence forces through sharing or exchanging the electrons between Cr(III) and bentonite. Cr(III) ions are difficult to desorb because the energy of making chemical bonds broken provided by simple mechanical agitation at a low temperature is not sufficient. Beyond 303 K, desorption rate decreases with increasing temperature. It can be found that desorption process is exothermic at a higher temperature. Above all, the desorption of Cr(III) from the adsorbed GMZ bentoniteis not significantly influenced by the temperature.

12.6.5   AGITATION

A study about the effect of shaking time, ionic strength, temperature and pH value on desorption of Cr(III) adsorbed onto GMZ bentonite was carried out by Yong Gui Chen et al. (2013). It is obvious from the results that the amount of desorption increases with increasing shaking time. The maximum desorption capacity was observed after 3 h, beyond which there was almost no further increase in desorption. According to the previous work [26], the adsorption reached equilibrium after 2 h. Obviously, desorption is slower than adsorption. Many studies also indicated that the adsorption rates are significantly higher than the desorption rates, such as, Co(II), Pb(II) and Cu(II) ions desorption from bentonite reported by Manohar et al. (2006) and Korkut et al. (2010), respectively.

12.7   CONCLUSION

This review summarizes the industrial applications of natural polysaccharides such as Chitin and Chitosan based materials in wastewater treatment. In addition, this review also opens up various factors affecting the adsorption and desorption processes. We expect that this chapter will provide insights on the use of these natural polysaccharides for researchers working to discover new materials with new properties for the valuable applications.

ACKNOWLEDGMENT

The authors are grateful to authorities of DKM College for Women and Thiruvalluvar University, Vellore, Tamil Nadu, India for the support. Thanks are also due to the editor for the opportunity to review such an innovating field.

KEYWORDS

- chitin
- chitosan
- factors affecting adsorption
- factors affecting desorption

REFERENCES

1. Abdus Salam, N., Adekola, F. A. The influence of pH and adsorbent concentration on adsorption of lead and zinc on a natural goethite, *African Journal of Science and Technology (AJST) Science and Engineering Series*. 2005, 6, 55–66.

2. Adepoju Bello, A. A., Ojomolade, O. O., Ayoola, G. A., Coker, H. A. B. Quantitative analysis of some toxic metals in domestic water obtained from Lagos metropolis. *The Nig. J. Pharm.* 2009, 42, 57–60.

3. Ahalya, N., Kanamadi, R. D., Ramachandra, T. V. Biosorption of chromium (VI) from aqueous solutions by the husk of Bengal gram (*Cicer arientinum*). *Electron. J. Biotechnol.* 2005, 8, 258–264.

4. Ahluwalia, S. S., Goyal, D. Microbial and plant derived biomass for removal of heavy metals from waste water, *Bioresour. Technol.* 2005, 98, 2243–2257.

5. Ajmal, M., Khan, A. H., Ahmad, S., Ahmad, A. Role of sawdust in the removal of copper (II) from industrial wastes. *Water Res.* 1998, 32, 3085–3091.

6. Aksu, Z., et al. The biosorption of copper (II) by C. *vulgaris* and *Zramigera. Environ Technol.* 1992, 13, 579–586.

7. Alves, N. M., Mano, J. F. Chitosan derivatives obtained by chemical modifications for biomedical and environmental applications. *International Journal of Biological Macro Molecules.* 2008, 43, 401–414.

8. Arica, M. Y., et al., Comparative biosorption of mercuric ions from aquatic systems by immobilized live and heat inactivated *Trarnetes versicolor* and *Pleurotus sajurcaju* (2003) (Received 4th September, 2003, accepted 15th November, 2003).

9. Arrascue, M. L., Garcia, H., Horna, M., Guibal, E. Gold sorption on chitosan derivatives. *Hydrometallurgy.* 2003, 71, 191–200.

10. Atia, A., Donia, A. M., Abou El Enein, S. A., Yousif, A. M. Studies on uptake behavior of copper (II) and lead (II) by amine chelating resins with different textural properties. *Sep Purif Technol.* 2003, 33, 295–301.

11. ATSDR, Toxicological Profile for Copper. U.S. Public Health Service. Agency for Toxic Substances, *Disease Registry*, 1990.

12. Azmal, M., Mohammad, A., Yousuf, R., Ahmed, A. Adsorption behavior of cadmium, zinc, nickel and lead from aqueous solutions by Mangifera indica seed shell. *Indian J. Environ. Health.* 1998, 40, 15–26.

13. Bakare Odunola, M. T. Determination of some metallic impurities present in soft drinks marketed in Nigeria. *The Nig. J. Pharm.* 2005, 4, 51–54.

14. Baysal, S. H., Önal, S., Özdemir, G. Biosorption of chromium, cadmium, and cobalt from aqueous solution by immobilized living cells of Chryzeomonas luteola TEM 05 *Preparative Biochemistry and Biotechnology. 2009*, 39, 419–428.

15. Beauvais, R. A., Alexandratos, S. D. Polymer supported Reagents for the Selective Complexation of Metal Ions. An Overview, *React. Funct. Polym.*, 1998, *36*, 113–123.

16. Becker, T., Schlaak, M., Strasdeit, H. Adsorption of nickel (II), zinc (II) and cadmium (II) by new chitosan derivatives. *Reactive & Functional Polymers.* 2000, 44, 289–298.

17. Benaissa, H., Benguella, B. Effect of anions and cations on cadmium sorption kinetics from aqueous solutions by chitin: experimental studies and modeling. *Environmental Pollution.* 2004, 130, 157–163.

18. Benguella, B., Benaissa, H. Cadmium removal from aqueous solutions by chitin: kinetic and equilibrium studies. *Water Research.* 2002, 36, 2463–2474.

19. Bhattacharyya, K. G., Gupta, S. S. Adsorption of a few heavy metals on natural and modified kaolinite and montmorillonite: A review. *Advances in Colloid and Interface Science.* 2008, 140, 114–131.

20. Blanco, A. B., Sanz, B., Llama, M. J., Serra, J. L. Biosorption of heavy metals to immobilized Phormidium laminosum biomass. *J. Biotechnol.* 1999, 69, 227–240.

21. Blazquez, G., Hernainz, F., Calero, M., Ruiz-Nunez, L. F. Removal of cadmium ions with olive stones: The effect of some parameters. *Proc. Biochem.* 2005, 40, 2649–2654.
22. Boehm, H. P. Carbon 1994, 32, 759.
23. Borba, C. E., Guirardello, R., Silva, E. A. Removal of nickel (II) ions from aqueous solution by biosorption in a fixed bed column. experimental and theoretical breakthrough. *Biochem Eng J.* 2006, 30, 184–191.
24. Bulut, Y., Akcay, G., Elma, D., Serhatli, I. E. Synthesis of clay-based superabsorbent composite and its sorption capability. *J. Hazar. Mater.* 2009, 171, 717.
25. Chao, A. C., Shyu, S. S., Lin, Y. C., Mi, F. L. Enzymatic grafting of carboxyl groups on to chitosan to confer o chitosan the property of a cationic dye adsorbent. *Bioresour Technol.* 2004, 91, 157–62.
26. Chen Yong Gui, He Yong; Ye Wei Min, Zhang Xiong Fei, Lin Cui Hua. Removal of chromium (III) from aqueous solutions by adsorption on bentonite from Gaomiaozi, China. *Environmental Earth Sciences.* 2012, 67, 1261–1268.
27. Chen, H. L., Wu, L. G., Tan, J., Zhu, C. L. PVA membrane filled bcyclodextrin for separation of isomeric xylenes by pervaporation. *Chem Eng J.* 2000, 78, 159–168.
28. Chern, J. M., Chien, Y. W. Competitive adsorption of benzoic acid and p-nitrophenol onto activated carbon: isotherm and breakthrough curves. *Water Res.* 2003, 37, 2347–56.
29. Cheryan, M., Rajagopalan, N. Membrane processing of oily streams. Wastewater treatment and waste reduction. *Journal of Membrane Science.* 1998, 151, 13–28.
30. Chiou, M. S., Li, H. Y. Equilibrium and kinetic modeling of adsorption of reactive dye on crosslinked chitosan beads. *J Hazardous Mat.* 2002, 93, 233–248.
31. Chojnacka, K., Chojnacki, A., Górecka, H. Biosorption of Cr3þ, Cd2þ and Cu2þ ions by blue-green algae Spirulina sp.: kinetics, equilibrium and the mechanism of the process. *Chemosphere.* 2005, 59, 75–84.
32. Choong Jeon; Kwang Ha Park. Adsorption and desorption characteristics of mercury (II) ions using aminated chitosan bead. *Water Research.* 2005, 39, 3938–3944.
33. Chu, K. H., Hashim, M. A. Desorption of Copper from Polyvinyl alcohol immobilized seaweed biomass. *Acta Biotechnol.* 2001, 21, 295–306.
34. Crini, G. Recent developments in polysaccharide-based materials used as adsorbents in wastewater treatment. *Prog. Polym. Sci.* 2005, 30, 38–70.
35. Crini, G. Studies on adsorption of dyes on beta cyclodextrin polymer. *Bioresour Technol.* 2003, 90, 193–8.
36. Crini, G., Janus, L., Morcellet, M., Torri, G., Morin, N. Sorption properties toward substituted phenolic derivative in water using macroporous polyamines containing b-cyclodextrin. *J Appl Polym* Sci. 1999, 73, 2903–2910.
37. Crini, G., Morcellet, M. Synthesis and applications of adsorbents containing cyclodextrins. *J Sep Sci.* 2002, 25, 789–813.
38. Crini, G., Morin, N., Rouland, J. C., Janus, L., Morcellet, M., Bertini, S. Adsorption de be'ta-naphtol sur des gels de cyclodextrine-carboxyme'thyl cellulose reticule's. *Eur Polym J* 2002, 38, 1095–1103.
39. Dastgheib, S. A., Rockstraw, D. A. Model for the Adsorption of Single Metal Ion Solutes in Aqueous Solution onto Activated Carbon Produced from Pecan Shells, *Carbon.* 2002, 40, 1843–1851.

40. Davis, T. A., Volesky, B., Vieira, R. H. S. F. Sargassum seaweed as biosorbent for heavy metals. *Water Res.* 2000, 34, 4270–4278.
41. Deans, J. R., Dixon, B. G. Uptake of Pb2+ and Cu2+ by novel biopolymers. *Water Res.* 1992, 26, 469–472.
42. Delval, F., Crini, G., Morin, N., Vebrel, J., Bertini, S., Torri, G. The sorption of several types of dye on crosslinked polysaccharides derivatives. *Dyes Pigments.* 2002, 53, 79–92.
43. Delval, F., Crini, G., Vebrel, J., Knorr, M., Sauvin, G., Conte, E. Starch-modified filters used for the removal of dyes from waste water. *Macromol Symp.* 2003, 203, 165–671.
44. Deng, S. B., Ting, Y. P. Fungal Biomass with Grafted Poly(acrylic acid) for Enhancement of Cu(II) and Cd(II) Biosorption, *Langmuir.* 2005, 21, 5940–5948.
45. Di Xu, Xiang Zhou, Xiangke Wang. Adsorption and desorption of Ni2+ on Na-montmorillonite: Effect of pH, ionic strength, fulvic acid, humic acid and addition sequences. *Applied Clay Science.* 2008, 39, 133–141.
46. Doina Hritcu, Gianina Dodi, Marcel Ionel Popa. Heavy Metal Ions Adsorption on Chitosan-Magnetite Microspheres. *International Review of Chemical Engineering.* 2012, 4, 364–368.
47. Dong-Wan Cho; Byong-Hun Jeon; Chul-Min Chon; Yongje Kim; Franklin W. Schwartz; Eung-Seok Lee d, Hocheol Song. A novel chitosan/clay/magnetite composite for adsorption of Cu(II) and As(V). *Chemical Engineering Journal.* 2012, 200–202, 654–662.
48. Eshel, H., Dahan, L., Dotan, A., Dodiuk, H., Kenig, S. Nanotailoring of Nanocomposite Hydrogels Containing POSS *Polym. Bull.* 2008, 61, 257.
49. Evans, J. R., David, W.G., MacRae, J. D., Amirbahman, A. Kinetics of Cadmium Uptake by Chitosan based Crab Shells, *Water Res.* 2002, 36, 3219–3226.
50. Fan, Y., Feng, Y.Q., Da, S. L. On-line selective solid-phase extraction of 4-nitrophenol with b-cyclodextrin bonded silica. *Anal Chim Acta.* 2003, 484, 145–53.
51. Fendorf, S. E. Surface reactions of chromium in soils and waters. *Geoderma.* 1995, 67, 55–71.
52. Fereidoun, H., Nourddin, M. S., Rreza, N. A., Mohsen, A., Ahmad, R., Pouria, H. The Effect of Long-Term Exposure to Particulate Pollution on the Lung Function of Teheranian and Zanjanian Students. *Pakistan Journal of Physiology.* 2007, 3, 1–5.
53. Findon, A., Mckay, G., Blair, H. S. (1993) Transport studies for the sorption of copper ions by chitosan. *J. of Environ. Sci. and Health.* 1993, 28, 173–185.
54. Finqueneisel, G., Zimny, T., Albiniak, A., Siemieniewska, T., Vogt, D., Weber, J. V. Cheap adsorbent – part 1 – active cokes from lignites and improvement of their adsorptive properties by mild oxidation. *Fuel.* 1998, 77, 549–556.
55. Fu, F. L., Wang, Q. Removal of heavy metal ions from wastewaters: a review. *J. Environ.* Manage. 2011, 92, 407–418.
56. Gadd, G. M. Biosorption: Critical review of scientific rationale, environmental importance and significance for pollution treatment. *J. Chem. Technol. Biotech.* 2009, 84(1), 13–28.
57. Galil, N., Rebhun, M. Primary chemical treatment minimizing dependence on bioprocess in small treatment plants. *Water Sci. Technol.* 1990, 22, 203–210.

58. Gang, S., Weixing, S. Sunflower stalk as adsorbents for the removal of metal ions from waste water. *Ind. Eng. Chem. Res.* 1998, 37, 1324–1328.

59. Garcia Miragaya, J., Page, A. L. Influence of ionic strength and inorganic complex formation on the sorption of trace amounts of Cd by montmorillonite. *Soil Sci. Soc. Am. J.* 1976, 48, 749–752.

60. Gardea-Torresdey, J. L., De la Rosa, G., Peralta-Videa, J. R. Use of phytofiltration technologies in the removal of heavy metals: a review. *Pure Appl. Chem.* 2004, 76, 801–813.

61. Garg, U. K., Kaur, M. P., Sud, D., Garg, V. K. Removal of hexavalent chromium from aqueous solution by adsorption on treated sugar cane bagasse using response surface methodological approach. *Desalination.* 2009, 249, 475–479.

62. Ghanem, S. A., Mikkelsen, D. S. Sorption of Zn on iron hydrous oxide. *Soil Sci.* 1988, 146, 15–21.

63. Ghoul, M., Bacquet, M., Morcellet, M. Uptake of heavy metals from synthetic aqueous solutions using modified PEI silica gels. *Water Res.* 2003, 37, 729–734.

64. Ghulam Mustafa; Rai S. Kookana; Balwant Singh Desorption of cadmium from goethite: Effects of pH, temperature and aging. *Chemosphere.* 2006, 64, 856–865.

65. Gotoh, T., Matsushima, K., Kikuchi, K. I. Preparation of alginate–chitosan hybrid gel beads and adsorption of divalent metal ions. *Chemosphere.* 2004, 55, 135–140.

66. Gregorio Crini; Recent developments in polysaccharide-based materials used as adsorbents in wastewater treatment. *Prog. Polym. Sci.* 2005, 30, 38–70.

67. Guibal, E. Interactions of metal ions with chitosan-based sorbents: a review. *Sep. Purif. Technol.* 2004, 38, 43–74.

68. Gupta, V. K., Lmran, A., Sain, V. K. Defluoridation of wastewaters using waste carbon slurry, *Water Res.* 2007, 41, 3307–3316.

69. Guven, O., Sen, M., Karadag, E., Saraydin, D. A review on the radiation synthesis of copolymeric hydrogels for adsorption and separation purposes. *Radiat Phys Chem.* 1999, 56, 381–386.

70. Hamai, S., Kikuchi, K. Room-temperature phosphorescence of 6-bromo-2-naphtol in poly(vinyl alcohol) films containing cyclodextrins. *J Photochem Photobiol A Chem.* 2003, 161, 61–68.

71. Harrison, R. M. Pollution, Causes, Cambridge. *Royal Society of Chemistry.* 1990, 63–83.

72. Hawke, D. J., Sotolongo, S., Millero, F. J. Uptake of Fe(II) and Mn(II) on chitin as a model organic phase. *Marine Chemistry.* 1991, 33, 201–212.

73. Hemalatha, R., Chitra, R., Xavier Raja Rathinam., Sudha, P. N., Synthesizing and characterization of chitosan graft co polymer: adsorption studies for Cu (II) and Cr (VI). *International journal of environmental sciences.* 2011, 2, 805–828.

74. Ho Cheon Lee; Young Gyu Jeong; Byung Gil Min; Won Seok Lyoo; Sang Cheol Lee. Preparation and acid dye adsorption behavior of polyurethane/chitosan composite foams. *Fibers and Polymers.* 2009, 10, 636–642.

75. Ho, Y. S. Removal of copper ions from aqueous solution by tree fern. *Water Research.* 2003, 37, 2323–2330.

76. Holan et al., Biosorption of cadmium by biomass of marine algae. *Biotechnol. Bioeng.* 1993, 41, 819–825.

77. Horikoshi, T., Nakajima, A., Sakaguchi, T. Studies on the Accumulation of Heavy Metal Elements in Biological Systems. Accumulation of Uranium by Microorganisms. *Eur. J. Appl. Microbiol. Biotechnol.*, 1981, 12, 90–96.

78. Horsfall, M., Spiff, A. I., Abia, A. A. Studies on the influence of Mercaptoacetic acid (MAA) modification of Cassava (Manihot Sculenta Cranz) Waste Biomass on the Adsorption of Cu2+ and Cd2+ from Aqueous Solution. *Bull. Korean chem. Soc.* 2004, 25, 310–320.

79. Hosea, M., Greene, B., McPherson, R., Henzl, M., Alexander, M. D., Darnall, D. Accumulation of elemental gold on the alga *Chlorella vulgaris. Inorganica Chimica Acta.* 1986, 123, 161–165.

80. Hu, C. Y., Lo, S. L., Kuan, W. H., Lee, Y. D. Removal of fluoride from semiconductor wastewater by electrocoagulation-flotation. *Water Res.* 2005, 39, 895–901.

81. Huang, C. P., Westman, D., Quirk, K., Huang, J. P. The removal of cadmium (II) from dilute aqueous solutions by fungal adsorbent. *Water Sci. Technol.* 1998, 20, 369–376.

82. Hubbe, M. A., Hasan, S. H., Docoste, J. J. cellulosic substrates for removal of pollutants from aqueous systems: A review. *Bioresources.* 2011, 6, 2167–2287.

83. Igwe, J. C., Abia, A. A. Maize cob and husk as Adsorbents for removal of cadmium, lead and zinc ions from wastewater. *The physical scientist.* 2003, 2, 210–215.

84. Inglezakis, V. J., Loizidou, M. D; Grigoropoulou, H. P. Ion exchange of Pb2+, Cu2+, Fe3+, and Cr3+ on natural clinoptilolite: selectivity determination and influence of acidity on metal uptake. *Journal of Colloid and Interface Science.* 2003, 261, 49–54.

85. Inoue, K., Baba, Y., Yoshizuka, K., Noguchi, H., Yoshizaki, M. Selectivity Series in the Adsorption of Metal Ions on a Resin. Crosslinking Copper (II) Complexed Chitosan. *Chem. Lett.* 1988, 1281–1284.

86. Jaman, H., Chakraborty, D., Saha, P. A study of the thermodynamic and kinetics of copper adsorption using chemically modified rice husk, *Clean.* 2009, 37, 704–711.

87. Jang, L. K. Biotechnol. Bioeng. 1993, 43, 183.

88. Janson, C. E., Kenson, R. E., Tucker, H. Treatment of Heavy Metals in Wastewaters, *Environ. Prog.* 1982, 1, 212–216.

89. Jessie Lue S; Peng, S. H. Polyurethane (PU) membrane preparation with and without hydroxypropyl b-cyclodextrin and their pervaporation characteristics. *J Membr Sci.* 2003, 222, 203–217.

90. Jha, I. N., Iyengar, L., Rao, A. V. S. P. Removal of cadmium using chitosan. *J. Environ. Eng.* 1998, 114, 962–974.

91. Jodra, Y., Mijangos, F. Cooperative Biosorption of Copper on Calcium Alginate Enclosing Iminodiacetic Type Resin, *Environ. Sci. Technol.* 2003, 37, 4362–4367.

92. Jung, M. W., Ahn, K. H., Lee, Y., Kim, K. P., Rhee, J. S., Park, J. T., et al. Adsorption characteristics of phenol and chlorophenols on granular activated carbons (CAC). *Microchem* J. 2001, 70, 123–131.

93. Kabata Pendias, A., Pendias, H. Trace Elements in Soils and Plants. *Boca Raton.* 1992, 365.

94. Kadirvelu, K., Namasivayam, C. (2003). Activated carbon from coconut coir pith as metal adsorbent: Adsorption of Cd (II) from aqueous solutions. *Adv. Environ. Res.* 2003, 7, 471–478.

95. Kalfat, R., Ben Ali, M., Mlika, R., Fekih-Romdhane, F., Jaffrezic Renault, N. Polysiloxane–gel matrices for ion sensitive membrane. *Int J Inorg Mat.* 2000, 2, 225–231.

96. Kapoor, A., Viraraghavan, T., Cullimore, D. R. Removal of heavy metals using the fungus Aspergillus niger. *Bioresour. Technol.* 1999, 70, 95–104.

97. Kapoor, A., Viraraghavan, T., Fungal biosorption—an alternative treatment option for heavy metal bearing wastewaters: a review. *Bioresour. Technol.* 1995, 53, 195.

98. Katarzyna Jaros; Władysław Kamiński; Jadwiga Albińska; Urszula Nowak; Removal of heavy metal ions: copper, zinc and chromium from water on chitosan beads. *Environment Protection Engineering.* 2005, 31, 154–162.

99. Khan, S. I. Dumping of Solid Waste: A Threat to Environment, *The Dawn.* Ed; [online] 2004, Retrieved from http://66.219.30.210/weekly/science/archive/040214/science13.htm

100. Khezami., Capart, R. Removal of chromium (VI) from aqueous solution by activated carbons. Kinetic and equilibrium studies. *Journal of Hazardous Materials,* 2005, 123, 223–231.

101. Kononova, O. N., Kholmogorov, A. G., Lukianov, A. N., Kachin, S. V., Pashkov, G. L., Kononov, Y. S. Sorption of Zn(II), Cu(II), Fe(III) on carbon adsorbents from manganese sulfate solutions. *Carbon.* 2001, 39, 383–387.

102. Korkut, O., Sayan, E., Lacin, O., Bayrak, B. Investigation of adsorption and ultrasound assisted desorption of lead (II) and copper (II) on local bentonite: A modeling study. *Desalination,* 2010, 259, 243–248.

103. Kost, J., Encyclopedia of Controlled Drug Delivery. Mathiowitz. E., Ed, Wiley: New York, 1999.

104. Ku, Y., Chiou, H. M. The adsorption of fluoride ion from aqueous solution by activated alumina. *Water Air Soil Pollut.* 2002, 133, 349–360.

105. Kumar, E., Bhatnagar, A., Ji, M., Jung, W., Lee, S. H., Kim, S.J., Lee G, Song H, Choi, J. Y., Yang, J. S., Jeon, B. H. Defluoridation from aqueous solutions by granular ferric hydroxide (GFH). *Water Res.* 2009, 43, 490–498.

106. Kunkoro, E. P., Roussy, J., Guibal, E. Mercury recovery by polymer-enhanced ultrafiltration: comparison of chitosan and poly(ethylenimine) used as macrolingand. *Sep. Sci. Technol.* 2005, 40, 659–684.

107. Kurita, K., Inoue, S. Preparation of iodo-chitins and graft copolymerization onto the derivatives. In: Chitin and Chitosan; Skjak-Braek, G., Anthonsen, T., Sandford, P., Ed., New York: Elsevier Applied Science; 1989; p. 365–372.

108. Kurniawan, T. A., Chan, G. Y. S. Lo, W. H., Babel, S. Comparisons of low-cost adsorbents for treating wastewaters laden with heavy metals. *Sci. Total Environ.* 2005, 366, 409–426.

109. Kurniawan, T. A., Chan, G. Y. S., Lo, W. H., Babel, S. Physico-chemical treatment techniques for wastewater laden with heavy metals. *Chem. Eng. J.* 2006, 118, 83–98.

110. Kusumocahyo, S. P., Kanamori, T., Sumaru, K., Iwatsubo, T., Shinbo, T. Pervaporation of xylene isomer mixture through cyclodextrins containing polyacrylic acid membranes. *J Membr Sci.* 2004, 231, 127–132.

111. Le Thuaut, P., Martel, B., Crini, G., Maschke, U., Coqueret, X., Morcellet, M. Grafting of cyclodextrins onto polypropylene nonwoven fabrics for the manufacture of reactive filters. I. Synthesis parameters. *J Appl Polym Sci.* 2000, 78, 2118–2125.

112. Lee, M. Y., Park, J. M., Yang, J. W. Micro precipitation of lead on the surface of crab shell particles. *Process Biochemistry.* 1997, 32, 671–677.

113. Lee, S. T., Mi, F. L., Shen, Y. J., Shyu, S. S. Equilibrium and kinetic studies of copper (II) ion uptake by chitosan tripolyphosphate chelating resin. *Polymer.* 2001, 42, 1879–1892.

114. Lerivrey, J., Dubois, B., Decock, P., Micera, J., Kozlowski, H. Formation of D-glucosamine complexes with Cu(II), Ni(II) and Co(II) ions. *Inorg. Chim. Acta.* 1986, 125, 187–190.
115. Lin, C. C., Lin, H. L. Remediation of soil contaminated with the heavy metal (Cd2+). *J. Hazard. Mater.* 2005, 122, 7–15.
116. Lin, L., Rhee, K.C., Koseoglu, S. S. Bench-scale membrane degumming of crude vegetable oil: process optimization. *Journal of Membrane Science.* 1997, 134, 101–108.
117. Linsen, B. G. Physical and Chemical Aspects of Adsorbents and Catalysts. Academic Press: London, 1970.
118. Lisa, N., Kanagaratnam, B., Trever, M. Biosorption of Zinc from aqueous solutions using biosolids. *Adv. Env. Res.* 2004, 8, 629–635.
119. Liu, X. D., Tokura, S., Haruki, M., Nishi, N., Sakairi, N. Surface modification of nonporous glass beads with chitosan and their adsorption property for transition metal ions. *Carbohydr Polym.* 2002, 49, 103–108.
120. Liu, X. D., Tokura, S., Nishi, N., Sakairi, N. A novel method for immobilization of chitosan onto non-porous glass beads through a 1,3-thiazolidine linker. *Polymer.* 2003, 44, 1021–1026.
121. Lv, L., He, J., Wei, M., Evans, D. G., Zhoua, Z. Treatment of high fluoride concentration water by MgAl-CO3 layered double hydroxides: Kinetic and equilibrium studies. *Water Res.* 2007, 41, 1534–1542.
122. Lv, L., Hou, M. P., Su, F., Zhao, X. S. Competitive adsorption of Pb2+, Cu2+ and Cd2+ ions on microporous titanosilicate ETS-10. *Journal of Colloid and Interface Science.* 2005, 287, 178–184.
123. Macura, R., Suder, B. J., Wightman, J. P. Interaction of heavy metals with chitin and chitosan III chromium. *J. Appl. Polym. Sci.* 1982, 27, 4827–4837.
124. Mah Yew Keong, K., Hidajat; Uddin, M. S. Surfactant enhanced electro-Kinetic remediation of contaminated soil. *Journal of the Institution of Engineers.* 2005, 45, 61–70.
125. Manohar, D. M., Noeline, B. F., Anirudhan, T. S. Adsorption performance of Al-pillared bentonite clay for the removal of cobalt(II) from aqueous phase. *Applied Clay Science.* 2006, 31, 194–206.
126. Marcovecchio, J. E., Botte, S. E., Freije, R. H. Heavy Metals, Major Metals, Trace Elements. In: Handbook of Water Analysis. Nollet, L. M., Ed., London: CRC Press: London, 2007, 275–311.
127. Martel, B., Le Thuaut, P., Crini, G., Morcellet, M., Naggi, A. M., Maschke, U., et al. Grafting of cyclodextrins onto polypropylene nonwoven fabrics for the manufacture of reactive filters. II. Characterization. *J Appl Polym Sci.* 2000, 78, 2166–2173.
128. Martel, B., Morcellet, M., Ruffin, D., Ducoroy, L., Weltrowski, M. Finishing of polyester fabrics cyclodextrins and polycarboxylic acids as crosslinking agents. *J Inclusion Phenom Macrocyclic Chem.* 2002, 44, 443–446.
129. McAfee, B. J., Gould, W. D., Nadeau, J. C., Da Costa, A. C. A. Biosorption of metal ions using chitosan, chitin, and biomass of Rhizopus oryzae. *Sep Sci Technol.* 2001, 36, 3207–3222.
130. Mckay, G., Blair, H. S., Findon, A. Equilibrium studies for the sorption of metal ions onto chitosan. *Ind. J. of Chem.* 1989, 28A, 356–360.
131. Mehta, S. K., Gaur, J. P. Use of algae for removing heavy metal ions from wastewater: progress and prospects. *Crit. Rev. Biotechnol.* 2005, 25, 113–152.

132. Mi, F. L., Shyu, S. S., Chen, C. T., Lai, J. Y. Adsorption of indomethacin onto chemically modified chitosan beads. *Polymer*. 2002, 43, 757–765.
133. Mishra, S. P., Singh, V. K., Tiwari, D. Radiotracer technique in adsorption study. Part XIV. Efficient removal of mercury from aqueous solutions by hydrous zirconium oxide, Appl. Radiat. Isot. 1996, 47, 15–21.
134. Mohammad Al-Anber; Removal of High-level Fe3 + from Aqueous Solution using
135. Muzzarelli, R. A. A.1973. Natural Chelating Polymers. Pergamon Press: New York, 1973, 83–227.
136. Naiya, T. K., Bhattacharya, A. K., Mandal, S.,S. K. Das, S. K. The sorption of Lead (II) Ions on Rice Husk Ash. *J. Hazard Mater*. 2009, 163, 1254–1264.
137. Namasivayam, C., Senthilkumar, S. Removal of Arsenic (V) from Aqueous solution using industrial Solid waste: Adsorption rates and Equilibrium studies. *Ind. Eng. Chem. Res*. 1998, 37, 4816–4822.
138. Naseem Zahra, Lead Removal from Water by Low Cost Adsorbents: A Review, *Pak. J. Anal. Environ. Chem*. 2012, 13, 1–8.
139. Nassar, M. M., Magdy, Y. H. Mass transfer during adsorption of basic dyes on clay in fixed bed. *Indian Chem. Eng*. 1999, 40, 27.
140. Nayak, D. P., Ponrathnam, S., Rajan, C. R. Macroporous copolymer matrix IV. Expanded by adsorption application. *J Chromatogr* A. 2001, 922, 63–76.
141. Ng, J. C. Y., Cheung, W. H., Mckay, G. Equilibrium studies of the sorption of Cu(II) ions onto chitosan. *J. Colloids Interface Sci*. 2002, 255, 64–74.
142. Nomanbhay, S. M., Palanisamy, K. Removal of heavy metal from industrial wastewater using chitosan coated oil palm shell charcoal. Electron. *J. Biotechnol*. 2005, 8, 43–53.
143. Nouri, S., Haghseresht, F., Lu, G. Q. M. Comparison of adsorption capacity of p-cresol 1 p-nitrophenol by activated carbon in single and double solute. *Adsorption*. 2002, 8, 215–223.
144. O'Connell, D. W., Birkinshaw, C., O'Dwyer, T. F., Heavy metal adsorbents pre-pared from the modification of cellulose: a review. *Bioresour. Technol*. 2008, 99, 6709–6724.
145. Okieimen, F. E., Maya, A. O., Oriakhi, C. O. Sorption of cadmium, lead and zinc ions on sulfur containing chemically modified cellulosic materials. *Intern. J. Environ. Anal. Chem*. 1988, 32, 23–27.
146. Onsøyen, E., Skaugrud, Ø., Metal recovery using chitosan. *J. Chem. Technol. and Biotechnol*. 1990, 49, 395–404.
147. Pan, B. J., Pan, B. C., Zhang, W. M., Lv, L., Zhang, Q. X., Zheng, S. R. Development of polymeric and polymer-based hybrid adsorbents for pollutants removal from waters. *Chem. Eng. J*. 2009, 151, 19–29.
148. Park, D., Yun, Y. S., Cho, H. Y., Park, J. M. Chromium Biosorption by Thermally Treated Biomass of the Brown Seaweed. *Ind. Eng. Chem. Res*. 2004, 43, 8226–8232.
149. Park, J., Comfort, S. D., Shea, P. J., Machacek, T. A. Remediating Munitions-Contaminated Soil with Zerovalent Iron and Cationic Surfactants. *J. Environ. Qual*. 2004, 33, 1305–1313
150. Patrickios, C. S., Georgiou, T. K. Covalent amphiphilic polymer networks. *Curr Opin Colloid Int Sci*. 2003,8, 76–85.
151. Peng, C., Wang, Y., Tang, Y. Synthesis of crosslinked chitosancrown ethers and evaluation of these products as adsorbents for metal ions. *J Appl Polym Sci*. 1998, 70, 501–506.

152. Pereira, M. F. R., Soares, S. F., Orfao, J. M. J., Figueiredo, J. L. Adsorption of dyes on activated carbons: influence of surface chemical groups. *Carbon.* 2003, 41, 811–821.
153. Phan, T. N. T., Bacquet, M., Morcellet, M. Synthesis and characterization of silica gels functionalized with monochlorotriazinyl b-cyclodextrin and their sorption capacities towards organic compounds. *J Inclusion Phenom Macrocylic Chem.* 2000, 38, 345–359.
154. Phan, T. N. T., Bacquet, M., Morcellet, M. The removal of organic pollutants from water using new silica-supported cyclodextrin derivatives. *React Funct Polym.* 2002, 52, 117–125.
155. Progressive Insurance, 2005. Pollution Impact on Human Health. Retrieved from http://www.progressiveic.com/n25feb05.htm
156. Puranik, P. R., Paknikar, K. M. Biosorption of lead and zinc from solutions using Streptoverticillium cinnamoneum waste biomass. *J. Biotechnol.* 1997, 55, 113–124.
157. Rajendra Dongre; Minakshi Thakur; Dinesh Ghugal; Jostna Meshram. Bromine pretreated chitosan for adsorption of lead (II) from water. *Bull. Mater. Sci.* 2012, 35, 875–884.
158. Ram Lokhande; Pravin, U., Singare; Deepali, S.. Pimple. Pollution in Water of Kasardi River Flowing along Taloja Industrial Area of Mumbai. *India World Environment.* 2011, 1, 6–13
159. Rangel-Mendez, J. R., Streat, M. Adsorption of cadmium by activated carbon cloth: influence of surface oxidation and solution pH. *Water Res.* 2002, 36, 1244–1252.
160. Ravi Divakaran; Antony Paul, A. J., Anoop, K. K., Alex Kuriakose, V. J., Rajesh. R. Adsorption of nickel (II) and chromium (VI) ions by chitin and chitosan from aqueous solutions containing both ions TIST. *Int. J. Sci. Tech. Res.* 2012, 1, 43–50
161. Ravi Kumar, M. N. V. A review of chitin and chitosan applications. *React Funct Polym.* 2000, 46, 1–27.
162. Ren, L., Miura, Y., Nishi, N., Tokura, S. Modification of chitin by ceric salt-initiated graft polymerization-preparation of poly(methylmethacrylate)-grafted chitin derivatives that swell in organic solvents. *Carbohydr Polym.* 1993, 21, 23–7.
163. Rhodes, R. C., Belasco, I. J., Pease, H. L. Determination of mobility and adsorption of agrochemicals in soils. *J. Agric. Food Chem.* 1970, 18, 524–528
164. Rivera-Utrilla, J., Bautista-Toledo, I., Ferro-Garcia, M. A., Moreno-Castilla, C. Bioadsorption of Pb(II) Cd(II), and Cr(VI) on activated carbon from aqueous solutions. *Carbon.* 2003, 41, 323–30.
165. Rorrer, G. L., Hsien, T. Y., Way, J. D. Synthesis of porous magnetic chitosan beads for removal of cadmium ions from waste water. *Ind. Eng. Chem. Res.* 1993, 32, 2170–2178.
166. Ruiz, M., Sastre, A. M., Guibal, E. Palladium sorption on glutaraldehyde crosslinked chitosan. *React Funct Polym.* 2000, 45, 155–173.
167. Russell, M. H., Recommended Approaches to Assess Pesticide Mobility in Soil" in Environmental Behavior of Agrochemicals; Roberts, T. R., Kearney, P. C., Ed., John Wiley & Sons Ltd., 1995.
168. Santos, I. R., Silva-Filho, E. V., Schaefer, C. E., Albuquerque- Filho, M. R., Campos, L. S. Heavy metals contamination in coastal sediments and soils near the Brazilian Antarctic Station, King George Island. *Mar. Poll. Bull.* 2005, 50, 85–194.

169. Saravanan, D., Hemalatha, R., Sudha, P. N. Synthesis and characterization of cross-linked chitin/bentonite polymer blend and adsorption studies of Cu (II) and Cr (VI) on chitin. *Der Pharma Chemica*, 2011, 3, 406–424.
170. Sekhar, K. C., Kamala, C. T. C., Hary, N. S., Sastry, A. R. K., Rao, T. N., Vairamani, M. Removal of lead from aqueous solutions using an immobilized biomaterial derived from a plant biomass. *J. Hazard. Mater.* 2004, 108, 111–117.
171. Selvaraj K, Chandramohan V and Pattabhi S. *Indian J. Environ. Protect.* 1998, 18, 641.
172. Setthamongkol, P., Salaenoi, J. Adsorption Capacity of Chitosan Beads in Toxic Solutions. *World Academy of Science, Engineering and Technology*, 2012, 69, 161–166.
173. Shahidi, F., Arachchi, J. K. M., Jeon, Y. Food application of chitin and chitosan. *Trends Food Sci. Technol.* 1999, 10, 37–51.
174. Shiftan, D., Ravenelle, F., Mateescu, MA., Marchessault, R. H. Change in the V B polymorph ratio and T1 relaxation of epichlorohydrin crosslinked high amylose starch excipient. Starch. 2000, 52, 186–95.
175. Somasekara Reddy, M. C. Removal of direct dye from an aqueous solutions with an adsorbent made from tamarind fruit shell an agricultural solid waste. *Journal of scientific and industrial research*. 2006, 65, 443–446.
176. Sorme, L., Lagerkvist, R. Sources of heavy metals in urban wastewater in Stockholm. *Sci Total Environ*. 2002, 298, 131–145.
177. Soundarrajan, M., Gomathi, T., Sudha, P. N. Understanding the Adsorption Efficiency of Chitosan Coated Carbon on Heavy Metal Removal. *International Journal of Scientific and Research Publications*. 2013, 3, 1–10.
178. Srivastava, S. K., Tygai, R., Pant, N., Pal, N. Studies on the removal of some toxic metal-ions. Part II. Removal of lead and cadmium by montmorillonite and kaolinite. *Environ. Technol. Lett.* 1989, 10, 275–282.
179. Steenkamp, G. C., Keizer, K., Neomagus, H. W. J. P., Krieg, H. M. Copper II removal from polluted water with alumina/chitosan composite membranes. *J Membr Sci.* 2002, 197, 147–56.
180. Stefan Dultz., Jong Hyok An., Beate Riebe. Organic cation exchanged montmorillonite and vermiculite as adsorbents for Cr(VI): Effect of layer charge on adsorption properties. *Applied Clay Science*. 2012, 67, 125–133.
181. Szygula, A., Ruiz, M., Guibal, E., Sastre, A. M. Removal of an Anionic Reactive Dye by Chitosan and its Regeneration. In waste management, water pollution, air pollution, indoor climate;(WWAI'08); 2008; 26–28.
182. Takeshi Gotoh, Keiei Matsushima, Ken-Ichi Kikuchi. Preparation of alginate–chitosan hybrid gel beads and adsorption of divalent metal ions. *Chemosphere*. 2004, 55, 135–140.
183. Tan, B., Hu, J., Zhang, P., Huang, D., Shavanov, V. N., Weiss, M., Knyazikhin, Y., Myneni, R. B. Validation of MODIS LAI product in croplands of Alpilles, *France. J. Geophys. Res.* 2004.
184. Tejedor-Tejedor, M. I., Anderson, M. C. Protonation of phosphate on the surface of goethite as studied by CIRFTIR and electrophoretic mobility. *Langmuir.* 1990, 6, 602–611.
185. Teker, M., Imamoglu, M., Saltabas, O. Adsorption of copper and cadmium ions by activated carbon from rice hulls. *Turkish J. Chem.* 1999, 23, 185–191.
186. Thilagan, J., Gopalakrishnan, S., Kannadasan, T. A comparative study on adsorption of copper (II) ions in aqueous solution by (A) chitosan blended with cellulose

and crosslinked by formaldehyde, (B)chitosan immobilized on red soil, (C) chitosan reinforced by banana stem fiber. *International Journal of Applied Engineering and Technology.* 2013, 3, 35–60.

187. Thirugnanasambandham, K., Sivakumar, V., Prakash Maran, J. Chitosan as an adsorbent to treat rice mill wastewater—Mechanism modeling and optimization. *Carbohydrate Polymers.* 2013, 97, 451–457.

188. Tien, C. *Adsorption Calculations and Modeling* Butterworth-Heinemann, Boston, 1994.

189. Tzu-Yang Hsien; Yu-Ling Liu. Desorption of Cadmium from Porous Chitosan Beads. In Advancing Desalination, Robert Y. Ning, Ed., InTech: Croatia, 2012; p 163–180.

190. Ucer, A., Uyanik, A., Aygun, S. F. Adsorption of Cu(II), Cd(II), Zn(II), Mn(II) and Fe(III) ions by tannic acid immobilized activated carbon. *Sep. Purif. Technol.* 2006, 47, 113–118.

191. Vanbenschoten, J. E., Reed, J. E., Matsumoto, M. R., McGarvey, P. J. Metal removal by soil washing for an iron-oxide coated sandy soil. *Water Environ. Res.* 1994, 66, 168–174.

192. Vanloon, G. W., Duffy, S. J. The Hydrosphere. In: Environmental Chemistry: A Global Perspective. 2nd Edn. Oxford University Press: New York, 2005; 197–211.

193. Varma, A. J., Deshpande, S. V., Kennedy, J. F. Metal complexation by chitosan and its derivatives: a review. *Carbohydrate Polymers.* 2004, 55, 77–93.

194. Vasudevan, P., Padmavathy, V., Dhingra, S.C. Kinetics of biosorption of cadmium on baker's yeast. *Bioresources Technology,* 2003, 89, 281–287.

195. Vijayaraghavan, K., Jegan, J., Palanivelu, K., Velan, M. Biosorption of copper, cobalt and nickel by marine green alga Ulva reticulata in a packed column. *Chemosphere.* 2005a, 60, 419–426.

196. Vijayaraghavan, K., Palanivelu, K., Velan, M. Biosorption of copper (II) and cobalt (II) from aqueous solutions by crab shell particles. *Bioresources Technology.* 2006, 97, 1411–1419.

197. Vijayaraghavan, K., Palanivelu, K., Velan, M. Crab shell-based biosorption technology for the treatment of nickel-bearing electroplating industrial effluents. *J. Hazard. Mater.* 2005b, 119, 251–254.

198. Volesky, B. Biosorbents for metal recovery. *Trends Biotechnol.* 1987, 5, 96–101.

199. Vymazal, J. Algae and Element Cycling in Wetlands. Lewis Pub., *Boca Raton.* 1995, 689.

200. Wan Ngah, W. S., Endud, C. S., Mayanar, R. Removal of copper (II) ions from aqueous solution onto chitosan and crosslinked chitosan beads. *React. Funct. Polym.* 2002, 50, 181–190.

201. Wan, M. W., Petrisor, I. G., Lai, H. T., Yen, T. F. Copper adsorption through chitosan immobilized on sand to demonstrate the feasibility for in situ soil decontamination. *Carbohydr Polym.* 2004, 55, 249–254.

202. Wang Min, Ling Xu., Peng Jing; Zhai Mao-lin; Li Jiu-qiang; Wei Gen-shuan. Adsorption and desorption of Sr(II) ions in the gels based on polysaccharide derivatives. *Journal of Hazardous Materials.* 2009, 171, 820–826.

203. Wang, Y. H., Lin, S. H Juang, R. S. Removal of heavy metal ions from aqueous solutions using various low-cost adsorbents. *J. Hazard. Mater.* 2003, 102, 291–302.

204. Wankasi, D., Horsfall, M. Jnr., Spiff, A. Desorption of Pb2+ and Cu2+ from Nipa palm (*Nypa fruticans Wurmb*) biomass. *African Journal of Biotechnology.* 2005, 4, 923–927.

205. Ware, G. Theory and Application. New York. 1983, 308.
206. Wasay, S. A., Parker, W., Van Geel, P. J., Barrington, S., Tokunaga, S. Arsenic Pollution of a Loam soil: Retention from and Decontamination. *Journal of Soil Contamination.* 2000, 9, 51–64.
207. Weber, W. J., Mc Ginley, P. M., Katz, L. E. Sorption phenomena in subsurface systems: concepts, models and effects on contaminant fate and transport, review paper. *Water Res.* 1991, 25, 499–528.
208. Wilczak, A. and Keinath, T.M. Kinetics of sorption and desorption of copper (II) and lead (II) on activated carbon. *Wat. Environ. Res.* 1993, 65, 238–244.
209. Xiao, F., Ju-Chang, H. H. Comparison of biosorbents with organic sorbents for removing Copper (II) from aqueous solutions. *J. Environ. Man.* 2009, 90, 3105–3109.
210. Yan, G., Viraraghavan, T. Effect of pretreatment on the bioadsorption of heavy metals on Mucor rouxi. *Water SA*, 2000, 26, 119–123.
211. Yang, T. C., Zall, R. R. Adsorption of metals by natural polymers generated from seafood processing wastes. *Ind Eng Chem Prod Res Dev.* 1984, 23, 168–172.
212. Yong-gui Chen; Yong, H. E., Wei-min, Y. E., Wang-Hua, S. U. I., Min-min Xiao. Effect of shaking time, ionic strength, temperature and pH value on desorption of Cr(III) adsorbed onto GMZ bentonite. *Trans. Nonferrous Met. Soc China.* 2013, 23, 3482–3489.
213. Yubin, T., Fangyan, C., Honglin, Z. *Adsorpt. Sci. Technol.*, 1998, 16, 595.
214. Yun, Y. S., Park, D., Park, J. M., Volesky, B. Biosorption of Trivalent Chromium on the Brown Seaweed Biomass, *Environ. Sci. Technol.* 2001, 35, 4353–4358.
215. Yushina, Y., Hasegawa, J. Process performance comparison of membrane introduced anaerobic digestion using food-industry waste-water. *Desalination,* 1994, 98, 413–421.
216. Zhang, A., Asakura, T., Uchiyama, G. The adsorption mechanism of uranium (VI) from seawater on a macroporous fibrous polymeric adsorbent containing amidoxime chelating functional group. *React Funct Polym.* 2003, 57, 67–76.
217. Zhang, X., Bai, R. Mechanisms and kinetics of humic acid adsorption onto chitosan-coated granules. *J Colloid Int Sci.* 2003, 264, 30–8.
218. Zhou, D., Zhang, L., Zhou, J., Guo, S. Cellulose/chitin beads for adsorption of heavy metals in aqueous solution. *Water Research.* 2004, 38, 2643–2650.
219. Zhou, J. L. Kiff, R. J. Uptake of copper from aqueous solutions by fungal biomass.*J.Chem.Technol.Biotechnol.*1991, 52, 317–330.
220. Zhou, J. L., Huag, P.L., Lin, R. G. Sorption and desorption of Cu and Cd by macroalgae and microalgae.. *Environ. Pollut.* 1998, 101, 6–75.
221. Zouboulis, A. I., Matis, K. A., Stalidis, G. A. Flotation methods and techniques in wastewater. Innovations in Flotation Technology. Mavros, P., Matis, K. A. (Eds.) Kluwer Academic: Dordrecht, NL, 1992, 96–104.

INDEX